P9-DUN-024

RACE, GENDER, AND SCIENCE

Anne Fausto-Sterling, *General Editor*

Feminism and Science
Nancy Tuana, Editor

The "Racial" Economy of Science: Toward a Democratic Future
Sandra Harding, Editor

The Less Noble Sex: Scientific, Religious, and Philosophical Conceptions of Woman's Nature
Nancy Tuana

Love, Power and Knowledge: Toward a Feminist Transformation of the Sciences
Hilary Rose

Women's Health—Missing from U.S. Medicine
Sue V. Rosser

Im/partial Science: Gender Ideology in Molecular Biology
Bonnie B. Spanier

Deviant Bodies

DEVIANT BODIES

Critical Perspectives on Difference in Science and Popular Culture

edited by

Jennifer Terry and Jacqueline Urla

Indiana University Press

BLOOMINGTON AND INDIANAPOLIS

The paper used in this publication meets the minimum requirements of American
National Standard for Information Sciences—Permanence of Paper for Printed
Library Materials, ANSI Z39.48-1984.

∞™

Manufactured in the United States of America

Library of Congress Cataloging-in-Publication Data

Deviant bodies : critical perspectives on difference in science and
popular culture / edited by Jennifer Terry and Jacqueline Urla.
p. cm. — (Race, gender, and science)
Includes bibliographical references and index.
ISBN 0-253-32898-5 (cl : alk. paper). — ISBN 0-253-20975-7 (pa : alk. paper)
1. Body, Human—Social aspects. 2. Body, Human—Public opinion.
3. Deviant behavior. 4. Social norms. 5. Social values.
I. Terry, Jennifer, date. II. Urla, Jacqueline. III. Series.
GN298.D49 1995
306.4—dc20 95-3260

1 2 3 4 5 00 99 98 97 96 95

Contents

Introduction: Mapping Embodied Deviance
Jacqueline Urla and Jennifer Terry 1

1

Gender, Race, and Nation: The Comparative Anatomy of "Hottentot" Women in Europe, 1815–1817
Anne Fausto-Sterling 19

2

Framed: The Deaf in the Harem
Nicholas Mirzoeff 49

3

Colonizing and Transforming the Criminal Tribesman: The Salvation Army in British India
Rachel J. Tolen 78

4

This Norm Which Is Not One: Reading the Female Body in Lombroso's Anthropology
David G. Horn 109

5

Anxious Slippages between "Us" and "Them": A Brief History of the Scientific Search for Homosexual Bodies
Jennifer Terry 129

6

The Destruction of "Lives Not Worth Living"
Robert N. Proctor 170

7

Domesticity in the Federal Indian Schools: The Power of Authority over Mind and Body
K. Tsianina Lomawaima 197

8
Nymphomania: The Historical Construction of Female Sexuality
Carol Groneman 219

9
Theatres of Madness
Susan Jahoda 251

10
The Anthropometry of Barbie: Unsettling Ideals of the Feminine Body in Popular Culture
Jacqueline Urla and Alan C. Swedlund 277

11
Regulated Passions: The Invention of Inhibited Sexual Desire and Sexual Addiction
Janice M. Irvine 314

12
Between Innocence and Safety: Epidemiologic and Popular Constructions of Young People's Need for Safe Sex
Cindy Patton 338

13
The Hen That Can't Lay an Egg (*Bu Xia Dan de Mu Ji*): Conceptions of Female Infertility in Modern China
Lisa Handwerker 358

14
The Media-ted Gene: Stories of Gender and Race
Dorothy Nelkin and M. Susan Lindee 387

Notes on Contributors 403
Name Index 407

Deviant Bodies

Introduction

Mapping Embodied Deviance

Jacqueline Urla and Jennifer Terry

Do petty thieves have a genetic propensity to shoplifting? Is homosexuality an effect of hormonal imbalances? Is prostitution the result of some neurochemical disorder? Are juvenile delinquents afflicted with bad genes and faulty biochemistry? Questions of this sort make the headlines of daily newspapers and pepper the feature stories of scientific publications with surprising frequency these days, revealing a tenacious anxiety and curiosity about the links between the body and "deviancy." Indeed, the notion that individuals identified as socially deviant are somatically different from "normal" people is a peculiarly recurring idea that is deeply rooted in Western scientific and popular thought, but one that takes many forms in relation to particular historical and political contexts. With early roots in Aristotelian comparative studies, the idea that moral character is rooted in the body has structured a wide variety of modern medical and scientific studies, and shapes the current conditions under which popular fictions circulate about the bodies of all kinds of people who are deemed to be in some way behaviorally aberrant or socially disruptive.

In nineteenth-century Europe and America, the belief that moral character and psychical features were fundamentally tied to biology came to the fore with a vengeance at a moment of heated debate about who would enjoy the privileges of legal and economic enfranchisement. Modern vernaculars of rationality, hygiene, and bureaucratic order made the sorting of different peoples an imperative of life scientists as well as of lawmakers and the police. Efforts to measure the ears of criminals, the clitorises of prostitutes, and the facial contours of "perverts" fueled a feverish desire to classify forms of deviance, to locate them in biology, and thus to police them in the larger social body. The somatic territorializing of deviance, since the nineteenth century, has been part and parcel of a larger effort to organize social relations according to categories denoting normality versus aberration, health versus pathology, and national security versus social danger.

In its late-twentieth-century form, the idea that deviance is a matter of somatic essence—having passed through a period of disrepute following revelations about the "excesses" of the Third Reich—has arrived again, through the back door, facilitated by moral discourses concerning addiction as well as techniques of genetic engineering, neuroanatomical imaging, and virology. In an age of "compulsive behavior," killer viruses, and dangerous genes, methods for finding a host of socially and scientifically menacing pathogens inside certain bodies are contributing to the construction of new, biologically demonized underclasses. The alliance between scientists, physicians, and the police that so distinctly characterized techniques of governing during the nineteenth century seems to have appeared again in the form of attempts to sort aspects of social deviance in terms of anatomical or physiological anomalies of the body. In this new wave, proposals to screen "incorrigible" inner-city children for genes for violence and projects to study the autopsied brains of HIV-infected homosexual men reveal the dynamics by which social conflict is being displaced again onto the body. Techniques for prevention and elimination of deviance in this moment enunciate an ethos of biological determinism whereby strategies of social reform from the 1960s are jettisoned as old-fashioned and ineffective ways of dealing with contemporary social problems. In the Reagan/Bush era, technoscientific biological determinism arrived to explain social injustice as natural and necessary. Thus, economic inequalities, the subordination of women and people of color, and the untimely HIV-related deaths of almost a million U.S. citizens were deemed epiphenomena of biology. It appears that a new voodoo science has emerged to rationalize brutality and undergird the banality of evil in a New World Order.

The collection of essays that follows is concerned both with the history of the idea that deviance is embodied and with the idea's relevance to how we as modern subjects understand ourselves and others today. As a whole, the book represents an inquiry into modern Western epistemology by examining the very idea of *embodied deviance,* which we define as the historically and culturally specific belief that deviant social behavior (however that is defined) manifests in the materiality of the body, as a cause or an effect, or perhaps as merely a suggestive trace. In short, embodied deviance is the term we give to the scientific and popular postulate that the bodies of subjects classified as deviant are essentially marked in some recognizable fashion. Each of the contributions to this volume examines particular discursive and practical connections that have been drawn between various kinds of deviancy and the body.

Our inquiry does not take place in a vacuum. A quick foray through any university bookstore, a scan of contemporary art exhibitions or scientific publications, and even a surf across the channels of afternoon talk shows reveal that, in the twilight of the twentieth century, we are in the midst of an explosion of scholarly and popular interest in the representational vicissitudes of the body. Curiously, at a time when the discrete object called the body is being theoretically deconstructed, technologically fragmented, and politically reterritorialized, it has become an object of intense intellectual fascination, a realm for futuristic fantasy, and, indeed, a grounding point of profound social anxiety. Scholarly and artistic production of the past few years presents to us a human body which, far from being a self-evident organic whole, is at best a nominal construct and a phantasmatic space, imagined very differently over time and across various cultural contexts. At this point, it is perhaps beyond dispute that representations of the body are a means for generating dynamic cultural meanings, structuring complex social relations, and establishing flows of power (Crary and Kwinter 1992; Feher, Naddaff, and Tazi 1989; Haraway 1989, 1991; Jordanova 1989; Kroker and Kroker 1987; Laqueur 1990; Gallagher and Laqueur 1987; Martin 1990; Russett 1989; Schiebinger 1987; Turner 1984).

Deviant Bodies is a tributary stream in this recent flood of interest in the body. The book makes its contribution to the contemporary scene through case-based analyses that look closely at instances of how the body has been figured discursively in relation to particular constructions of deviance. In framing the volume, we as its editors were guided by Michel Foucault's important observation that the scientific and lay desire for authoritative Truth has brought the practice of representing bodies together with modes of regulation, containment, incitement, and resistance. And from Foucault we take the now almost commonplace axiom that the modern life sciences and medicine—and, indeed, popular perceptions to which they give rise—have not merely observed and reported on bodies; they *construct* bodies through particular investigatory techniques and culturally lodged research goals. Bodies do not exist in terms of an a priori essence, anterior to techniques and practices that are imposed upon them. They are neither transhistorical sets of needs and desires nor natural objects preexisting cultural (and, indeed, scientific) representation. They are effects, products, or symptoms of specific techniques and regulatory practices. In short, bodies are points on which and from which the disciplinary power of scientific investigations and their popular appropriations is exercised (Feher 1987). Knowable only through culture and history, they are not in any simple way natural or ever free of relations of power. What is at stake in *Deviant Bodies* is not merely that what we know to be bodies are always representations; what matters is that scientific and popular modes of representing bodies are never innocent but always tie bodies to larger systems of knowledge production and, indeed, to social and material inequality (Foucault 1979, 1980).

Thanks to the epistemological and productively disruptive interventions of poststructuralism, feminist theory, and more recently antiracist and queer theory, it is no longer viable to talk about "the human body" as a single, universal entity. Indeed, as it turns out, to do so is to comply with an agenda of domination through reduction, a logic and practice characteristic of the enlightened humanist tradition. Within the modern Western framework, to talk about a generic human body has meant to talk in terms of Self and Other and, moreover, of elite Subjects and subordinated Objects. The project of *Deviant Bodies* is to take on this binaristic facet of Western thought as it materializes in more oppressive forms—that is, when and where divisions between health and pathology, normalcy and deviance, purity and adulteration are used to map a complex and often conflictual set of local and global social relations. Our collective focus in this volume is to analyze how constructions of bodies are used to encode and enact these relations.

Fictions of Oneness are now in deep crisis, effaced by the weighty duplicity of liberal democracy and the emergence of a radical politics of multiplicity over the course of the late nineteenth and twentieth centuries. *Deviant Bodies* is dedicated to furthering productive disruptions, refusing the terms of reduction by which important differences and multiplicities are turned into binary face-offs between privileged Subjects and marginalized Objects. In this way, the present anthology is aligned with other efforts to unsettle the false sense of Oneness and the Self-centeredness of humanism and positivism that heretofore relied upon ideas of a single, generic human body to generate hypocritical fictions of unity, identity, truth, and authenticity. In other words, this volume seeks to show how the ideal human body has been cast implicitly in the image of the robust, European, heterosexual gentleman, an ideal defined by its contradistinction to a potpourri of "deviant" types. The chapters herein—together and separately—subject the modern fiction of Man to critical scrutiny by revealing that its sacred Oneness is based upon its distinction from a host of Others. Our strategy, then, is to deconstruct a particular bodily fiction that has been deployed for the maintenance of Man: the idea that individuals who deviate from that ideal are morally and socially inferior, and that their social or moral disruptiveness is always somehow embodied. Thus, our volume is framed by a set of questions: How are bodies marked in relation to systems of social stratification based on race, gender, age, economic class, and sexuality? How do bodies do the work of supporting unequal relations of power? Which bodies in particular incite suspicion, scrutiny, desire, and controversy in clashes of power? And how are representations of bodies deployed in resistant discourses and social movements that oppose domination?

The contributors to this anthology generally concur that the construction of deviance is always already also a process of constructing some model of the "normal" body. Because the normal body is an unmarked figure that gains its meaning mainly in residual contrast to various deviant bodies, the book focuses on constructions of the bodies of subordinated Others to reveal this dependent relationship. Thus, together, the chapters stress that "deviant bodies" have been used implicitly and intricately to shore up notions of what is normal and what is not. To this end, a number of the assembled cases trace the construction of particular "deviant bodies," including the homosexual body (Terry), the deaf body (Mirzoeff), the "sex-addicted" body (Irvine), the HIV-infected body (Patton), the infertile body (Handwerker), and the criminal body (Horn). In addition to these case studies, several other chapters reveal that whole categories of people—women, Jews, Native Americans, blacks—have been seen to be fundamentally deviant, not by virtue of particular symptoms they manifest, but simply because of their subordinate location in systems for distinguishing gender, ethnicity, class, and race (Fausto-Sterling, Groneman, Urla/Swedlund, Nelkin/Lindee, Lomawaima, Proctor, Jahoda, Tolen). Throughout the book, the specter of the normal body, be it a white, heterosexual, healthy, or male body, is always simultaneously present—even if in shadow form—in discourses of deviance.

Deviant Bodies embraces the challenge to explore and struggle with the dynamics of difference, taking a deconstructive approach to map how complex systems for making social distinctions gain their authority and power by assuming that all cultural and psychical differences among people reside fundamentally in their bodies. Thus, in many of the book's chapters, authors explore how techniques for finding deviance articulate and reflect the complex classificatory systems by which racial, socioeconomic, gender, and sexual distinctions are produced. Treating bodies as discursive constructs and not as given natural objects, the volume's contributors explore various techniques used by experts to isolate parts of the body and read them as signs of morality, intelligence, and social worth. Thus, the cases presented here suggest that representational practices focused on bodies operate as mechanisms of power for negotiating conflicts and differences in particular historical and cultural contexts.

Contemporary political struggles linking power, knowledge, and bodies further motivate and inform the critical analysis of this book. We are concerned with the myriad tangible effects of the pernicious idea that deviance is a matter of an individual's biological essence. In many instances, how people live in their bodies and in social relations is profoundly structured by this notion. Thus the volume as a whole asks not only how bodies of socially marginalized people have been constructed through authoritative discourses and scientific practices, but also how these constructions articulate and structure power relations in society under the powerful sign of Science. Today, bodies function as powerful semiotic fields for staging passionate political demands.

The inquiry we invite in *Deviant Bodies* is concerned with analyzing the power associated with bodily constructs, and thus the book benefits from political activism centered on the intersection between the body and medico-scientific authority. Certainly, in our contemporary cultural context, bodies have become sites of political struggles precisely over representation and over the meaning of what is normal and what is not. Through their motions, habits, behaviors, and significations, bodies have been territorialized, inscribed, contained, and dispersed in relation to high-stakes political positionings about what should be permitted and what should be forbidden in issues as disparate as abortion, reproduction, homosexuality, genetic screening, drug consumption, crime, and disease control. The power of staking political claims through the body—whether from the left or the right—rests in the assumption that bodies, like property, are real material objects whose disposition is of great concern to society as a whole. In this way, bodies—like all objects that acquire the status of the Real through elaborate processes that represent them as material—are condensations and displacements of social relations. They do indeed enjoy the status of the Real and the material, and thus become a powerful source of medical evidence and scientific knowledge as much as they become grounds for articulating political claims.

In formulating this volume, we found it crucial to ask in both general and specific terms why the body has been so central to scientific and popular constructions of deviance. Its privileged status as a source of evidence and incontrovertible truth is rooted in a modern scientific assumption that bodies are fundamentally authentic and material, rather than ethereal or imagined. Within the crosscutting traditions of Enlightenment science and humanism, the body has been understood to be given by nature, and thus to be real and objective, capable of overriding even the most abstruse attempts of an individual to disguise his or her true self. Palpable and visible, the body's contours, anatomical features, processes, movements, and expressions are taken to be straightforward, accurate indications of an individual's essence and character. Paradoxically, even as they are understood to be self-evident, bodies continue to require expert interpretation. Thus they become surfaces onto which physicians, scientists, and lay people can inscribe and project powerful cultural meanings and moral prohibitions.

The modern bodies we imagine today are in many ways the legacies of techniques of measurement, visualization, and classification that grow out of the powerful domains of scientific empiricism and medical treatment. Perhaps not surprisingly, many of this book's authors focus their attention on the discourses of science and medicine, including comparative anatomy, sexology, gynecology, physical anthropology, genetics, epidemiology, and psychiatry as they have constituted policies of eugenics, public health, education, and criminology. The chapters that follow reject the assumption that science is a domain separate from culture or outside of history. Each illustrates that there is an enormous amount of traffic between the domains of social life and scientific research, noting how scientific questions emerge in relation to social anxieties about race, class, gender, and sexuality, though never in any simple or straightforward, unidirectional fashion. It is not our intention to depict science as some master conspiratorial apparatus that determines social relations, or to see science as simply reflective of these relations. In fact, the contingencies, mistakes, blind spots, inconsistencies, and unintended effects that characterize scientific research are often starkly revealed by the project of analyzing how certain bodies have been probed for signs of deviance. Efforts to mark and maintain the boundary between what is "normal" and what is "deviant" can give rise to many unanticipated consequences. Terry's chapter on the history of scientific quests for discerning the homosexual body illustrates the point that unforeseen and troublesome outcomes can actually arise from the careful, microscopic search for somatic distinctions between "normal" and "sex variant" bodies. In an attempt to draw a line of demarcation, scientists investigating homosexuality in the 1930s inadvertently produced enough evidence of variation to show that there was no firm or clear boundary between the bodies of normal and "perverse" people. In this and many other cases, meticulous techniques of quantification and classification eroded the very possibility of fixing the distinctions that researchers presumed were present. Efforts to classify homosexuals vis-à-vis certain somatic features were originally undertaken to allay the anxiety surrounding this problem of urban hygiene, but in the end these efforts actually fueled social and scientific anxiety even further, precisely because science produced no such clear demarcation.

The relentless search for signs of deviance, alongside attempts to single out and align certain individuals with morbid typologies, even in the face of contrary evidence, is part of the larger history of modern life sciences and their preoccupation with naming and classifying diverse things that make up the world. During the late eighteenth and nineteenth centuries, a great deal of scientific effort was devoted to tracing variations among species using new classificatory techniques from natural history, comparative anatomy, and evolutionary biology. Modern biological sciences were elaborated through a European zeal to render human diversity in biological terms, and bodies generally became territories for siting all sorts of cultural differences (Gould 1981; Stocking 1968, 1987). Phenotypic variations, perceived often to be the causes of cultural diversity, were charted on the basis of objectifying, measuring, and classifying the bodies of various peoples of color whom white anthropologists and biologists encountered in their fieldwork in Africa, Asia, and the Americas. Thus, human cultural diversity was equated with biological difference, and human beings were eventually categorized in hierarchically arranged groups corresponding to aspects of geography, race, nationality, gender, and socioeconomic class. In turn, lay appropriations of this schema underscored the idea that European gentlemen were biologically and culturally superior.

Fausto-Sterling's historical account of French scientific dissections of Sarah Bartmann, otherwise known as the "Hottentot Venus," gives us a particularly vivid example of this typologizing process. By focusing on the early-nineteenth-century writings of renowned anatomists Georges Cuvier and Henri de Blainville, her analysis shows how African women's bodies figured in the imagination of European male scientists and travelers during a period of French colonial expansion and capitalist development. Like the newly traversed lands and seas, women on the African continent were represented in European texts as wild and unruly. Their bodies, like the geographies of Africa and the "New World," were areas to be explored, exploited, dissected, and tamed. Representations of these wild territories, in turn, became central to European gentlemanly commentary on proper femininity that posed African women in contradistinction to women in "civilized" societies. Fausto-Sterling points to the ways in which anatomists' techniques of classifying and measuring African women's bodies were embedded in a masculinist desire for epistemological mastery over nature, "savages," and femininity. In contrast to previous scholarly analyses of the Hottentot Venus, she focuses critical scrutiny on the psychical and material investments of the male scientists rather than on the body of Bartmann herself. In noting that the "primitive" female is best understood as a product of the European imagination, Fausto-Sterling argues that scientific representations of Bartmann's body speak volumes about the scientists' desire to make clear-cut distinctions between men and women and between racial types, in order to underscore European male superiority.

Turning from Fausto-Sterling's work to that of Robert Proctor, we see that in the latter half of the nineteenth century the objectifying gaze of scientists expanded from a primary focus on racial typologies in colonial contexts to construct and study a group of "internal others" who existed within European and American society and were believed to be threatening to the social order. In his chapter on Nazi medicine, Proctor reminds us that some of the most egregious typologizing techniques were deployed within the boundaries of Europe in state-sanctioned campaigns of mass murder. Instead of considering Nazi medicine an extreme or exceptional case of pseudoscience, Proctor analyzes the rational classification of "lives not worth living," a category flexible enough to encompass an array of people deemed to be physically or morally degenerate. In the Nazi case, the classification of "deviant bodies" became central to a nationalist strategy of genocide, but had much in common with assumptions held by American and British scientists from the first half of the twentieth century.

Throughout the history of modern scientific inquiries about embodied deviance we find recurring slippages between conceptions of "difference," "deviance," and "pathology." Both in the colonial context and within the West, phenotypic and cultural variations among humans were seldom understood by scientists to be benign or insignificant. Especially during the nineteenth century, deficiency came to be inferred from difference, and deviant behavior, in the sense of variation from the norm, was frequently taken to be a pathological matter of concern to physicians as well as to governing officials. Nicholas Mirzoeff's chapter, focusing on the meanings associated with deafness, is particularly interesting in this regard. Mirzoeff looks at the influence of Western speculations about the Turkish harem on the modern medicalization of the deaf, noting that in the eighteenth and nineteenth centuries deafness came to be associated with the despotic and clandestine world of the Oriental harem. European gentleman travelers who visited Istanbul in an effort to pierce the veil of the mysterious East returned with stories about deafness and the common use of sign language among harem eunuchs. A partial result was that, even as deafness was constructed in modern forms as a physical deficiency, it was also associated with despotic secrecy, dangerous sensuality, and irrationality. Taking a broad interdisciplinary approach, Mirzoeff examines French literature, art, and scientific writings to explore a key moment of transition from orientalist to medicalized constructions of the deaf. He finds a matrix of ideas linking race, crime, and sexual deviance with deafness. Mirzoeff points to the persistence of this matrix in the medical construction of deafness following the French Revolution of 1789, but he also notes that modern ideas about treating disorders served to turn the deaf into targets for special training and management. Sign language, in turn, as the only visible symptom of this disorder, came to be seen as an irrational, lewd, and socially threatening practice that had to be repressed for the good of the deaf and for society at large.

In tracing the history of the idea of embodied deviance, we find that classificatory practices at the heart of this notion depended theoretically and pragmatically on making deviance visible. Thus, in many cases, the search for signs of deviance privileged sight above the other senses and measurements over other kinds of evidence (Foucault 1975; Sekula 1986; Watney 1990; Green 1984; Marshall 1990). Techniques for studying and measuring deviant bodies traveled back and forth between the Euro-American metropole and the colonial periphery. Rachel Tolen's chapter on Salvation Army moral reform campaigns in British India analyzes the complex interrelationships between efforts to discipline unruly bodies in England and in its star colony. As Tolen notes, "under colonial rule, the constitution of the notion of a criminal caste drew on prevailing [British] discourses about crime, class, and work, as well as on British notions about the nature of Indian society." Moreover, Tolen, by illustrating how the disciplining of bodies was at the center of British colonial domination, effectively argues that reform efforts to make Indian tribespeople productive and compliant were key to the overall expansion of the British Empire. Considered alongside other chapters in this volume, Tolen's piece suggests that methods used to classify "criminal tribesmen" in India, like those devised for describing African "savages," were integrally linked to the regulatory apparatuses of hospitals, museums, government-run boarding schools for Native Americans, and even bawdy theaters in Europe and the United States. Thus, colonial and missionary schools in Africa, India, and parts of Asia shared a diagnostic and classificatory logic with prisons, clinics, and insane asylums in the metropole. This was precisely the desire to isolate categories of Others, whether they be prostitutes, homosexuals, nymphomaniacs, imbeciles, tribal rioters, or disease carriers. Each type was believed to have bodily signs of deviance that could be made visible.

The power of empirical observation lies primarily in its ability to render information visible, thus offering a means for controlling deviance through the clinical gaze. In this way, empiricism can have much in common with the penological surveillance of bodies, when both share techniques of watching suspiciously with an eye to control and regulation. In prisons, schools, hospitals, factories, and families, bodily activities are isolated through the careful gaze of experts. Timing, form, expression, gait, and gesture are means for breaking the body down into elements that can be disciplined as well as interpreted for their classificatory significance. Thus the body is fragmented and territorialized into zones from which degeneracy, perversion, savagery, and deviance are presumed to emanate. The legitimacy of surveillance in both the broad social context and much of scientific practice is dependent upon a rationale of protecting the social body from potentially dangerous forces, and of bestowing upon individuals the responsibility of health and hygiene. Modern surveillance demands the inspection of individuals in order to judge their status, analyze their deficits, and evaluate their function. This has been played out to a large extent through scientific investigations, which, in claiming to be therapeutic and ameliorative, produce an array of varying and contradictory definitions of what could count as deviant.

Looking into places not readily visible to the layperson in everyday life, science transforms the body into the province of specialists who alone can decode its many signs. Thus, scientific surveillance of deviance moves into ever more embedded and microscopic zones, where experts survey everything from pelvic angles to facial expressions, measure a cornucopia of bodily dimensions from femur bones to scrota, and ultimately transgress the threshold of the skin to look for deviance in all manner of bodily fluids and genetic properties. To paraphrase cultural critic Stuart Marshall, this is where markers of deviance are "to be captured reverberating throughout the body at every level of detail and magnification" (1990: 30). In the process of this moral and somatic surveillance, the soul of the deviant is both exteriorized to the body's surface and annexed to its various recesses where only men of science can peer.

The scopic regimes associated with looking for somatic markings of deviance position the expert simultaneously as objective scientist, informed interpreter, and voyeur. Voyeuristic investments of experts combine the libidinal pleasures of viewing with the authoritative power of science to dramatize the aphorism that "images don't lie." As Marshall points out, this kind of scientific surveillance requires that the expert's gaze be "clarified and intensified in order that the disease should give up its hidden secrets into the domain of the visual" (1990: 25).

Within a larger matrix of power, scopic regimes enact oppressive gender dynamics. More often than not, they involve men looking at women's bodies, and thus, through the penetrating clinical gaze, scopic techniques become tools for gaining scientific mastery over troublesome women. In her historical analysis of nymphomania, Carol Groneman notes that women's bodies were scrutinized for signs of "excessive" sexual desire by male doctors concerned not only with establishing medical authority, but also with policing white, middle-class women's desire during a period of turbulent social transformation in the United States. Thus the doctors' desire for scientific mastery over female sexuality—especially by rendering visible the signs of nymphomania—can be seen as a means to guard against the uncertainties of changing gender relations in the late Victorian period. Groneman argues that efforts to pathologize women's sexual desire were themselves symptoms of male dread and discomfort over women's changing roles and their demands for equal rights. Hence, nymphomania, like hysteria and neurasthenia, functioned to turn women's nonconforming behavior into medical problems, placing them under the dominion of male experts. Groneman's analysis not only reveals the constructed and changing nature of what were considered female sexual disorders, but also offers a picture of how women's bodies became phantasmatic spaces, where the fantasies and anxieties of male physicians were registered in the name of symptoms and diseases. In addition to naming some women as inherently pathological, the medicalization of female sexual desire served to put all women on notice about what was acceptable and what was forbidden, this time replacing the language of sin and virtue with that of disease and hygiene.

Many of the cases presented in *Deviant Bodies* suggest that scientific efforts to find deviance situate, implicate, and perhaps even transform the bodies of investigating experts. The idea that science is objective usually obscures the fact that scientists too have bodies, and those bodies are integrally involved in establishing norms and deviations. Those who speak in the omniscient language of science, medicine, and truth are deeply enmeshed in the process by which deviance is supposed to be discovered. Even though his postures, perceptions, and desires are disguised by the protocols of clinical observation, a scientist's gesture in taking measurements places his body in the frame of observation and representation. Moreover, any deviant body is always constructed in close proximity to the observing expert's body, even as the latter is never subjected to the bright, blinding gaze he is trained to focus on others. To draw attention to the fact that experts' bodies are present in the clinical scene could be a step toward destabilizing the power of omniscient science. In this way, artist Susan Jahoda displays the misogynist terror provoked by female sexual desire in her photographic essay highlighting the mechanisms of constraint and punishment to which women's bodies, and thus their sexualities, have been subjected. In these images, male experts' bodies are absent, and yet their dread and aversion is perceptible throughout. Hiding behind the comfort of scientific objectivity, the doctors' bodies appear to escape representation, but Jahoda's juxtaposition of text with images enacts a deconstruction through which male physicians' bodies become palpable in their quivering, terrified states.

Readers will note the preponderance of attention to women's bodies in this volume. Indeed, women's bodies have presented a particularly troubling case to scientists: for women, like men, are seen to be human but have been regarded historically as belonging to a biologically inferior class. Western metaphysical thought has bound Woman to the destiny of biology through her capacity to give birth. Particularly since the emergence of modern materialist science, Woman has been aligned with nature, irrationality, and the body, in contrast to Man, the bastion of culture and reason. Woman is symbolized by her body, Man is symbolized by his mind. With the Cartesian ascent of rationality came a denigration of the body, alongside a masculinist assault on nature and a scientific repugnance toward things feminine. Thus, it is possible to interpret the modern scientific scrutiny of women's bodies as a manifestation of a philosophical tradition concerned with establishing, again and again, the superiority of Man.

Several chapters in this volume highlight the ways women's bodies have been interpreted to be inherently inferior to men and their bodies. David Horn, analyzing Cesare Lombroso's nineteenth-century study of female criminal offenders' bodies, suggests that the relationship between the normal and the deviant female body is always vexed, since female bodies are perceived to be inherently deviant in relation to a male norm of the human body. Horn argues that the scientific effort to mark off a physically distinct type called the "female offender" from normal women was part of a reconfiguration of "crime," away from an earlier concern with classification of illegal acts toward a "typology of criminals." But it was also an effort to mark women who broke the law as particularly deviant and socially dangerous. The failure to find definitive anatomical signs of female criminality led not to its abandonment as a scientific goal, but to the extension of suspicion toward all women. The idea that deviance could be masquerading under the guise of a "normal" woman served to construct all women as potentially dangerous. As Horn's cleverly titled chapter suggests, it was practically impossible to imagine any female as normal within the terms of Lombroso's criminal anthropology.

Urla and Swedlund approach the theme of feminine deviance from a rather different angle. Taking as their point of departure that female bodies are perceived as inherently defective, they turn their attention to the analysis of one of the most popular icons of the post–World War II period: the Barbie doll. Through their tongue-in-cheek anthropometry experiment, these anthropologists turn the tables on science, away from the scrutiny of women's bodies, to focus on the ideals by which we come to measure ourselves and others. Based on their anthropometric calculations, the authors note that few, if any, women will ever measure up to the ideal form of Barbie. Thus, because Barbie's body is statistically deviant compared to most women, the doll has functioned to underscore the predicament of femininity: no female body is ever appropriate. Yet Urla and Swedlund's aim is not simply to argue that Barbie dolls are oppressive. Instead, their cultural explication of Barbie plays upon the varied and contradictory meanings of the thin-body ideal, noting how it is linked more generally to the construction of femininity in late consumer capitalism. By looking at a genre of satire called Barbie Noire, Urla and Swedlund analyze how the female body has increasingly become, like Barbie herself, a form of cultural plastic, a ground for displaying contested notions of gender identity.

In addition to the trials and tribulations of beauty, women's reproductive capacity historically has been the focus of studies linking women with deviance and pathology. Lisa Handwerker's chapter offers a way of looking at the persistence of this tendency in contemporary China. Based on original ethnographic fieldwork in Beijing, Handwerker analyzes the paradoxical predicament of female infertility in the context of China's policy restricting families to the birth of only one child. Offering an interesting cross-cultural and contemporary analysis of embodied deviance, Handwerker explores the social stigmatization of the childless or infertile woman. She notes that even as Chinese women are admonished to use birth control and avail themselves of abortion, those who fail to reproduce are seen as selfish, overly modern, and bourgeois. Likewise, having too many children is commonly perceived as a form of indulgence and immorality. In one of the more interesting aspects of her chapter, Handwerker notes how infertile Chinese women are combining both traditional Chinese medicine and high-tech Western biomedicine to find a solution to their source of shame. While each medical system has its own distinct vision of the body, both, she argues, construe childless women as defects to be corrected.

Clearly, the mechanisms of power associated with discerning the "deviant" from the "normal" constrain those placed in the former category while privileging those in the latter. But this anthology is not intended to present a narrative of miserable abjection and crude victimization at the hands of those classified as "normal." In many instances, authoritative practices of discerning the deviant from the normal not only shape the subjectivity of those being observed, but also give rise to forms of resistance. In her study of sex-segregated American Indian boarding schools, Tsianina Lomawaima focuses on the mechanisms of bodily surveillance and discipline exercised through the education of Native American girls during the first half of the twentieth century. Lomawaima shows how, in the boarding schools' attempt to make good subservient Christian women out of Indian girls, "the regimentation of the external body was the essential sign of a new life, of a successful transformation" into a middle-class Victorian notion of domesticated femininity. She notes that "the acute, piercing focus on Indian girls' attire, comportment, posture, and hairstyles betrays a deep-seated, racially defined perception of Indian people's physical bodies as 'uncivilized.' " Thus, strategies to reform Native Americans at the microscopic level brought racialist biology into the service of a governmental plan of social containment.

But the process of "civilizing" these girls was anything but smooth, thanks to their astute mischief. Through her lively oral histories of Native American women, Lomawaima gives us a rare insight into the ways girls surreptitiously defied the rules of comportment and dress that structured their training to become compliant subjects. Their seemingly ordinary schoolgirl antics of hiding forbidden bloomers under their official school uniforms functioned as a way of eluding the control that school authorities sought to exercise over their dress and, indeed, over their moral character.

Scientific and medical discourses permeate the realm of popular culture, where they carry particular kinds of authority and appeal. The promise of science to help individuals understand themselves and the world in which they live is offered up through feature stories on television, in newspapers and magazines, and in the information-cluttered marketplace. Knowing one's origins, one's environment, one's proficiencies, and one's weaknesses constitute the modern technologies of the self, which, to a large extent, are animated by scientific advice and expertise in the public sphere. At least in the United States, science is positioned to offer Joe and Jane SixPack knowledge about themselves that is understood to be necessary for their vitality, happiness, and authenticity. In her chapter exploring scientific sexology and its popular appropriations, Janice Irvine notes that medical diagnoses of sexual disorders and social perceptions of deviance result from complex negotiations between the perceptions of afflicted patients, the trained perspective of experts, and the larger context of cultural meanings. Even as recently named sexual disorders figure the afflicted individual as deviant, Irvine argues that medical discourses enjoy great popular appeal because they authorize and legitimate subjective perceptions of being sexually abnormal. She notes that recent historical events centering on sexual expression—either to encourage or repress it—contribute to an interesting trend toward self-diagnosis, whereby individuals name themselves as either diseased or deficient. Irvine concludes by making a call for progressive movements to analyze critically the new disorders of desire in a way that explores how and why individuals find these diagnoses meaningful, useful, and perhaps even reassuring. Moreover, the sexual anxieties and discomfort reported by self-diagnosing individuals are problems, Irvine notes, in the body politic around the meanings given to sexuality and the anxious warnings about avoiding risks. They are anxieties of a cultural place and a historical moment affecting many more than those who are "afflicted" by them, even as they function to localize disorders in individual bodies.

Cindy Patton's chapter stresses the deadly consequences of presuming certain bodies are innately deviant. In her discussion of the cultural contradictions that surround HIV-positive youth, Patton argues that epidemiological and popular strategies of risk reduction are shot through with distinctions and prejudices that defeat any real possibility of disease control. Basing her study on a survey of media coverage on AIDS among youths, Patton notes that certain types of young people are constructed as inherently deviant due to their presumed demographic risk for contracting the virus. Specifically, she finds a basic distinction between, on the one hand, "normal" teens, besieged by hormones, but innocent and outside the range of real risk, and, on the other hand, both gay teens, who are at risk for infection, and youth of color who are presumed to be already infected because of their innate promiscuity and atavism. These distinctions, Patton notes, are mapped onto geographies of the putatively disease-free, white, middle-class suburb and its Other, the germ-ridden inner city. Linked to the particular environment of the ghetto, youth of color are figured as irrational and primitive, while white gay teens are described as deviant but recuperable. The consequence of this classification process, Patton suggests, is not only the further neglect of populations in which HIV is rampant, but also the defeat of measures to prevent the disease in so-called normal populations.

Finally, Dorothy Nelkin and M. Susan Lindee bring us to the brave new world of genetic research in the present. We live in a time when the human genome project and high-tech representations of the body capture the popular imagination, promising to rid the social body of disease-causing agents and deleterious genes. To a curious and anxious public, the gene has become a cultural icon, capable of explaining anything and everything, from epidemic disease to economic inequality. As Nelkin and Lindee argue, the magical sign of the gene is now being used to shore up long-standing lies about the innate inferiority of African Americans and women. By focusing on the persistent stereotypes that structure scientific research and popular perceptions of gender and race, Nelkin and Lindee note that, in spite of their futuristic appeal and high-tech authority, contemporary genetic explanations of "natural" and inherent abilities grow out of and promote established agendas of social stratification. The authors leave us with a word of warning about how even the most technically complex forms of scientific research—those surrounding the magical genes—can be read as the "just so" stories of a putatively democratic society plagued by the hypocrisies of racism and sexism.

Deviant Bodies represents an effort both to efface "normal" and "healthy" bodies as fictions of science, and to document the subversiveness and resistances of "deviant" bodies. In demonstrating that the notion of normalcy is parasitic to the construction of deviance, we recognize that systems of binary classification work in particularly oppressive ways to subjugate those classified as deviant. Yet we would like to leave readers with a reminder that no one is free from the powerful and haunting effects of discipline and subjectification that binary classifications impose. Because this volume appears at a time rife with political questions about the significance of cultural diversity, we invite readers to approach the following chapters as part of a critical commentary on the present zeal to map difference and deviance onto bodies. We hope that the analysis presented here will underscore the point that the costs of perceiving social problems as matters of somatic essence are, in a word, devastating.

References

Crary, Jonathan, and Sanford Kwinter. *Incorporations.* New York: Zone Books, 1992.

Feher, Michel. "Of Bodies and Technologies." In *Discussions in Contemporary Culture. Number One.* Ed. Hal Foster. Seattle: Bay Press, 1987.

Feher, Michel, Ramona Naddaff, and Nadia Tazi, eds. *Fragments for a History of the Human Body,* 3 vols. New York: Zone Books; Cambridge: distributed by MIT Press, 1989.

Foucault, Michel. *The Birth of the Clinic: An Archaeology of Medical Perception.* Trans. A. M. Sheridan Smith. New York: Vintage Books, 1975.

———. *Discipline and Punish: The Birth of the Prison* Trans. Alan Sheridan. New York: Vintage Books, 1979.

———. *The History of Sexuality:* Vol. 1, *An Introduction.* Trans. Robert Hurley. New York: Vintage Books, 1980.

Gallagher, Catherine, and Thomas Laqueur, eds. *The Making of the Modern Body: Sexuality and Society in the Nineteenth Century.* Berkeley: University of California Press, 1987.

Gould, Stephen Jay. *The Mismeasure of Man.* New York: Norton, 1981.

Green, David. "Classified Subjects—Photography and Anthropology: The Technology of Power." *Ten* 8, no. 14 (1984).

Haraway, Donna. *Primate Visions: Gender, Race, and Nature in the World of Modern Science.* New York: Routledge, 1989.

———. *Simians, Cyborgs, and Women: The Reinvention of Nature.* New York: Routledge, 1991.

Jordanova, Ludmilla. *Sexual Visions: Images of Gender in Science and Medicine between the Eighteenth and Twentieth Centuries.* Madison: University of Wisconsin, 1989.

Kroker, Arthur, and Marilouise Kroker, eds. *Body Invaders: Panic Sex in America.* New York: St. Martin's Press, 1987.

Laqueur, Thomas. *Making Sex: Body and Gender from the Greeks to Freud.* Cambridge: Harvard University Press, 1990.

Marshall, Stuart. "Picturing Deviancy." In *Ecstatic Antibodies: Resisting the AIDS Mythology.* Ed. Tessa Boffin and Sunil Gupta. London: Rivers Oram, 1990.

Martin, Emily. "Science and Women's Bodies: Forms of Anthropological Knowledge." In *Body/Politics: Women and the Discourses of Science.* Ed. Mary Jacobus, Evelyn Fox Keller, and Sally Shuttleworth. New York: Routledge, 1990.

Russett, Cynthia Eagle. *Sexual Science: The Victorian Construction of Womanhood.* Cambridge: Harvard University Press, 1989.

Schiebinger, Londa. "Skeletons in the Closet: The First Illustrations of the Female

Skeleton in Eighteenth-Century Anatomy." In *The Making of the Modern Body.* Ed. Catherine Gallagher and Thomas Laqueur. Berkeley: University of California Press, 1987.

Sekula, Allan. "The Body and the Archive." *October* 39 (Winter 1986): 3–64.

Stocking, George W., Jr. *Race, Culture, and Evolution: Essays in the History of Anthropology.* London: Collier-Macmillan, 1968.

Stocking, George W., Jr. *Victorian Anthropology.* New York: Free Press, 1987.

Turner, Bryan S. *The Body and Society: Explorations in Social Theory.* New York: Basil Blackwell, 1984.

Watney, Simon. "Representing AIDS." In *Ecstatic Antibodies: Resisting the AIDS Mythology.* Ed. Tessa Boffin and Sunil Gupta. London: Rivers Oram, 1990.

1

Gender, Race, and Nation

The Comparative Anatomy of "Hottentot" Women in Europe, 1815–1817

Anne Fausto-Sterling

A *note about language use:* Writing about nineteenth-century studies of race presents the modern writer with a problem: how to be faithful to the language usage of earlier periods without offending contemporary sensibilities. In this chapter I have chosen to capitalize words designating a race or a people. At the same time, I will use the appellations of the period about which I write. Hence I will render the French word *Negre* as Negro. Some nineteenth-century words, especially "Hottentot," "primitive," and "savage," contain meanings that we know today as deeply racist. I will use these words without quotation marks when it seems obvious that they refer to nineteenth- rather than twentieth-century usage.

A note about illustrations: This chapter is unillustrated for a reason. The obvious illustrations might include drawings and political cartoons of Sarah Bartmann or illustrations of her genitalia. Including such visual material would continue to state the question as a matter of science and to focus us visually on Bartmann as a deviant. Who could avoid looking to see if she really was different? I would have had to counter such illustrations with an additional discussion of the social construction of visual imagery. But this essay is meant to focus on the scientists who used Bartmann. Thus an appropriate illustration might be the architectual layout of the French Museum, where Cuvier worked, or something of that order. Failing to have in hand a drawing that keeps us focused on the construction and constructors of scientific knowledge, I felt it would be better to have none at all. Readers who are dying to see an image of Bartmann may, of course, return to any of the original sources cited.

Introduction

In 1816 Saartje Bartman, a South African woman whose original name is unknown and whose Dutch name had been anglicized to Sarah Bartmann, died in Paris. Depending upon the account, her death was caused by smallpox, pleurisy, or alchohol poisoning (Cuvier 1817; Lindfors 1983; Gray 1979). Georges Cuvier (1769–1832), one of the "fathers" of modern biology, claimed her body in the interests of science, offering a detailed account of its examination to the members of the French Museum of Natural History. Although now removed, as recently as the early 1980s a cast of her body along with her actual skeleton could be found on display in case #33 in the Musée de l'Homme in Paris; her preserved brain and a wax mold of her genitalia are stored in one of the museum's back rooms (Lindfors 1983; Gould 1985; Kirby 1953).[1]

During the last several years Bartmann's story has been retold by a number of writers (Altick 1978; Edwards and Walvin, 1983; Gilman, 1985).[2] These new accounts are significant. Just as during the nineteenth century she became a vehicle for the redefinition of our concepts of race, gender, and sexuality, her present recasting occurs in an era in which the bonds of empire have broken apart, and the fabric of the cultural systems of the nations of the North Atlantic has come under critical scrutiny. In this article I once again tell the tale, focusing not on Bartmann but on the scientists who so relentlessly probed her body. During the period 1814–70 there were at least seven scientific descriptions of the bodies of women of color done in the tradition of classical comparative anatomy. What was the importance of these dissections to the scientists who did them and the society that supported them? What social, cultural, and personal work did these scientific forays accomplish, and how did they accomplish it? Why did the anatomical descriptions of women of color seem to be of such importance to biologists of the nineteenth century?

The colonial expansions of the eighteenth and nineteenth centuries shaped European science; Cuvier's dissection of Bartmann was a natural extension of that shaping. (By "natural" I mean that it seemed unexceptional to the scientists of that era; it appeared to be not merely *good* science; it was forward-looking.) But a close reading of the original scientific publications reveals the insecurity and angst about race and gender experienced by individual researchers and the European culture at large. These articles show how the French scientific elite of the early nineteenth century tried to lay their own fears to rest. That they did so at the expense of so many others is no small matter.

Constructing the Hottentot before 1800

Several of the African women who ended up on the comparative anatomists' dissecting tables were called Hottentots or, sometimes, Bushwomen. Yet the peoples whom the early Dutch explorers named Hottentot had been extinct as a coherent cultural group since the late 1600s (Elphick 1977). Initially I thought written and visual descriptions would help me figure out these women's "true" race; I quickly discovered, however, that even the depictions of something so seemingly objective as skin color varied so widely that I now believe that questions of racial origin are like will-o'-the-wisps. Human racial difference, while in some sense obvious and therefore "real," is in another sense pure fabrication, a story written about the social relations of a particular historical time and then mapped onto available bodies.

As early as the sixteenth century, European travelers circling the world reported on the peoples they encountered. The earliest European engravings of nonwhites presented idyllic scenes. A depiction by Theodor de Bry from 1590, for example, shows Adam and Eve in the garden, with Native Americans farming peacefully in the background. The de Bry family images of the New World, however, transformed with time into savage and monstrous ones containing scenes of cannibalism and other horrors (Bucher 1981). Similarly, a representation of the Hottentots from 1595 (Raven-Hart 1967) shows two classically Greek-looking men standing in the foreground, with animals and a pastoral scene behind. A representation from 1627, however, tells a different story. A man and woman with yellow brown skin stand in the foreground. The man's hair is tied in little topknots; his stature is stocky and less Adonis-like than before, and he looks angry. The woman, naked except for a loincloth holds the entrails of an animal in her hand. One of her breasts is slung backwards over her shoulder, and from it a child, clinging to her back, suckles. As we shall see, the drawings of explorers discussed here in turn became the working background (the cited literature) of the racial studies of the early nineteenth century, which are presented in a format designed to connote scientific certainty.

The Adamic visions of newly discovered lands brought with them a darker side. Amerigo Vespucci, whose feminized first name became that of the New World, wrote that the women went about "naked and libidinous; yet they have bodies which are tolerably beautiful" (Tiffany and Adams 1985: 64). Vespucci's innocents lived to be 150 years old, and giving birth caused them no inconvenience. Despite being so at one with nature, Vespucci found Native American women immoral. They had special knowledge of how to enlarge their lovers' sex organs, induce miscarriages, and control their own fertility (Tiffany and Adams 1985). The early explorers linked the metaphor of the innocent virgin (both the women and the virgin land) with that of the wildly libidinous female. As one recent commentator puts it:

> Colonial discourse oscillates between these two master tropes, alternately positing the colonized 'other' as blissfully ignorant, pure and welcoming as well as an uncontrollable, savage, wild native whose chaotic, hysterical presence requires the imposition of the law, i.e., suppression of resistance. (Shohat 1991: 55)

From the start of the scientific revolution, scientists viewed the earth or nature as female, a territory to be explored, exploited, and controlled (Merchant 1980). Newly discovered lands were personified as female, and it seems unsurprising that the women of these nations became the locus of scientific inquiry. Identifying foreign lands as female helped to naturalize their rape and exploitation, but the appearance on the scene of "wild women" raised troubling questions about the status of European women. Hence, it also became important to differentiate the "savage" land/woman from the civilized female of Europe. The Hottentot in particular fascinated and preoccupied the nineteenth-century scientist/explorer—the comparative anatomist who explored the body as well as the earth. But just who were the Hottentots?

In 1652 the Dutch established a refreshment station at the Cape of Good Hope, which not long after became a colonial settlement. The people whom they first and most frequently encountered there were pastoral nomads, short of stature, with light brown skin, and speaking a language with unusual clicks. The Dutch called these people Hottentots, although in the indigenous language they were called Khoikhoi, which means "men of men." Within sixty years after the Dutch settlement, the Khoikhoi, as an organized, independent culture, were extinct, ravaged by smallpox and the encroachment of the Dutch. Individual descendants of the Khoikhoi continued to exist, and European references to Hottentots may have referred to such people. Nevertheless, nineteenth-century European scientists wrote about Hottentots, even though the racial/cultural group that late-20th-century anthropologists believe to merit that name had been extinct for at least three-quarters of a century. Furthermore, in the eighteenth and nineteenth centuries Europeans often used the word "Hottentot" interchangeably with the word "Bushman."[3] The Bushmen, or Khoisan, or hunter-gatherer Khoi, were (and are) a physically similar but culturally distinct people who lived contiguously with the Khoikhoi (Elphick 1977; Guenther 1980). They speak a linguistically related language and have been the object/subject of a long tradition of cultural readings by Euro-Americans (Haraway 1989; Lewin

1988; Lee 1992). In this chapter I look at studies with both the word "Bushman/ Bushwoman" and the word "Hottentot" in the titles. Cuvier, for example, argued vehemently that Sarah Bartmann was a Bushwoman and not a Hottentot. The importance of the distinction in his mind will become apparent as the story unfolds.

Constructing the Hottentot in the French Museum of Natural History

The encounters between women from southern Africa and the great men of European science began in the second decade of the nineteenth century when Henri de Blainville (1777–1850) and Georges Cuvier met Bartmann and described her for scientific circles, both when she was alive and after her death (Cuvier 1817; de Blainville 1816). We know a lot about these men who were so needful of exploring non-European bodies. Cuvier, a French Protestant, weathered the French Revolution in the countryside. He came to Paris in 1795 and quickly became the chair of anatomy of animals at the Museum of Natural History (Appel 1987; Flourens 1845). Cuvier's meteoric rise gave him considerable control over the future of French zoology. In short order he became secretary of the Académie des Sciences, an organization whose weekly meetings attracted the best scientists of the city, professor at the museum and the Collège de France, and member of the Council of the University. Henri de Blainville started out under Cuvier's patronage. He completed medical school in 1808 and became an adjunct professor at the Faculté des Sciences, while also teaching some of Cuvier's courses at the museum. But by 1816, the year his publication on Sarah Bartmann appeared, he had broken with Cuvier. After obtaining a new patron, he managed, in 1825, to enter the Académie and eventually succeeded Cuvier, in 1832, as chair of comparative anatomy.

Cuvier and de Blainville worked at the Musée d'Histoire Naturelle, founded in 1793 by the Revolutionary Convention. It contained ever-growing collections and with its "magnificent facilities for research became the world center for the study of the life sciences" (Appel 1987: 11). Work done in France from 1793–1830 established the study of comparative anatomy, paleontology, morphology, and what many see as the structure of modern zoological taxonomy. Cuvier and de Blainville used the museum's extraordinary collections to write their key works. Here we see one of the direct links to the earlier periods of exploration. During prior centuries private collectors of great wealth amassed large cabinets filled with curiosities—cultural artifacts and strange animals and plants. It was these collections that enabled the eighteenth-century classifiers to begin their work.

Bruno Latour identifies this process of collection as a move that simultaneously established the power of Western science and domesticated the "savage" by making "the wilderness known in advance, predictable" (Latour 1987: 218). He connects scientific knowledge to a process of accumulation, a recurring cycle of voyages to distant places in which the ships returned laden with new maps, native plants, and sometimes even the natives themselves. Explorers deposited these mobile information bits at centers, such as museums or the private collections that preceded them. Scientists possessed unique knowledge merely by working at these locations, which enabled them literally to place the world before their eyes without ever leaving their place of employ. Latour writes: "[T]hus the history of science is in large part the history of the mobilization of anything that can be made to move and shipped back home for this universal census" (Latour 1987: 225). Cuvier literally lived, "for nearly forty years, surrounded by the objects which engrossed so great a portion of his thoughts" (James 1830: 9). His house on the museum grounds connected directly to the anatomy museum and contained a suite of rooms, each of which held material on a particular subject. As he worked, he moved (along with his stove) from one room to the next, gathering his comparative information, transported from around the world to the comfort of his own home (Coleman 1964).

As centers of science acquired collections, however, they faced the prospect of becoming overwhelmed by the sheer volume of things collected. In order to manage the flood of information, scientists had to distill or summarize it. Cuvier, de Blainville, and others approached the inundation by developing coherent systems of animal classification. Thus the project of classification comprised one aspect of domesticating distant lands. The project extended from the most primitive and strange of animals and plants to the most complex and familiar. The history of classification must be read in this fashion; the attention paid by famous scientists to human anatomy cannot be painted on a separate canvas as if it were an odd or aberrant happening within the otherwise pure and noble history of biology.

During the French Revolution the cabinets of the wealthy who fled the conflict, as well as those from territories that France invaded, became part of the museum's collections. The cabinet of the Stadholder of Holland, for example, provided material for several of Cuvier's early papers. Appel describes the wealth of collected material:

> . . . in 1822, the Cabinet contained 1500 mammals belonging to over 500 species, 1800 reptiles belonging to over 700 species, 5000 fishes from over 2000 species, 25,000 arthropods . . . and an unspecified number of molluscs. . . . (Appel 1987: 35–36)

Cuvier's own comparative anatomy cabinet contained still more. He championed the idea that, in order to classify the animals, one must move beyond their mere surface similarities. Instead, one must gather facts and measurements from all of the internal parts. Without such comparative information, he believed, accurate classification of the animals became impossible. By 1822, among the 11,486 preparations in Cuvier's possession were a large number of human skeletons and skulls of different ages and races.

The human material did not innocently fall his way. In fact he had complained unbelievingly "that there is not yet, in any work, a detailed comparison of the skeletons of a Negro and a white" (Stocking 1982: 29). Wishing to bring the science of anatomy out of the realm of travelers' descriptions, Cuvier offered explicit instructions on how to procure human skeletons. He believed skulls to be the most important evidence, and he urged travelers to nab bodies whenever they observed a battle involving "savages." They must then "boil the bones in a solution of soda or caustic potash and rid them of their flesh in a matter of several hours" (Stocking 1982: 30). He also suggested methods of preserving skulls with flesh still intact, so that one could examine their facial forms.

As we shall see, Egyptian mummies—both animal and human—supplied another significant source that Cuvier used to develop and defend his theories of animal classification. These he obtained from the travels of his mentor-turned-colleague, and eventual archenemy, Étienne Geoffroy Saint-Hilaire. Geoffroy Saint-Hilaire spent several years in Egypt as part of the young general Napoleon Bonaparte's expedition. Cuvier declined the opportunity, writing that the real science could be done most efficiently by staying at home in the museum, where he had a worldwide collection of research objects at his fingertips (Outram 1984).[4] In 1798 Bonaparte took with him the Commission of Science and the Arts, which included many famous French intellectuals. During his years in Egypt, Geoffroy Saint-Hilaire collected large numbers of animals and, of particular importance to this story, several human and animal mummies. By 1800, British armies had defeated the French in Egypt; the capitulation agreement stipulated that the British were to receive all of the notes and collections obtained by the French savants while in Egypt. But in a heroic moment, Geoffroy Saint-Hilaire refused. In the end he kept everything but the Rosetta stone, which now resides in the British Museum (Appel 1987). Once again we see how the fortunes of modern European science intertwined with the vicissitudes of colonial expansion.

Cuvier and de Blainville used the technologies of dissection and comparative anatomy to create classifications. These reflected both their scientific and their religious accounts of the world, and it is from and through these that their views on race, gender, and nation emerge. In the eighteenth century the idea of biologically differing races remained undeveloped. When Linnaeus listed varieties of men in his *Systema Naturae* (1758), he emphasized that the differences between them appeared because of environment. There were, of course, crosscurrents. Proponents of the Great Chain of Being placed Hottentots and Negroes on a continuum linking orangutans and humans. Nevertheless, "eighteenth-century writers did not conceptualize human diversity in rigidly hereditarian or strictly physical terms. . . . " (Stocking 1990: 18).

Cuvier divided the animal world into four branches: the vertebrates, the articulates, the molluscs, and the radiates. He used the structure of the nervous system to assign animals to one of these four categories. As one of his successors and hagiographers wrote, "the nervous system is in effect the entire animal, and all the other systems are only there to serve and maintain it. It is the unity and the multiplicity of forms of the nervous system which defines the unity and multiplicity of the animal kingdom" (Flourens 1845: 98).[5] Cuvier expected to find similarities in structure within each branch of the animal world. He insisted, however, that the four branches themselves existed independently of one another. Despite similarities between animals within each of his branches, he believed that God had created each individual species (which he defined as animals that could have fertile matings). As tempting as the interrelatedness was to many of his contemporaries, Cuvier did not believe that one organism evolved into another. There were no missing links, only gaps put there purposely by the Creator. "What law is there," he asked, "which would force the Creator to form unnecessarily useless organisms simply in order to fill gaps in a scale?" (Appel 1987: 137).[6]

Cuvier's emphasis on the nervous system makes it obvious why he would consider the skull, which houses the brain, to be of utmost importance in assigning animals to particular categories. It takes on additional significance if one remembers that, unlike present-day taxonomists, Cuvier did not believe in evolution. At least in theory, he did not build the complex from the primitive, although his treatment of the human races turns out to be more than a little ambiguous in this regard. Instead he took the most complex as the model from which he derived all other structures. Because humans have the most intricate nervous system, they became the model to which all other systems compared. In each of his *Leçons d'anatomie comparée*, he began with human structures and developed those of other animals by comparison (Coleman 1964). In this sense, his entire zoological system was homocentric.

Cuvier's beliefs about human difference mirror the transition from an eighteenth-century emphasis on differences in levels of "civilization" to the nineteenth-century construction of race. His work on Sarah Bartmann embodies the contradictions such a transition inevitably brings. In 1790, for example, he scolded a friend for believing that Negroes and orangutans could have fertile matings and for thinking that Negroes' mental abilities could be explained by some alleged peculiarity in brain structure (Stocking 1982). By 1817, however, in his work on Sarah Bartmann, he brandished the skull of an Egyptian mummy, exclaiming that its structure proved that Egyptian civilization had been created by whites from whom present-day Europeans had descended (Cuvier 1817).[7]

Cuvier believed in theory that all humans came from a single creation, a view we today call monogeny. He delineated three races: Caucasians, Ethiopians (Negroes), and Mongolians. Despite uniting the three races under the banner of humanity (because they could interbreed), he found them to contain distinct physical differences, especially in the overall structure and shape of the head. One could not miss the invisible capabilities he read from the facial structures:

> It is not for nothing that the Caucasian race had gained dominion over the world and made the most rapid progress in the sciences while the Negroes are still sunken in slavery and the pleasures of the senses and the Chinese [lost] in the [obscurities] of a monosyllabic and hieroglyphic language. The shape of their head relates them somewhat more than us to the animals. (Coleman 1964: 166)

Cuvier, it is worth noting, was opposed to slavery. His was "a beneficent but haughty paternalism. . . . " (Coleman 1964: 167). In practice, however, his brother Fredéric, writing "under the authority of the administration of the Museum" (i.e., brother Georges), would include Georges Cuvier's description of Sarah Bartmann as the only example of the human species listed in his *Natural History of the Mammals* (Geoffroy Saint-Hilaire and Cuvier 1824: title page). Accompanying the article were two dramatic illustrations similar in size, style, and presentation to those offered for each of the forty-one species of monkeys and numerous other animals described in detail. The Hottentots' inclusion as the only humans in a book otherwise devoted to mammalian diversity suggests quite clearly Cuvier's ambivalence about monogeny and the separate creation of each species. Clearly, his religious belief system conflicted with his role in supporting European domination of more distant lands. Perhaps this internal conflict generated some of the urgency he felt about performing human dissections.

Other scientists of this period also linked human females with apes. While they differentiated white males from higher primates, using characteristics such as language, reason, and high culture, scholars used various forms of sexual anatomy—breasts, the presence of a hymen, the structure of the vaginal canal, and the placement of the urethral opening—to distinguish females from animals. Naturalists wrote that the breasts of female apes were flabby and pendulous—like those in the travelers' accounts of Hottentots (Schiebinger 1993). Cuvier's description of Sarah Bartmann repeats such "observations." The Hottentot worked as a double trope. As a woman of color, she served as a primitive primitive: she was both a female and a racial link to nature—two for the price of one.

Although Cuvier believed that the human races had probably developed separately for several thousand years, there were others, who we today call polygenists, who argued that the races were actually separate species (Stepan 1982). Presentations such as those in the *Natural History of the Mammals* provided fuel for the fire of polygeny. Cuvier's system of zoological classification, his focus on the nervous system, and his idea that species were created separately laid the foundations for the nineteenth-century concepts of race (Stocking 1982, 1990; Stepan 1982).

In Search of Sarah Bartmann

In contrast to what we know about her examiners, little about Bartmann is certain. What we do know comes from reading beneath the surface of newspaper reports, court proceedings, and scientific articles. We have nothing directly from her own hand. A historical record that has preserved a wealth of traces of the history of European men of science has left us only glimpses of the subjects they described. Hence, from the very outset, our knowledge of Sarah Bartmann is a construction, an effort to read between the lines of historical markings written from the viewpoint of a dominant culture. Even the most elementary information seems difficult to obtain. Cuvier wrote that she was twenty-six when they met and twenty-eight when she died, yet the inscription in the museum case that holds her body says that she was thirty-eight (Kirby 1949). She is said to have had two children by an African man, but de Blainville (1816) says that she had one child. One source says that the single child was dead by the time Bartmann arrived in Europe. According to some accounts, she was the daughter of a drover who had been killed by Bushmen. According to others, she was herself a Bushwoman (Altick 1978; Cuvier 1817). One London newspaper referred to her as "a Hottentot of a mixed race," while a twentieth-century writer wrote that he was "inclined to the view that she was a Bushwoman who possessed a certain proportion of alien blood" (Kirby 1949: 61).

Some sources state that Bartmann was taken in as a servant girl by a Boer family named Cezar. In 1810 Peter Cezar arranged to bring her to London, where he put her on exhibition in the Egyptian Hall of Picadilly Circus.[8] She appeared on a platform raised two feet off the ground. A "keeper" ordered her to walk, sit, and stand, and when she sometimes refused to obey him, he threatened her. The whole "performance" so horrified some that abolitionists brought

Cezar to court, charging that he held her in involuntary servitude. During the court hearing on November 24, 1810, the following claims emerged: The abolitionists charged that she was "clandestinely inveigled" from the Cape of Good Hope without the permission of the British governor, who was understood to be the guardian of the Hottentot nation "by reason of their general imbecile state" (Kirby 1953: 61). In his defense, her exhibitor presented a contractual agreement written in Dutch, possibly after the start of the court proceedings. In it Bartmann "agreed" (no mention is made of a signature, and I have not examined the original), in exchange for twelve guineas per year, to perform domestic duties for her master and to be viewed in public in England and Ireland "just as she was." The court did not issue a writ of habeas corpus because—according to secondhand accounts—Bartmann testified in Dutch that she was not sexually abused, that she came to London of her own free will in order to earn money, and that she liked London and even had two "black boys" to serve her, but that she would like some warmer clothes. Her exhibition continued and a year later, on December 7, 1811, she was baptized in Manchester, "Sarah Bartmann a female Hottentot of the Cape of Good Hope born on the Borders of Caffraria" (Kirby 1953: 61). At some point prior to 1814, she ended up in Paris, and in March of 1815 a panel of zoologists and physiologists examined her for three days in the Jardin du Roi. During this time an artist painted the nude that appears in Geoffroy Saint-Hilaire and Cuvier's tome (1824). In December of 1815 she died in Paris, apparently of smallpox, but helped along by a misdiagnosis of pleurisy and, according to Cuvier, by her own indulgence in strong drink.

Why was Bartmann's exhibition so popular? Prior to the nineteenth century there was a small population of people of color living in Great Britain. They included slaves, escaped slaves, and the children of freedmen sent to England for an education. Strikingly, the vast majority of the nonwhite population in England was male. Thus, even though people of color lived in England in 1800, a nonwhite female was an unusual sight (Walvin 1973). This, however, is an insufficient explanation. We must also place Bartmann's experiences in at least two other contexts: the London entertainment scene and the evolving belief systems about sex, gender, and sexuality.

The shows of London and those that traveled about the countryside were popular forms of amusement. They displayed talking pigs, animal monsters, and human oddities—the Fattest Man on Earth, the Living Skeleton, fire-eaters, midgets, and giants. Bartmann's exhibition exemplifies an early version of ethnographic displays that became more complex during the nineteenth century. After her show closed, "the Venus of South America" appeared next. Tono Maria, a Botocudo Indian from Brazil, publically displayed the scars (104 to be exact) she bore as punishment for adulterous acts. In time, the shows became more and more elaborate. In 1822 an entire grouping of Laplanders shown in the Egyptian Hall drew 58,000 visitors over a period of a few months. Then followed Eskimos and, subsequently, a "family grouping" of Zulus, all supposedly providing live demonstrations of their "native" behaviors. Such displays[9] may be seen as a living, nineteenth-century version of the early-twentieth-century museum diorama, the sort that riveted my attention in the American Museum of Natural History when I was a child. The dioramas, while supposedly providing scientifically accurate presentations of peoples of the world, instead offer a Euro-American vision of gender arrangements and the primitive that serves to set the supposedly "civilized" viewer apart, while at the same time offering the reassurance that women have always cooked and served, and men have always hunted (Haraway 1989).

Sometimes the shows of exotic people of color involved complete fabrication. A Zulu warrior might really be a black citizen of London, hired to play the part. One of the best documented examples of such "creativity" was the performer "Zip the What-is-it," hired and shown by P. T. Barnum. In one handbill, Zip was described as having been "captured by a party of adventurers while they were in search of the Gorilla. While exploring the river Gambia . . . they fell in with a race of beings never before discovered . . . in a PERFECTLY NUDE STATE, roving among the trees . . . in a manner common to the Monkey and the Orang Outang" (Lindfors 1983: 96). As it turns out, Zip was really William Henry Johnson, an African American from Bridgeport, Connecticut. He made what he found to be good money, and in exchange kept mum about his identity. Interviewed in 1926, at the age of 84, while still employed at Coney Island, he is reported to have said, "Well, we fooled 'em a long time, didn't we?" (Lindfors 1983: 98).

The London (and in fact European) show scene during the nineteenth century became a vehicle for creating visions of the nonwhite world.[10] As the century progressed, these visions "grew less representative of the African peoples they . . . were meant to portray. . . . Black Africa was presented as an exotic realm beyond the looking glass, a fantasy world populated by grotesque monsters—fat-arsed females, bloodthirsty warriors, pre-verbal pinheads, midgets and geeks" (Lindfors 1983: 100). From this vision Britain's "civilizing colonial mission" drew great strength. And it is also from this vision, this reflection of the other, that Europe's self-image derived; the presentation of the exotic requires a definition of the normal. It is this borderline between normal and abnormal that Bartmann's presentation helped to define for the Euro-American woman.

Bartmann's display linked the notion of the wild or savage female with one of dangerous or uncontrollable sexuality. At the "performance's" opening, she appeared caged, rocking back and forth to emphasize her supposedly wild and potentially dangerous nature. The *London Times* reported, "She is dressed in a colour as nearly resembling her skin as possible. The dress is contrived to exhibit the entire frame of her body, and spectators are even invited to examine the peculiarities of her form" (Kirby 1949: 58). One eyewitness recounted with horror the poking and pushing Bartmann endured, as people tried to see for themselves whether her buttocks were the real thing. Prurient interest in Bartmann became explicit in the rude street ballads and equally prurient cartoons that focused on her steatopygous backside.[11]

According to the *Oxford English Dictionary*, the term *steatopygia* (from the roots for fat and buttocks) was used as early as 1822 in a traveler's account of South Africa, but the observer said the "condition" was not characteristic of all Hottentots nor was it, for that matter, characteristic of any particular people. Later in the century, what had been essentially a curiosity found its way into medical textbooks as an abnormality. According to Gilman, by the middle of the nineteenth century the buttocks had become a clear symbol of female sexuality; and the intense interest in the backside, a displacement for fascination with the genitalia. Gilman concludes, "Female sexuality is linked to the image of the buttocks, and the quintessential buttocks are those of the hottentot" (Gilman 1985: 210).[12] Female sexuality may not have been the only thing at stake in all of the focus on Bartmann's backside. In this same historical period, a new sexual discourse on sodomy also developed. Male prostitutes, often dressed as women, walked the streets of London (Trumbach 1991), and certainly at a later date the enlarged buttocks became associated with female prostitution (Gilman 1985). Until more historical work is done, possible relationships between cultural constructions of the sodomitical body and those of the steatopygous African woman will remain a matter of speculation.

Bartmann's story does not end in England. Her presentation in Paris evoked a great stir as well. There was a lively market in prints showing her in full profile; crowds went to see her perform. And she became the subject of satirical cartoons filled with not particularly subtle sexual innuendo. The French male's sexual interest in the exotic even became part of a one-act vaudeville play in which the male protagonist declares that he will love only an exotic woman. His good, white, middle-class cousin, in love with him, but unable to attract his attention, disguises herself as the Hottentot Venus, with whom he falls in love, making the appropriate mating, even after the fraud is revealed. (The full story has many more twists and turns, but this is the "Cliff Notes plot" [Lindfors 1983: 100].)

Of all the retellings of Bartmann's story, only Gould's attempts to give some insight into Bartmann's own feelings. We can never see her except through the eyes of the white men who described her. From them we can glean the following:

first, for all her "savageness," she spoke English, Dutch, and a little French. Cuvier found her to have a lively, intelligent mind, an excellent memory, and a good ear for music. The question of her own complicity in and resistance to her exploitation is a very modern one. The evidence is scant. During her "performances" "she frequently heaved deep sighs; seemed anxious and uneasy; grew sullen, when she was ordered to play on some rude instrument of music" (Altick 1978: 270). Writing in the third person, de Blainville, who examined her in the Jardin du Roi, reported the following:

> Sarah appears good, sweet and timid, very easy to manage when one pleases her, cantankerous and stubborn in the contrary case. She appears to have a sense of modesty or at least we had a very difficult time convincing her to allow herself to be seen nude, and she scarcely wished to remove for even a moment the handkerchief with which she hid her organs of generation. . . . [H]er moods were very changeable; when one believed her to be tranquil and well-occupied with something, suddenly a desire to do something else would be born in her. Without being angry, she would easily strike someone. . . . [S]he took a dislike to M. de Blainville, probably because he came too near to her, and pestered her in order to obtain material for his description; although she loved money, she refused what he offered her in an effort to make her more docile. . . . She appeared to love to sleep: she preferred meat, especially chicken and rabbit, loved (alcoholic) spirits even more and didn't smoke, but chewed tobacco. (de Blainville 1816: 189)

In this passage, de Blainville expressed the same conflicts evinced two centuries earlier by Vespucci. He found her to be modest, good, sweet, and timid (like any modern, "civilized" Frenchwoman), but he could not reconcile this observation with what seemed to him to be the remnants of some irrational wildness (including habits such as chewing tobacco), which were out of line for any female he would wish to call civilized.

It is also worth comparing de Blainville's language to that used by Geoffroy Saint-Hilaire and F. Cuvier in the *Natural History of the Mammals*. In the section describing *Cynocephalus* monkeys (which follows immediately on the heels of Sarah Bartmann's description), they write that "one can see them pass in an instant from affection to hostility, from anger to love, from indifference to rage, without any apparent cause for their sudden changes" (Geoffroy Saint-Hilaire and Cuvier 1824: 2). They write further that the monkeys are "very lascivious, always disposed to couple, and very different from other animals, the females receive the males even after conception" (Geoffroy Saint-Hilaire and Cuvier 1824: 3). Clearly, de Blainville's language echoes through this passage framing the scientists' concerns about human animality and sexuality.

Constructing the (Nonwhite) Female

Although a theater attraction and the object of a legal dispute about slavery in England, it was in Paris, before and after her death, that Bartmann entered into the scientific accounting of race and gender. This part of the story takes us from Sarah's meeting with scientists in the Jardin du Roi to her death, preservation, and dissection by Georges Cuvier—and to other scientific and medical dissections of nonwhites in the period from 1815 to, at least, the 1870s.[13]

The printed version of de Blainville's report to the Société Philomatique de Paris (given orally in December of 1815 and appearing in the Society's proceedings in 1816) offers two purposes for the publication. The first is "a detailed comparison of this woman [Sarah Bartmann] with the lowest race of humans, the Negro race, and with the highest race of monkeys, the orangutan," and the second was to provide "the most complete account possible of the anomaly of her reproductive organs" (de Blainville 1816: 183). De Blainville accomplished his first purpose more completely than his second. On more than four occasions in this short paper he differentiates Bartmann from "Negroes," and throughout the article suggests the similarity of various body structures to those of the orangutan.

De Blainville began with an overall description of Sarah Bartmann's body shape and head. He then systematically described her cranium (one paragraph), her ears (two long, detailed paragraphs), her eyes (one paragraph), and other aspects of her face (five paragraphs, including one each devoted to her nose, teeth, and lips). In terms of printed space, her facial structure was the most important aspect. The final segment of his paper includes brief accounts (one paragraph each) of her neck, trunk, and breasts. In addition, he briefly described her legs, arms, and joints, devoting a full paragraph complete with measurements, to her steatopygous buttocks.

De Blainville's attempts to get a good look at her pudendum, especially at the "hottentot apron," which Cuvier finally succeeded in describing only after her death, were foiled by her modesty (see above). Despite this, de Blainville offers three full paragraphs of description. He verbally sketches the pubis, mentioning its sparse hair covering, and lamenting that, from a frontal view, one cannot see the vaginal labia majora, but that, when she leaned over or when one watched from behind as she walked, one could see hanging appendages that were probably the sought-after elongated labia minora.

De Blainville's ambivalences emerge clearly in the written text. He placed Bartmann among other females by reporting that she menstruated regularly, "like other women," but noted that she wasn't really like white women because her periodic flow "appear[ed] less abundant" (de Blainville 1816: 183). (Debates about menstruation from the turn of the eighteenth century considered menstruation a measure of full humanity; the heavier the flow, the higher one's place in nature [Schiebinger 1993].) Although the person who showed her in Paris claimed that she had a highly aggressive sexual appetite—one day even throwing herself on top of a man she desired—de Blainville doubted the truth of the specific incident. Not to have her too closely linked to European women, however, he also suggests that the modesty he observed might have resulted from her presence for some years among Europeans, conceding that, even after so many years, "it is possible that there still remained something of the original" (de Blainville 1816: 183). Finally, de Blainville suggests "that the extraordinary organization which this woman offers" (de Blainville 1816: 189) is probably natural to her race, rather than being pathological. In support of his contention, he cites travelers who found the same peculiarities—of jaws, buttocks, and labia—among "natives" living in their home environments. Hence, he finishes with the assertion of natural racial difference.

In de Blainville's text different parts of the body carried specific meanings. To compare the Negro and the orangutan, he spent paragraphs on detailed descriptions of the head, face, jaws, and lips. He used these to link Hottentots to orangs, writing that the general form of the head and the details of its various parts, taken together, make clear that Hottentots more closely resemble orangs than they do Negroes. He repeatedly invoked Pieter Camper's facial angle (Gould 1981; Russett 1989), the shape and placement of the jaws, and—in somewhat excruciating detail—the arrangement and structure of the ears. These passages evoke the tradition of physiognomy elaborated by Lavater (1775–78), whose work, widely translated into French and other languages, offered a basis for Gall's phrenology and a method of using the face to read the internal workings of animals. Of humans Lavater wrote:

> The intellectual life . . . would reside in the head and have the eye for its center . . . the forehead, to the eyebrows, [will] be a mirror . . . of the understandings; the nose and cheeks the image of the moral and sensitive life; the mouth and chin the image of the animal life. . . . (Graham 1979: 48)

When de Blainville and then Cuvier offered detailed comparisons between Sarah Bartmann's cheeks and nose and those of Caucasians, they set forth more than a set of dry descriptions. Her "moral and sensitive life" lay evident upon the surface of her face.[14]

It is to the description of the genitalia that de Blainville turns to place Bartmann among women. Here he balances his belief in the civilizing effects of Europe against a scarcely hidden savage libido. The gender norms of white women appear as a backdrop for the consideration of "savage" sexuality. Although he gave detailed descriptions of most of her exterior, de Blainville did not succeed in fully examining Bartmann's genitalia. Where he failed on the living woman, Cuvier suceeded after her death. Clearly a full account of this "primitive woman's" genitalia was essential to putting her finally in her appropriate place. By exposing them to what passed for scientific scrutiny, Cuvier provided the means to control the previously uncontrollable. Triumphantly, he opened his presentation to the French Academy with the following: "There is nothing more celebrated in natural history than the Hottentot apron, and at the same time there is nothing which has been the object of such great argumentation" (Cuvier 1817: 259). Cuvier set the stage to settle the arguments once and for all.

Twentieth-century scientific reports open with an introduction that uses previously published journal articles to provide background and justification for the report to follow. In Cuvier's piece we see the transition to this modern format from an older, more anecdotal style. Rather than relying on official scientific publications, however, Cuvier relied on travelers' accounts of the apron and the steatopygia. In later works, although these anecdotal, eyewitness testimonials fade from sight, they remain the source for knowledge incorporated into a more "objective" scientific literature. (Sexologists William Masters and Virginia E. Johnson, for example, in their scientifically dispassionate work on the *Human Sexual Response,* include a claim that African women elongate their vaginal labia by physical manipulation; their cited source is a decidedly unscientific (by modern standards) compendium of female physical oddities that dates from the 1930s but draws on nineteenth-century literature of the sort discussed here. [Masters and Johnson 1966: 58].)

To set the stage for his revelations about the Hottentot apron, Cuvier first needed to provide a racial identity for his cadaver (which he referred to throughout the article as "my Bushwoman"). Travelers' accounts indicated that Bushmen were a people who lived much deeper in "the interior of lands" than did Hottentots. The apron and enlarged buttocks were peculiarly theirs, disappearing when they interbred with true Hottentots. Cuvier believed that the confusion between Bushmen and Hottentots explained the inconsistent nature of travelers' reports, since some voyagers to the Cape of Good Hope claimed sightings of the Hottentot apron, while others did not. Nevertheless, he had to admit that many people did not believe in the existence of a Bushman nation. Cuvier threw his weight behind what he believed to be the accumulation of evidence: that there existed "beings almost entirely savage who infested certain parts of the Cape colony . . . who built a sort of nest in the tufts of the brush; they originated from a race from the interior of Africa and were equally distinct from the

Kaffir and the Hottentot" (Cuvier 1817: 261). Cuvier believed that the Bushman social structure had degenerated, so that eventually "they knew neither government nor proprieties; they scarcely organized themselves into families and then only when passion excited them. . . . They subsisted only by robbery and hunting, lived only in caves and covered their bodies with the skins of animals they had killed" (Cuvier 1817: 261). By naming Bartmann as a Bushwoman, Cuvier created her as the most primitive of all humans—a female exemplar of a degenerate, barely human race. Despite his lack of belief in evolution, he constructed her as the missing link between humans and apes.

To the modern reader, several noteworthy aspects emerge from these introductory passages. First, Cuvier melds the vision of an interior or hidden Africa with the hidden or interior genitalia of the Hottentot Venus. This becomes even clearer in subsequent passages in which, like de Blainville, he complains that when he examined her as a living nude in the Jardin du Roi in 1815 she "carefully hid her apron either between her thighs or more deeply" (Cuvier 1817: 265). Second, he connected a hidden (and hypothetical) people from the deep African interior with an animal-like primitiveness. The passage about making nests from brush tufts evokes monkey and ape behaviors (chimps sleep each night in nests they weave from tree branches). Cuvier's goal in this paper was to render visible the hidden African nations and the hidden genitalia. By exposing them he hoped to disempower, to use observation to bring these unknown elements under scientific control. In the remainder of the account, Cuvier devoted himself simultaneously to the tasks of racial and sexual localization. Where among humans did these interior people belong, and what did their women conceal in their body cavities?

In his presentation to the members of the Museum of Natural History, Cuvier moved from a description of the exterior, living, and never quite controllable Bartmann (for he needed her permission to examine her hidden parts) to the compliant cadaver laid out before him, now unable at last to resist his deepest probings. In both life and death Sarah Bartmann was a vessel of contradictions. He found that her "sudden and capricious" movements resembled those of a monkey, while her lips protruded like those of an orangutan. Yet he noted that she spoke several languages, had a good ear for music, and possessed a good memory. Nevertheless, Cuvier's vision of the savage emerged: belts and necklaces of glass beads "and other savage attires" pleased her, but more than anything she had developed an insatiable taste for "l'eau-de-vie" (Cuvier 1817: 263).

For fully one-fifth of the paper we read of her exterior. Cuvier paints what he clearly found to be a picture gruesome in its contradictory aspects. Only four and a half feet tall, she had enormous hips and buttocks, but otherwise normal body parts. Her shoulders and back were graceful, the protrusion of her chest not excessive, her arms slender and well made, her hands charming, and her feet pretty. But her physiognomy—her face—repelled him. In the jutting of the jaw, the oblique angle of her incisors, and the shortness of her chin, she looked like a Negro. In the enormity of her cheeks, the flatness of the base of her nose, and her narrow eye slits, she resembled a Mongol. Her ears, he felt, resembled those of several different kinds of monkeys. When finally, in the spring of 1815, she agreed to pose nude for a painting, Cuvier reported the truth of the stories about the enormity of her protruding buttocks and breasts—enormous hanging masses[15]—and her barely pilous pubis.

When she died, on December 29, 1815, the police prefect gave Cuvier permission to take the body to the museum, where his first task became to find and describe her hidden vaginal appendages. For a page and a half the reader learns of the appearance, folded and unfolded, of the vaginal lips, of their angle of joining, the measurements of their length (more than four inches—although Blumenbach reportedly had drawings of others whose apron extended for up to eight inches) and thickness, and the manner in which they cover the vulval opening. These he compared to analogous parts in European women, pointing out the considerable variation and stating that in general the inner vaginal lips are more developed in women from warmer climates. The variation in vaginal development had, indeed, been recognized by French anatomists, but a mere ten years earlier, medical writers failed to connect differences in vaginal structures to either southern races or nonwhite women. In a straightforward account of "over-development" of vaginal lips, Dr. M. Baillie, a British physician and member of the Royal Society of Medicine of London (whose book was translated into French 1807), wrote matter-of-factly of this variation, listing it among a number of genital anomalies, but not connected to non-European women (Baillie 1807). As Gilman (1985) points out, however, by the middle of the nineteenth century elongated labia had taken their place in medical textbooks alongside accounts of enlarged clitorises, both described as genital abnormalities, rather than as part of a wide range of "normal" human variation.

Cuvier acknowledged the great variation in length of the inner vaginal lips found even among European women. But nothing, he felt, compared to those of "negresses" and "abyssynians," whose lips grew so large that they became uncomfortable, obliging their destruction by an operation carried out on young girls at about the same age that Abyssinian boys were circumcised. As an aside that served to establish a norm for vaginal structure and a warning to those whose bodies did not conform, we learn that the Portuguese Jesuits tried in the sixteenth century to outlaw this practice, believing that it was a holdover from ancient Judaism. But the now Catholic girls could no longer find husbands because the men wouldn't put up with such "a disgusting deformity" (Cuvier

1817: 267), and finally, with the authorization of the Pope, a permission was made possible by a surgeon's verification that the elongated lips were natural rather than the result of manipulation, and the ancient custom resumed.

Cuvier contrasts the vaginal lips of Bushwomen with those of monkeys, the near invisibility of which provided no evidence to link them to these primitive humans. But the steatopygia was another matter. Bartmann's buttocks, Cuvier believed, bore a striking resemblance to the genital swellings of female mandrills and baboons, which grow to "monstrous proportions" at certain times in their lives. Cuvier wanted to know whether the pelvic bone had developed any peculiar structures as a result of carting around such a heavy load. To answer the question, he made use of his well-established method of comparative anatomy, placing side by side the pelvises of "his bushwoman", those of "negresses," and those of different white women. In considering Bartmann's small overall size, Cuvier found her pelvis to be proportionally smaller and less flared, the anterior ridge of one of the bones thicker and more curved in back, and the ischial symphysis thicker. "All these characters, in an almost unnoticeable fashion, resemble one another in Negro women, and female Bushwomen and monkeys" (Cuvier 1817: 269). Just as the differences themselves were practically imperceptible, amidst a welter of measurement and description, Cuvier imperceptibly separated the tamed and manageable European woman from the wild and previously unknown African.

But something worried Cuvier. In his collection he had also a skeleton of a woman from the Canary Islands. She came from a group called the Guanche (extinct since shortly after the Spanish settlement), a people who inhabited the islands before the Spanish and who, by all accounts, were Caucasians. An astonished Cuvier reported to his colleagues that he found the most marked of Bartmann's characters not in the skeleton of Negro women but in that of the Canary Islander. Since he had too few complete skeletons to assess the reliability of these similarities, he turned finally to more abundant material. In the last part of his account, he compares the head and skull (which "one has always used to classify nations" [Cuvier 1817: 270]) of "our Bushwoman" with those of others in his collection.

Bartmann's skull, he wrote, mixed together the features of the Negro and the Mongol, but, chiefly, Cuvier declared that he "had never seen a human head more similar to those of monkeys" (Cuvier 1817: 271). After offering more detailed comparisons of various bones in the skull, Cuvier returned in the last few pages of his paper to the problem that concerned him at the outset—did the Bushmen really exist as a legitimate people, and just how far into the interior of Africa did they extend? Here he relied once more on travelers' reports. Although modern voyagers did not report such people in northern Africa, Herodotus and others described a group that seemed in stature and skin color to resemble the Bushmen. According to some sources, these people invaded Abyssinia, although the evidence in Cuvier's view was too prescientific to rely on. But he could be sure of one thing: Neither the

> Bushmen, nor any race of Negros, gave birth to the celebrated people who established civilization in ancient Egypt and from whom one could say that the entire world had inherited the principles of law, science and perhaps even religion. (Cuvier 1817: 273)

At least one modern author suggested that the ancient Egyptians were Negroes with wooly hair, but Cuvier could be sure that this, too, was in error. All he needed to do was compare the skulls of ancient Egyptians with those of the pretender races. One can picture him, as he spoke, dramatically producing from beneath his dissecting table the skulls of Egyptian mummies, those very same ones brought back by Geoffroy Saint-Hilaire from the Napoleonic incursion into Egypt.

Cuvier studied the skulls of more than fifty mummies. These, he pointed out, had the same skin color and large cranial capacity as modern Europeans. They provided further evidence for "that cruel law that seems to have condemned to eternal inferiority those races with depressed and compressed crania" (Cuvier 1817: 273). And finally, he presented to his museum colleagues the skull of the Canary Islander whose skeleton had so troublingly resembled Bartmann's. This too "announced a Caucasian origin" (Cuvier 1817: 274), which is the phrase that concludes his report. In this last section of his paper we watch him struggle with his data. First, he realized that he had a Caucasian skeleton that looked identical to Bartmann's. If he could not explain this away (what modern scientists call eliminating outliers—data points that don't neatly fit an expected graph line), his thesis that Bushmen represented a primitive form of humanity was in trouble. But that wasn't all that worried him: If his thesis was in trouble, so too was the claim of European superiority on which European and American colonization, enslavement, and disenfranchisement so depended. Thus, he went to considerable trouble to explain away the Guanche skeleton; ultimately he succeeded by using the scientific spoils of colonial expansion—the Egyptian mummies captured during Napoleon's Egyptian campaign.

Conclusion

This chapter places the scientific study of nonwhite women in several contexts. The investigations were, to be sure, part of the history of biology and, especially, a component of the movement to catalogue and classify all the living creatures of the earth. But this movement was in turn embedded in the process of European capitalist expansion. Not only did traders and conquerers, by collecting from around the world, create the need for a classification project, they also required the project to justify continued expansion, colonialism, and slavery. Further entangling the matter, the vast capital used to build the museums and house the collections came from the economic exploitation of non-European goods—both human and otherwise. This entire essay has been an argument against a narrowly constructed historiography of science; instead, I more broadly socialize the history of Euro-American biology in the first quarter of the nineteenth century by exposing its intersections with gender, race, and nation.

If one looks at the process less globally, one sees Cuvier and de Blainville as significant actors in a period of scientific change. From the perspective of the history of Euro-American biology, parochially extracted from its role in world expansion, one can say that the biologists of this period, and Cuvier in particular, made enormous scientific progress with the "discovery of the great information content of the internal anatomy of the invertebrates" (Mayr 1982: 183). According to this view, Cuvier "discovered" the importance of the nervous system as a way to organize animals. But "Cuvier's vision of the animal world was deeply coloured by that of the human society in which he was forced to make his way" (Outram 1984: 65). Far from reflecting some underlying natural system, Cuvier's use of the nervous system in his classification schemes had a homocentric starting point. The ideas formed a meshwork. Cuvier gave the focus on the nervous system and brain (obtained from his conviction that classification should proceed from the most complex—in this case human—structure to the simplest) the status of scientific fact by developing a reasonably coherent story about how the structure of the nervous system enabled him to classify all animals. Once scientists agreed on the validity of Cuvier's animal classification scheme, it fed back on the question of human classification. It seemed only "natural" to focus on the structure of the brain (as reflected in cranial and facial characteristics) to obtain evidence about the relative standing of the human races.

Sarah Bartmann's story is shocking to modern sensibilities. The racism of the period seems obvious—even laughable. But in the rush to create distance between nineteenth-century racist science and our modern, putatively less racist selves, even highly sophisticated scholars often lose sight of an important point. The loss becomes evident when I am asked (as I frequently am) what the *real* truth about Bartmann was. Just how big were those forbidden parts? The question reflects an ongoing belief in the possibility of an objective science. It suggests that, now that we have escaped all that silly racism of the nineteenth century, we ought to be able to get out our measuring tapes and find the real truth about other people's bodies. In this essay I argue that Bartmann's bodily differences were constructed using the social and scientific paradigms available at the time. The historical record tells us nothing about her agency; we can only know how Europeans framed and read her. Were she somehow magically alive today, contemporary biologists or anthropologists might frame and read her differently, but it would be a framing and reading, nevertheless. One contemporary difference might be that the varying worldwide liberation movements could offer her a context in which to contest the constructions of Euro-American science. In fact we see such contestations regularly in debates over such questions as brain size, race, and IQ (Maddock 1992; Schluter and Lynn 1992; Becker, Rushton, and Ankney 1992), brain shape and gender, and genetics and homosexuality (Fausto-Sterling 1992).

In *Playing in the Dark,* Toni Morrison (1992) makes her intellectual project "an effort to avert the critical gaze from the racial object to the racial subject; from the described and imagined to the describers and imaginers. . . . " (Morrison 1992: 90). By analogy I look at the fears and anxieties of the scientists, rather than worrying about the (in)accuracies of their descriptions of Sarah Bartmann and other people of color. To quote further from Morrison:

> The fabrication of an Africanist persona is reflexive; an extraordinary meditation on the self; a powerful exploration of the fears and desires that reside in the writerly conscious. It is an astonishing revelation of longing, of terror, of perplexity, of shame, of magnaminity. It requires hard work NOT to see this. (Morrison 1992: 17)

For our purposes we need only substitute the word "scientific" for the word "writerly." What can we glean of the fears, desires, longings, and terrors that perfuse the works we've just considered? And how are race, gender, and nationality woven into the story? In the accompanying chart I have listed some of the paired contradictions that emerge from my reading of Cuvier and de Blainville.

The simultaneous anxiety about European women and the savage Other is especially clear in de Blainville's account. He identified Bartmann as a woman because she menstruated. But she also drank, smoked, and was alleged to be sexually aggressive—all masculine characteristics. And if Bartmann, a woman, could behave thus, why not French women? Furthermore, the soap opera dramas about Bartmann that played in contemporary Paris suggested that French men, despite their "civilization," actually desired such women; civilization kept the European woman under control, decreasing the danger of rebellion, but thwarting male desire. Minute scientific observation converted the desire into a form of voyeurism, while at the same time confining it to a socially acceptable location.

conquest	resistance
human	animal
surface	interior
tame	wild
sexually modest	libidinous
civilized	savage
compliant	angry
ruler	subject
powerlessness	hidden power
male	female
white	nonwhite
colonizer	colonized

Cuvier most clearly concerned himself with establishing the priority of European nationhood; he wished to control the hidden secrets of Africa and the woman by exposing them to scientific daylight. The French Revolution had frightened him, and certainly the prospect of resistance from other peoples must have seemed terrifying (Outram 1984; Appel 1987). Hence, he delved beneath the surface, bringing the interior to light; he extracted the hidden genitalia and defined the hidden Hottentot. Lying on his dissection table, the wild Bartmann became tame, the savage civilized. By exposing the clandestine power, the ruler prevailed. But one need only look at the list of anxieties glossed from the scientific literature to know how uneasy lay the head that wore a crown.[16]

Notes

Acknowledgments: This paper was written with the financial support of the National Science Foundation, Fellowship #DIR-9112556 from the Program in History and Philosophy of Science. I would like to thank Evelynn Hammonds, Joan Richards, Gregg Mitman, and Londa Schiebinger, as well as the editors of this volume, for reading and commenting on recent drafts of this paper. Londa Schiebinger also kindly shared with me drafts of chapters of her book *Nature's Body: Gender in the Making of Modern Science* (Beacon 1993).

1. In 1992 the Musée de l'Homme had removed the remnants of the Bartmann exhibit. In its place was a modern one entitled "All relatives, all different," celebrating human genetic diversity. Discussion of Bartmann could still be found in a part of the exhibit devoted to the history of scientific racism.

2. There is also a book of poetry featuring the Venus Hottentot in the title poem: Elizabeth Alexander, *The Venus Hottentot* (Charlottesville: University Press of Virginia), 1990.

3. The Dutch word for Bushman is *bosjeman,* which translates as "little man of the forest." This is also the translated meaning of the Malay word *orangutan.*

4. This is in perfect accord with Latour's account of how scientific knowledge is constructed.

5. All translations from works cited in the original are mine.

6. In fact, de Blainville's break with Cuvier came over just this question. He devised a different classificatory system based on external, rather than internal characters, but he linked his divisions by creating intermediate groupings.

7. The question of the racial origins of European thought has been raised in our own era by the work of Martin Bernal (1987).

8. The detailed ins and outs of her sale and repurchase may be found in the references in note 11.

9. In contrast to the family groupings of Laps, Eskimos, and Zulus, the displays of Bartmann, Tono Maria, and Zip made no attempt to present a working culture.

10. Nonwhites were not the only "others" constructed. I plan to address the use of "freaks" in the construction of the Other in a book-length account of the construction of race and gender by biologists, anthropologists, and sociologists.

11. All the details cited here may be found in Altick (1978), Edwards and Walvin (1983), Gould (1985), Kirby (1949, 1953), and Lindfors (1983). Remarkably, prurient interest in the figure of the Hottentot continues to this day. Gould (1985) discusses a 1982 cover of the French magazine *Photo* that features a naked woman named "Carolina, La Vénus hottentote de Saint-Domingue." In the copy of the Geoffroy Saint-Hilaire and Cuvier held by the Brown Library, the frontal drawing of Bartmann (which exhibited her breasts in full form) has been razored out. The mutilation was first noticed by librarians in 1968. This is not the first time I have encountered such mutilation of material of this sort.

12. Although the bustle was not invented until 1869, various fashions in the eighteenth and nineteenth centuries accentuated the backside of middle- and upper-class white women (Batterberry and Batterberry 1977). The relationship between these fashions and scientific accounts of the body has yet to be detailed.

13. There were at least seven articles, falling into three chronological groupings, published in scientific journals in England, France, and Germany. The first two, by Henri de Blainville and Georges Cuvier, exclusively on Sarah Bartmann, were published in 1816 and 1817, respectively. The second group, containing two by German biologists, appeared in the 1830s. The first of these was written by Johannes Müller (1801–58) (Müller 1834), a physiologist and comparative embryologist, while the second, written by Frederick Tiedemann (1781–1861) (Tiedemann 1836), Professor of Anatomy and Physiology at the University of Heidelberg and Foreign Member of the Royal Society of London, appeared in 1836. Müller's article is about a Hottentot woman who died in Germany, and is in the same scientific style as the French papers. Tiedemann's work, on the other hand, represents a scientific departure. Although Bartmann's is among a wide variety of brains obtained from museum collections, it is not the focus of the article. From a scientific point of view, Tiedemann's study represents a transition from a period in which scientists offered detailed examinations of the outside of the body, while focusing on a single individual and describing all body parts. Tiedemann awarded priority to one organ—the brain. A comparison of the brains of Europeans, Negroes, and orangutans convinced him

that there was no difference among the humans. He used his results to condemn the practice of slavery. His method, though, is primitive compared to the approach of the scientists working in the 1860s (Marshall 1864; Flower and Murie 1867), whose work provides a useful contrast to the changing scientific and political times. In this paper I will consider the first two exemplars, reserving detailed examination of the other works for a future occasion.

14. Outram (1984) documents Cuvier's dispute with Franz Joseph Gall over the scientific nature of phrenology. But Cuvier clearly believed in the principle that the face could be read for deeper meaning.

15. In the seventeenth century, breasts—as natural and social objects—had undergone a transformation, as male social commentators launched a successful campaign to do away with wetnursing and reestablish the breast as an object that connected women to nature through the act of nursing. For middle- and upper-class white women, doing the right thing with the right kind of breasts hooked them into a growing cult of domesticity, which exploded as the nineteenth-century ideal for gender relationships for the middle and upper classes in Europe and America. This naturalization of motherhood worked hand in glove with the desexualization of white women (Schiebinger 1993; Perry 1991). Perry cites Thomas Laqueur (1986) as explaining "this cultural reconsideration of the nature of women's sexuality as part of a process . . . committed to sweeping clean all *socially* determined differences among people" (Perry 1991: 212), instead relocalizing difference in the biological body. No part of the body escaped unscathed from this process.

16. In one of the lovely ironies of history, Cuvier himself was dissected when he died (in 1832), and his brain and head measurements were taken. In a ranking of 115 men of note, Cuvier's brain weight came in third (Turgenev's was first). The French as a group ranked behind Americans and the British. The author of this 1908 paper concluded that "the brains of men devoted to the higher intellectual occupations, such as the mathematical sciences . . . [or] those of men who have devised original lines of research [Cuvier] and those of forceful characters, like Ben Butler and Daniel Webster, are generally heavier still. The results are fully in accord with biological truths" (Spitzka 1908: 215). In a second, larger sample, Spitzka included four women—mathematician Sonya Kovaleskaya, physician Caroline Winslow, actress Marie Bittner and educator and orator Madame Leblais—who ranked 134th–137th, in brain weight.

References

Altick, Richard D.
1978 *The Shows of London.* Cambridge: The Belknap Press of Harvard University.

Appel, Toby A.
1987 *The Cuvier-Geoffroy Debate: French Biology in the Decades before Darwin.* Oxford: Oxford University Press.

Baillie, Mathieu
1807 *Anatomie pathologique des organes les plus importans du corps humain.* Paris: Crochard.

Batterberry, Michael, and Ariane Batterberry
1977 *Mirror Mirror: A Social History of Fashion.* New York: Holt, Rinehart and Winston.

Becker, Brent A., J. Philippe Rushton, and C. Davison Ankney
1992 "Differences in Brain Size," *Nature* 358: 532.

Bernal, Martin
1987–91 *Black Athena: The Afroasiatic Roots of Classical Civilization,* vols. 1 and 2. New Brunswick, N.J.: Rutgers University.

Bucher, Bernadette
1981 *Icon and Conquest: A Structural Analysis of the Illustrations of de Bry's* Great Voyages. Trans. Basia Miller Gulati. Chicago: University of Chicago Press.

Coleman, William
1964 *Georges Cuvier, Zoologist: A Study in the History of Evolution Theory.* Cambridge: Harvard University Press.

Cuvier, Georges
1817 "Faites sur le cadavre d'une femme connue à Paris et à Londres sous le nom de Vénus Hottentotte." *Mémoires du Musée nationale d'histoire naturelle* 3: 259–74.

de Blainville, Henri
1816 "Sur une femme de la race hottentote." *Bulletin du Société philomatique de Paris,* pp. 183–90.

Edwards, Paul, and James Walvin
1983 *Black Personalities in the Era of the Slave Trade.* Baton Rouge: Louisiana State University Press.

Elphick, Richard
1977 *Kraal and Castle: Khoikhoi and the Founding of White South Africa.* New Haven: Yale University Press.

Fausto-Sterling, Anne
1992 *Myths of Gender: Biological Theories about Women and Men.* 2d ed. New York: Basic Books.

Figlio, Karl M.
1976 "The Metaphor of Organization: An Historiographical Perspective on the Bio-Medical Sciences of the Early Nineteenth Century." *History of Science* 14: 17–53.

Flourens, P.
1845 *Cuvier. Histoire de ses travaux.* 2d ed. rev. and corr. Paris: Paulin.

Flower, W. H., and James Murie
1867 "Account of the Dissection of a Bushwoman." *Journal of Anatomy and Physiology* 1: 189–208.

Geoffroy Saint-Hilaire, Etienne, and Fredéric Cuvier
1824 *Histoire Naturelle des Mammifères,* vols. 1 and 2. Paris: A. Belin.

Gilman, Sander L.
1985 "Black Bodies, White Bodies: Toward an Iconography of Female Sexuality in Late 19th-Century Art, Medicine and Literature." *Critical Inquiry* 12: 204–42.

Gould, Stephen Jay
1985 "The Hottentot Venus." In Stephen Jay Gould, *The Flamingo's Smile: Reflections in Natural History*, pp. 291–305. New York: Norton.
1981 *The Mismeasure of Man.* New York: Norton.

Graham, John
1979 *Lavater's Essays on Physiognomy: A Study in the History of Ideas.* Berne: Peter Lang.

Gray, Stephen
1979 *Southern African Literature: An Introduction.* New York: Barnes and Noble.

Guenther, Mathias Georg
1980 "From 'Brutal Savage' to 'Harmless People': Notes on the Changing Western Image of the Bushmen." *Paideuma* 26: 124–40.

Haraway, Donna
1989 *Primate Visions: Gender, Race, and Nature in the World of Modern Science.* New York: Routledge.

James, John Angell
1830 *Memoir of Clementine Cuvier, Daughter of Baron Cuvier.* New York: American Tract Society.

Kirby, Percival R.
1949 "The Hottentot Venus." *Africana News and Notes* 6: 55–62.
1953 "More about the Hottentot Venus." *Africana News and Notes* 10: 124–34.

Laqueur, Thomas
1986 "Orgasm, Generation, and the Politics of Reproductive Biology." *Representations* 14: 1–41.

Latour, Bruno
1987 *Science in Action: How to Follow Scientists and Engineers through Society.* Milton Keynes: Open University Press.

Lavater, J. C.
1775–78 *Physiognomische Fragmente zur Beförderung der Menschenkenntnis und Menschenliebe.* Leipzig: Weidmanns Erben und Reiche, H. Steiner und Companie.

Lee, Richard B.
1992 "Art, Science, or Politics? The Crisis in Hunter-Gatherer Studies." *American Anthropologist* 94(1): 31–54.

Lewin, Roger
1988 "New Views Emerge on Hunters and Gatherers." *Science* 240: 1146–48.

Lindfors, Bernth
1983 "The Hottentot Venus and Other African Attractions in Nineteenth-Century England." *Australasian Drama Studies* 1: 83–104.

Linnaeus (Carl von Linne)
1758 *Caroli Linnaei Systema Naturae. Regnum Animale.* 10th ed. Stockholm.

Maddock, John
1992 "How to Publish the Unpalatable?" *Nature* 358: 187.

Marshall, John
1864 "On the Brain of a Bushwoman; and on the Brains of Two Idiots of European Descent." *Philosophical Transactions of the Royal Society of London*, pp. 501–58.

Masters, William H., and Virginia E. Johnson
1966 *Human Sexual Response.* Boston: Little, Brown.

Mayr, Ernst
1982 *The Growth of Biological Thought: Diversity, Evolution, and Inheritance.* Cambridge: Belknap Press of Harvard University.

Merchant, Carolyn
1980 *The Death of Nature: Women, Ecology, and the Scientific Revolution.* San Francisco: Harper and Row.

Morrison, Toni
1992 *Playing in the Dark: Whiteness and the Literary Imagination.* Cambridge: Harvard University Press.

Müller, Johannes
1834 "Ueber die äusseren Geslechtstheile der Buschmänninnen." *Archiv für Anatomie, Physiologie und Wissenschaftliche Medicin*, pp. 319–45.

Outram, Dorinda
1984 *Georges Cuvier: Vocation, Science, and Authority in Post-revolutionary France.* Manchester: Manchester University Press.

Perry, Ruth
1991 "Colonizing the Breast: Sexuality and Maternity in Eighteenth-Century England." *Journal of the History of Sexuality* 2: 204–34.

Raven-Hart, Rowland
1967 *Before Van Riebeeck: Callers at South Africa from 1488 to 1652.* Cape Town: C. Struik.

Russett, Cynthia Eagle
1989 *Sexual Science: The Victorian Construction of Womanhood.* Cambridge: Harvard University Press.

Schiebinger, Londa
1993 *Nature's Body: Gender in the Making of Modern Science.* Boston: Beacon Press.

Schluter, Dolph, and Richard Lynn
1992 "Brain Size Differences." *Nature* 359: 181.

Shohat, Ella
1991 "Imaging Terra Incognita: The Disciplinary Gaze of the Empire." *Public Culture* 3(2): 41–70.

Spitzka, Edward Anthony
1908 "A Study of the Brains of Six Eminent Scientists and Scholars Belonging to the American Anthropometric Society, together with a Description of the Skull of Professor E. D. Cope." *American Philosophical Society Transactions* 21: 175–308.

Stepan, Nancy
1982 *The Idea of Race in Science: Great Britain, 1800–1960.* Hamden, Conn.: Archon.

Stocking, George W., Jr.
1982 *Race, Culture, and Evolution: Essays in the History of Anthropology.* Chicago: University of Chicago.
1987 *Victorian Anthropology.* New York: Free Press.

Tiedemann, Frederick
1836 "On the Brain of a Negro, Compared with That of the European and the Orang-outang." *Philosophical Transactions of the Royal Society of London* pp. 497–558.

Tiffany, Sharon W., and Kathleen J. Adams
1985 *The Wild Woman: An Inquiry into the Anthropology of an Idea.* Cambridge, Mass.: Schenkman.

Trumbach, Randolf
1991 "Sex, Gender and Sexual Identity in Modern Culture: Male Sodomy and Female Prostitution in Enlightenment London." *Journal of the History of Sexuality* 2: 187–203.

Walvin, James
1973 *Black and White; The Negro and English Society, 1555–1945.* London: Allen Lane and Penguin.

2

Framed

The Deaf in the Harem

Nicholas Mirzoeff

THE HAREM HAS been a remarkably dense site of Otherness for Western culture since the sixteenth century. In the sultan's harem in Istanbul, the Western gaze found an array of sexually available women, eunuchs (who were often black), dwarves, same-sex encounters by both men and women—and the deaf. By definition, the harem had not been seen by those who were re-creating it. In Arabic, the word *harim* means sacred, inviolable place and, by extension, the women of the household. No traveler could describe the sultan's harem from his or her own experience, as Corneille Le Bruyn attested: "It may be said that not a Man has ever seen the Face of one of the Sultanas belonging to the Seraglio during her stay there."[1] Even Lady Mary Montagu was able to join the harem women only in the baths and did not see the harem itself. The harem descriptions we have were based on private households, usually in North Africa, rather than the imperial harem in Istanbul itself. Nonetheless, many travelers and historians held very precise views on the subject, for, as Ali Behad has argued, the harem was for the Occident "nothing but a phantasm, a purely fictional construction onto which Europe's own sexual repressions, erotic fantasies, and desire of domination were projected."[2] In this sense, the harem had ever existed in Western eyes only as a phantasmic space, a place of deviancy and pathology, with uses both at home and abroad.

Deaf servants were privileged by the sultans, and their sign language became used by the entire court. However, in twentieth-century reconstructions of the harem, the deaf have disappeared. Previously, the deaf were included in travelers' accounts, histories of the Ottoman Empire, literature, and art depicting this zone of the forbidden. This chapter examines how these reports were transformed from being one of many exotic details in the harem into a defining framework for deafness itself, involving the sign language of the deaf, sexual deviance, race, and gender. Through the metonym of the harem, deafness became visible as a pathological symptom in nineteenth-century France. The medical elusiveness of deafness, combined with the subtle cultural politics of the deaf themselves, required this seemingly extravagant maneuver to make deafness different. By the late nineteenth century, however, the systematic application first of anthropology and then of medico-psychology to deafness and the deaf rendered such strategies unnecessary, and the deaf disappeared from public debate and, incidentally, from the myth of the harem.

Although a full account requires an investigation of the institutionalization of phonocentrism—that is, the assumption that human intelligence is fully manifested only through the voice and that any other signifying system, such as writing or sign language, is both secondary and inferior—my focus in this chapter is on the centrality of the visual.[3] For if phonocentrism mandated that the deaf be considered pathological, pathology demanded a visible symptom of the disease. The normative and disciplinary regimes described by Michel Foucault as controlling the boundaries of the normal and the pathological were above all scopic and spatial regimes. In the Panopticon and other such institutions, the inmates were controlled by the surveying gaze, not by linguistic oppositions, and, as Foucault observed, "visibility is a trap." To be seen was in certain senses to be controlled, and, in the Panopticon, "one is totally seen, without ever seeing; in the central tower one sees everything without ever being seen."[4] Within the frame of that gaze, a range of different registers of difference could be accommodated. In certain cases, such as the presumed opposition between white and black, the sign of difference was clearly visible, but deafness was, by its nature, difficult to see. It could be made visible within one of its best-known environments, the harem. Although by no means the sole register of the difference of deafness, the harem constituted one space in which its multiple signs of difference and deviance could be visualized. Initially, they were condemned by association and propinquity, until the more "rational" schemes of anthropology and degeneracy were able to account for the connection. In order to unravel the construction of such complex moments, disciplinary affiliations, whether from art history, the history of medicine, anthropology, or literature, must be set aside, or these nuances will be missed, leaving us with seemingly inexplicable prejudice and arbitrariness.

The Deaf in the Harem

Let's begin with an anecdote. Early in his career, the celebrated mime Jean-Baptiste Deburau (1796–1846) visited the sultan's court in Istanbul as a high-wire performer in a circus troupe.[5] The visitors performed for the women of the harem but were separated from them by a gauze curtain that prevented the circus members from seeing their audience. But Deburau's act involved a somersault on the high wire, and, in the course of his leap, he was able to catch a glimpse of the hidden women of the harem. Deburau never spoke of what he saw, for his reputation as a mime depended on his voluntary silence.[6] This image of the mute masculine gaze, with literally no ground beneath its feet, glimpsing the unseen while upside-down in mid-air, epitomizes the allure and inaccessibility of the harem, not to mention its power to disrupt. In untangling this intricate moment, I shall consider, firstly, the Western awareness of the deaf in the harem since the sixteenth century and the paradoxical framing of deafness as a visual phenomenon. Secondly, I will discuss how, in the nineteenth century, the new distinction between the normal and the pathological transformed both deafness and its locus, the harem, into symptoms of degeneracy. The invisibility of deafness remained intact throughout, rendered visible only by its symptom, sign language, and by being framed in the context of the deviant harem.

The first description of the deaf in the harem was made by Domenico Hierosolimitano, an Italian doctor employed by the sultan Murad III in the 1580s.[7] But it was the Englishman Paul Rycaut who first published a detailed description of the work of the deaf in the harem in 1668:

> There is a sort of Attendant to make up the *Ottoman* Court, called *Bizebani* or *Mutes;* men naturally born deaf, and so consequently for want of receiving the sound of words are dumb: These are in number about 40 who by night are lodged amongst the Pages in the two Chambers [of the seraglio], but in the day time have their stations before the *Mosque* belonging to the Pages, where they learn and perfect themselves in the language of the *Mutes,* which is made up of several signs in which they can discourse and fully express themselves; not only to signify their sense in familiar questions, but to recount stories, understand the Fables of their own Religion, the laws and Precepts of the *Alchoran,* the name of *Mahomet* and what else may be capable of being expressed by the Tongue. . . . This language of the *Mutes* is so much in fashion at the Ottoman Court that none almost but can deliver his sense in it, and is of much use to those who attend the Presence of the Grand Signior, before whom it is not reverent or seemly so much as to whisper.[8]

Rycaut also informed his readers that sign language was widely used by members of court involved in what he termed "the doctrine of Platonick love," a homoerotics in which he believed men from the rank of student to the Grand

Sultan himself were busily engaged. Within this imaginary harem, sign language thus suppressed difference of rank between courtiers, and even between the sultan and his courtiers. In both factual and fictional accounts of the Orient, the distinction between the seraglio, or palace, and the harem itself became blurred beyond recognition, so that the entire Ottoman court was envisaged as the harem and a place of danger. This confusion makes it impossible to determine from such accounts whether the deaf were palace servants or actually worked in the harem. But, irrespective of the actual state of affairs in Istanbul, the phantasm of the harem created an association of sign language with sexual lasciviousness, immorality, and racial Otherness in the eyes of the hearing in the West.

When the French botanist Joseph Pitton de Tournefort visited the sultan in the early eighteenth century on orders of Louis XIV, he reported in similar terms, inflected by his scientific training and courtly *politesse:*

> There is a remarkable variety of reasoning animal, the Mutes of the Seraglio. In order not to disturb the Prince's rest, they have invented a language amongst themselves, whose characters are only expressed by signs; and these signs are as intelligible at night as in the day by the touching of certain parts of their bodies. This language is so well received in the Seraglio that those who wish to flatter and who are close to the Prince, learn it with great care; for it would be lacking in the respect which is due to him to speak out loud in his presence.[9]

Tournefort found the deaf in the harem no more than "reasoning animals," even more alien than the eunuchs, and discreetly alluded to the erotic dimension of sign language. His use of the term *prince* to refer to the sultan blurred the distinction between East and West, initiating two centuries of French fascination and interaction with the Orient. Immediately following his section on the deaf in the harem, Tournefort lamented: "If only M. Descartes and M. Gassendi had traveled to Constantinople, as they had wished, how many excellent reflections would they not have made on the morality and politics of the Turks?"[10] As if to compensate for this loss, Rycaut and Tournefort's description of the deaf in the harem was part of its legend for over two hundred years.

The notoriety of these deaf servants was enhanced by their reputed service as executioners. Armed with bowstrings, the "mutes" were feared assassins. In Racine's *Bajazet* (1672), Roxana imagines their assault in a fit of jealous rage:

> See the hands of the mutes prepare his doom.
> See they come armed with those fatal bowstrings
> That squeeze the life out of betrayers like him.[11]

Tournefort confirmed the existence of these deaf guards, who protected the sultan in public audiences and used their bowstrings in the execution of "the Grandees of the Empire."[12] The perception of the deaf was thus highly colored by this titillating mixture of Eros and Thanatos in the most exotic of settings, which also provided a space in which sign language could be imagined in action.

Apart from the sultan, all those who had access to the harem were in some way mutilated, in Western eyes, whether it be the eunuchs, dwarves, or the deaf. The repeated accounts of the dangers associated with violating the codes of the harem "thematize," in Ali Behad's phrase, "as erotic the transgressive crossing of the harem's limits."[13] The eunuchs, for example, were slaves, often taken by force from Africa. In this particularly brutal episode of the slave trade, the process of castration was often fatal, especially to adult men. But, in the Western imagination, eunuchs have always been connected to sexuality, owing to their physical proximity to the forbidden women of the harem. As the sultan's executioners, the deaf servants were connected to this erotic, for it was their bowstrings that awaited the detected transgressor. Indeed, Tournefort reported that a dwarf born deaf and then made into a eunuch "forms the most beautiful of creatures in the eyes and judgment of the Turks."[14] Once again, the empirical truth of these statements is not important in terms of their reception in the West, where it was axiomatic that the sultan's absolute power could be turned on any intruder. It was thus logical that it should be inflicted on his servants.

In Rousseau's *Essay on the Origins of Language* (1749), he made use of sign language in its Western and Oriental settings to buttress his argument that:

> if the only needs we ever experienced were physical, we should most likely never have been able to speak; we would fully express our meanings by the language of gestures alone. . . . Without fear of jealousy, the secrets of Oriental gallantry are passed across the more strictly guarded harems in the epistolary language of salaams. The mutes of great nobles understand each other, and understand everything that is said to them by means of signs, just as well as one can understand anything said in discourse.[15]

Rousseau moved from a discussion of the presumed original gesture language to the harem and then back to sign language in eighteenth-century France in the space of a few lines, precisely because his argument was not original but had been formed and refined in a century of Orientalist discourse. Indeed, as soon as the abbé Charles-Michel de l'Epée (1712–89) published his system of methodical signs in 1776, he was embroiled in a dispute with the German oralist Samuel Heinecke, who accused him of teaching the deaf the language of the harem.

Signs, Symptoms, and Citizens

Since Plato, the hearing have conceived of the deaf and their sign language in visual terms.[16] The modern history of the deaf began with a much mythologized encounter between the Abbé de l'Epée and two deaf sisters in a poor district of Paris. Epée had a rather unsuccessful career, first as a Jansenist priest and later as a barrister, until, in his forties, he observed the women signing and realized that they were not simply miming but conversing, which provided him with an unsuspected opportunity to save their souls for the Church. Epée's heroic status as the founder of deaf education has obscured the peculiarity of his moment of visual recognition, which is his only real claim to originality. His methodical manual alphabet had been devised in sixteenth-century Spain, while the "natural" signs of the deaf, as he called them, formed the rest of his program. He himself considered that the deaf were suddenly made visible, describing them as "men similar to us, but reduced in some ways to the condition of beasts, as long as no-one worked to free them from the gloomy shadows in which they were enslaved."[17] In England, the first charitable institution for the deaf was established in London's East End in 1792, and strikingly similar visual terms were used to describe the deaf, who were described as being in "an obscurity little short of death itself."[18] Unlike the English, however, Epée understood the value of sign language, which he saw as a variant of the visual arts: "Like painting, the art of methodical signs is a silent language which speaks only to the eyes."[19] Epée believed deaf sign to be a visible demonstration of the Enlightenment, proving both Condillac's theory of the conventional sign (that is to say, that there is no necessary relation between a word, gesture, or mark and that which it describes) and the possibility of a universal language.

The visibility of the deaf via their sign language was accompanied by the invisibility of the condition itself. Until the nineteenth century, doctors could neither explain nor cure what was regarded as a permanent state of misfortune ordained by God and whose causes, prognosis, and development were mysterious. However, the state of being deaf was transformed into the pathological condition of deafness by the introduction of that which Georges Canguilhem defined as "normalization": "Between 1759, the date of the first appearance of the word *normal,* and 1834, the date of the first appearance of the word *normalized,* a normative class conquered the power to identify . . . the function of social norms, whose content it determined, with the use that that class made of them."[20] If hearing was henceforward regarded as the "normal" human condition, deafness was, by extension, pathological. The pathology of deafness led to the deaf being defined as "savages" and their sign language as a similarly "primitive" means of communication. Even active supporters of the deaf gave credence to such ideas. In a sermon given in 1824, the American Thomas Gallaudet—the founder of the eponymous university—described the uneducated deaf as "heathen": "I only crave a cup of consolation, for the Deaf and Dumb, from the same fountain at which the Hindoo, the African and the Savage, is beginning to draw the waters of eternal life."[21] One part of the nineteenth-century "civilizing mission" of colonialism was thus to rescue the deaf from their pathological condition by restoring them to (normal) speech. In his article on physiognomy for the *Dictionnaire des sciences médicales* (1820), J.-J. Virey (1755–1847) asserted that, unless people can be seen to speak, they have no physiognomy to interpret. His example of this condition was the veiled Oriental woman, but his stricture applied equally to the signing deaf. Virey thus connected deafness to Oriental (harem) women and transformed the visual symptom of sign language into a sign of illegibility.[22]

Two tasks now awaited the would-be doctor of the deaf: a medical definition of deafness and a cure. It was Jean-Marc-Gaspard Itard (1774–1838), the doctor at the Paris Institute for the Deaf, who addressed these questions and discovered the difficulties involved, after his first medico-educational endeavor with the famous Wild Boy of Aveyron. These new requirements entailed a need to make the problem of deafness visible and thus treatable. As Canguilhem notes: "To see an entity is already to foresee an action."[23] It seemed obvious that a person who was profoundly deaf and someone with perfect hearing formed medical opposites. But at what point did the person with some residual hearing become known as deaf? On 26 September 1807, a boy named Lefebvre was admitted to the Institute for the Deaf. Unfortunately, his behavior was soon found "intolerable" and a bad moral influence on the rest of the pupils, and a reexamination of his right of entry to the institute was ordered.[24] The doctor's report from his native Le Havre claimed that Lefebvre could hear "completely," although he could not talk. The boy was admitted, as the institute prided itself on its ability to render the mute speaking, but Itard now focused his investigation on the organ of hearing. If Lefebvre could hear, then his mutism was not a consequence of deafness, and the administration would feel justified in expelling him. But, as Itard reported on 30 January 1808, "it was extremely difficult to determine with regard to a being who still had no means of communicating his ideas or rather his sensations, and who has anyway a small susceptibility for attention, to what point he could hear [*entendre*]." Itard determined that he looked like "a being disgraced by nature," a circumlocution for deaf, and placed him under "surveillance." He discovered that Lefebvre would repeat the word *oui* whenever it was spoken to him, even if it was inaudible, and concluded that "his intellectual faculties are no more susceptible of improvement than his moral faculties."

From this point forward, deafness was cross-referenced with the new moral terminology of mental illness devised by Philippe Pinel. The double meaning of the French verb *entendre,* meaning both to hear and to understand, inevitably intertwined discussions of the ability to hear and to reason. Lefebvre was excluded from the institute because he was classified an "idiot," someone whose reasoning *esprit* could not be revived under any circumstances. Pinel described the idiot as suffering from dementia, leading to an incurable "automatic existence," although it was the most common of mental illnesses.[25] For Pinel, deafness was a common symptom of idiocy. His classification recalled Enlightenment definitions of the uneducated deaf either as automata or as the presensory statue imagined by Condillac, and thus seemed a logical complement to their treatment. Itard simply argued that idiocy was not accidental but inherent in many cases of deafness, and thereby arrived at a very convenient form of diagnosis. If the patient showed signs of improvement in understanding and intelligence, the disease was simply deafness; if not, deafness compounded with idiocy. Diagnosis thus depended upon the results of treatment, not upon the invisible and unmeasurable deficiency of hearing it was supposed to address. Although the categorization of deafness had changed, its invisibility as a condition remained intact.

Itard was diligent in seeking a cure, even if his diagnoses were uncertain. He attempted the use of injections, electricity, astringents, and even hot irons in his efforts to overcome the "blockage" in the ears he believed to be impeding hearing.[26] The deaf professor Pierre Pelissier later wrote angrily: "All those of my brothers in misfortune who were *operated* upon had their intellectual faculties deranged and they continue to feel the effects."[27] Itard came to realize the deleterious effects of his actions and later apologized for them: "One can regard a cure, properly speaking, for congenital deafness as impossible, so rare is it and so great is the number of deaf people who have been uselessly tormented to render them a sense which nature has pitilessly denied them. I accept the share of this reproach which adheres to me."[28] But he did not repent of his diagnostic procedure, arguing that the deaf were not susceptible to mania or delirium because of "the inactivity of this faculty of the intelligence [*entendement*]."[29] A person without *entendement* was either deaf or an idiot in Itard's view, the difference being that it was possible to do something for the former, but not for the latter. It is less surprising in light of this diagnostic connection that the deaf and the insane were dealt with by the same department of the Ministry of the Interior throughout the nineteenth century.[30]

Although the chances of curing deafness were known to be very low, Itard's medical successor at the institute, Prosper Ménière, argued in 1853: "The deaf believe that they are our equals in all respects. We should be generous and not destroy that illusion. But whatever they believe, deafness is an infirmity and we should repair it whether the person who has it is disturbed by it or not."[31] He spoke at a meeting of the Academy of Medicine, which discussed a Second Republic (1848–52) initiative to cure deafness using the methods of one Dr. Blanchet. Blanchet had revived Itard's discredited belief that everyone has some hearing, even if very little, and that the faculty could be improved with practice, just as muscles can be improved by exercise. The Republican government had been keen to adopt his therapy, for it held the promise of eliminating deafness as a category and restoring the deaf to oral society. Despite Itard's recantation of his own method, a section of the Academy also supported the proposal. Sign language was now taken to be a symptom of the deaf patient's immoral refusal to be cured. One Dr. Bouvier reported—after one week's study—that it was "a sort of primitive language; its expressions resemble the first language of a child, that of peoples little advanced in civilization."[32] Sign language could have no place in the advanced civilization of the West.

In the early nineteenth century, a new word, *surdi-mutité,* had been coined to express the condition of being a deaf-mute, for mutism was then believed to be an inescapable consequence of deafness.[33] We might translate it as "deafy-mutism" to keep the strange sound of the neologism. For the first time, it was possible to describe the "deaf-mute" (to use the nineteenth-century term) as a person afflicted with a reversible condition that one could define, isolate, and hence cure. Deafness as a handicap, as opposed to being a deaf person, is thus a recent construction that properly belongs to the modern period. It was energetically resisted by the deaf activists in Paris. The deaf journalist and activist Henri Gaillard (b. 1866), editor of the *Journal des Sourds-Muets,* focused his energies on convincing politicians of the Republican case for deaf political rights: "For a long time we have not ceased to proclaim that the Deaf-Mutes differ in no respect from the Hearing-Talking."[34] A left-leaning Parisian, Gaillard did not oppose the teaching of lipreading and of speech, both skills that he possessed, but argued that sign language was the essential preliminary stage in deaf education.

It is clear that this claimed equality was in fact a very unequal one, with the predominant power lying on the side of the hearing majority. For as the hearing look at the deaf, the deaf look back and disrupt or confirm the image produced. The result is what I shall call the "silent screen" of deafness, which depicts neither the deaf themselves nor the view of the deaf seen by the hearing but the product of the interaction of the two looking at each other. It takes two people to see deafness, one with hearing and one without. As such, it can never be stable or essential, but is always and already a cultural construction in need of renewal, and an image in need of focus and definition. The screen is the product of the intersection of two gazes that forms a certain space for perception, making it possible (in this case) to see the deaf within a category known as deafness. This notion of the screen is derived from that proposed by Jacques Lacan in his *Four Fundamental Concepts of Psychoanalysis*. For Lacan, "the screen here is the locus of mediation" between the gaze and the subject of representation.[35] Thus, as I look at the subject of representation, a third element comes into being on the screen. Deafness cannot be seen only by the hearing or only by the deaf: both see each other in "seeing" deafness.

But, as I have suggested, these views are by no means equivalent. The production of deafness as a visual category by the hearing requires that this blurred picture be clarified and that the images coalesce in a legible fashion. Central to this process is the framing of the visual image. Lacan's notion of the screen does not suggest how the image itself is produced or what defines and forms the screen. In visualizing deafness, a screen must be created and defined, which requires that it be framed: that is, have defined borders and parameters. Just as the visualization of deafness is a historically constructed process, so the frame is not 'natural' to the picture. In his reading of this question, which has given a new impetus to the theorizing of the visual space of representation, Jacques Derrida has insistently asked: "Now where does this frame come from? Who supplies it? Who constructs it? Where is it imported from?"[36] Derrida's questions indicate that the creation of *surdi-mutité*, of deafness as a handicap, cannot be fully explained by the overarching opposition of the normal and the pathological. The more precise question is why deafness was imported into that frame—and by what means? Frames are of little use with no picture inside, but every picture needs a frame. Such framing was essential if the invisibility of deafness was to become visible.[37]

Pathology and the Harem

The long-standing myth of the harem was refigured in the nineteenth century from a textual construction into a visualized space of cultural, sexual, and political deviancy. French Orientalism became as much a visual discourse as a written one. The Ottoman Empire had always been regarded as cruel and despotic, but texts such as Montesquieu's *Lettres Persanes* (1721), which made great play with the harem, could also envisage learning from the East. In the nineteenth century, the harem became seen as a prime example of the backward, unchanging nature of the Orient, which required the colonial presence as a catalyst for change. Imperial possessions were accessible to the majority of the colonizing population only through visual images, which needed to convey at once the desirability and necessity of colonialism. The imperative to visualize the harem in particular contained a complex series of conscious and unconscious motivations. It rendered the harem, and especially the women within, "accessible, credible, and profitable," to use Malek Alloulah's phrase.[38] The profit from such representations was actualized in the libidinal economy, either in fantasy or, for those with the money to travel, in reality. For sexuality was pathologized at the same time as deafness, creating categories of deviance and repression that found one outlet in the phantasm of the Orient.

The pathologizing of the harem accompanied and reinforced its use as a metonymic justification for imperial rule. Harry Boyle, Oriental secretary to the notorious Lord Cromer, held that as a consequence of British rule in Egypt, "[t]he unwholesome—and frequently degrading—associations of the old harem life [should] give place to the healthy and elevating influence of a generation of mothers, keenly alive to their responsibilities as regards the moral training and welfare of their children."[39] Imperialism was thus the unlikely protector of women and morality in the "backward" races of the East, who were held to be inferior to the Western races. Such polygenetic theories of race, that is, the belief that there were several distinct varieties of the human species, were especially prevalent in France.[40] For example, Hippolyte Taine, one of the leading French philosophers and critics in the later nineteenth century, argued that not only were there distinguishable races of people but that "[a] civilization forms a body, and its parts are connected with each other like the parts of an organic body." Each individual body was both a representative of a racial type and an exemplar of social organization. The different races and their cultures were located upon an invariable scale, with pride of place unsurprisingly reserved by Taine for the Aryan races. He held that all societies depended upon obedience, but that obedience was organized in very different ways: "If the sentiment of obedience is merely fear, you will find, as in most Oriental states, a brutal despotism, exaggerated punishment, oppression of the subject, servility of manners, insecurity of property, an impoverished production, the slavery of women and the customs of the harem."[41]

When the polygenetic theory of human history gave way to the monogenetic theory of evolution, many of the features of the previous system were simply retained. In the words of one left-wing Republican, the deaf "type" was particularly low, being "like *homo alalus,* like the pre-historic man without words, but even more backward because he does not hear, he passes amongst his fellow men, for him the equivalent of shadows, without hearing them and without understanding them: everything human is foreign to him."[42] The individual's "given" psychological characteristics were thus used to read off existing systems of political, gender, and economic organization. The shortcomings of any particular race in these matters were thus innate and could not be altered, except by outside intervention from a superior type. Let us then consider how a deaf person is represented under such a scheme. Taine's polygenetic corporal metaphor turns signing space into "Oriental" space. When a deaf person uses sign language, a space is created from the head to the midriff, across the shoulders and up to a foot in front of the body, which is semantic from the point of view of the language user, as sign language uses a spatial grammar.[43] But for Itard, Bouvier, or Taine, who did not understand sign, this reading was impossible. They were convinced that spoken Indo-European languages were the only means of access to abstract thought and rational constructions such as grammar. Taine created a scale of language running from Chinese, which he placed at the bottom, up to the languages of the Aryan races: "In this interval between the particular representation and the universal conception are found the germs of the greatest human differences."[44] Gestural sign was classified as primitive, closely related to Chinese, and incapable of abstraction—in a word, as Oriental. The use of sign language in the harem by deaf and hearing people was a mutually reinforcing indicator of pathology, which could be overcome only by the determined intervention of Western speech.

This newly deviant frame of the harem created a space in which Romantic artists could, and perhaps should, flout all the norms of realistic visual representation with impunity. If the normal visual space was France, then the pathological visual space was the harem. The veils and sign language of the harem were illegible barriers to understanding, in contrast to the pure speech and physiognomy of the West. Eugène Delacroix's painting *Greece on the Ruins of Missolonghi* (Bordeaux: Musée des Beaux Arts, 1826) represented Greece as a young woman with luminescent white skin, whose shift conveniently falls apart to reveal her tortured bosom. She is contrasted to the smaller figure of a black standard-bearer behind, as the critic Boutard remarked: "In the distance a negro raising the crescent over the ruins of Greece, evokes the idea of the country of Miltiades and of Pericles now reduced to be the slave of a slave."[45] In other words, the white woman was enslaved into the Oriental harem under the control of a black eunuch, himself a slave to the despotic sultan. A year later, Delacroix envisaged the consequences of such enslavement in his *The Death of Sardanapalus* (Paris: Musée du Louvre), his famous *succès de scandale* at the Salon of 1827. He depicted Byron's character Sardanapalus as an Oriental sultan using his "eunuchs and officers of the palace," to quote Delacroix's description in the Salon catalogue, to put his other servants to death while his capital falls to an invading army.[46] Prominent in the foreground is a bearded man—hence not a eunuch—putting a woman to death by the sword, evoking the famous deaf executioners of the seraglio. As recently as 1820, the recollections of M. d'Ohsson, the Swedish ambassador to the Court of Constantinople, had been published in France, emphasizing the importance of the Ottomans' deaf servants and their use of sign language: "The mutes . . . express themselves by rapid gestures and this language is known to the courtiers, women of the Harem and the Sultan himself, who often only uses a sign of his hand to give orders to those who surround him."[47] Sign language connected the Others of the harem into a homogenous Otherness. If a viewer of the *Sardanapalus* perceived the executioners to be deaf, it was now within a space of pathology and deviation. Both of Delacroix's pictures render a very unconventional sense of pictorial space. In the *Greece*, the disparity in size between the figures is far too great to be accounted for by the recession of perspective, and the *Sardanapalus* has no consistent spatial dimension—the sultan's bed slopes dramatically to the right corner, but in the left foreground an African man struggles with a horse that appears to be sinking below the floor, as if into hell. The breakdown of norms of pictorial representation was itself indicative of the densely interwoven strands of deviance and pathology represented by the harem. A history of Orientalist painting, a school so popular it gained its own annual Salon in 1893, could be written in terms of a visualization of pathology, deviance, and the physical consequences of such abnormality.[48]

The harem served to make the degenerative, physical stigma of race, sexuality, and even deafness visible to the Western gaze. These stigmata were the marks of the West itself. For if the mirror was the symbol of Classical representation, modernity presented the "silent screen" that refused to speak in answer to the observer's questioning of identity. As spectatorship hybridized, the body became an opaque medium. As Edward Said has cogently observed, Orientalism's deepest desire was to "make the [mute] Orient speak." The Oriental body at home and abroad could be categorized only with the aid of clarifying, pure sound. Gérard de Nerval attested to this resistance when he wrote from Egypt to Théophile Gautier of their dream of the real Orient: "Think of it no more! That Cairo lies beneath the ashes and dirt, . . . dust-laden and dumb."[49] What could not be discovered in the open streets, which, as de Nerval stressed, could not be seen as a whole, could be made apparent by enframing. Such procedures were particularly necessary in the effort to make deafness visible. For in a picture, everyone is deaf and mute, and there is no visible sign of the difference of deafness. Like the Orient in which they were framed, the deaf had to be classified and "demutized" in order to restore them to civilization.

This combination of medical, racial, and political opprobrium made the Orient in general, and the harem in particular, into what Said has called: "a political vision of reality whose structure promoted the difference between the familiar (Europe, the West, us) and the strange (the Orient, the East, 'them')".[50] Yet the clarity of this theoretical distinction was less easily maintained in practice. The British traveler Adolphus Slade gave witness to such dilemmas in his 1833 account of the harem:

> We delayed a few minutes to converse with two regular mutes; they were boys about 14 years old, very genteel and good-looking, whereby we were completely undeceived in regard of their species, having previously understood that a mute was a kind of animal between a dwarf and a monkey. The little urchins were exceedingly amused, and laughed and conversed about us with great rapidity, making most expressive language about us with their eyes and fingers. Their quick wit is proverbial in Turkey, and in the secret deliberations of the seraglio, where they alone are allowed to be present as domestics, nothing escapes their intelligence.[51]

Slade had accepted the polygenist theory that the deaf constituted a separate species, but, having met them, he seems to have been struck by their physical charms in marked contrast to the "abhorrent ugliness" of the eunuchs. Rather than be repelled by the deviance of deafness, Slade found himself attracted to them in a fashion that, while he still perceived them as Other, was complicated by a bond of erotic attraction. The harem was, of course, a classic location for such operations of transgressive desire, which disrupted the neatness of the binary opposition between East and West by connecting the pathological deaf Other to the normal Western male: the frame was a bad fit.

Mimicry and Originality

The pathologization of deafness presented the deaf themselves with an overriding imperative to demonstrate the "normality" of their perceptions and ideas. In 1834, the deaf community created by the Institute for the Deaf in Paris founded the *Central Society for the Deaf,* which organized annual banquets to commemorate Epée and promoted the work of such deaf artists as Fréderic Peyson, Léopold Loustau, and Joseph de Widerkehr. To be a deaf artist was, in nineteenth-century terms, virtually an oxymoron. This belief was strikingly expressed by Baron de Gérando in his magisterial two-volume work on deaf education: "One hears of several deaf people who have become distinguished painters; they still exist today; but although they are skilled in execution, in everything which relates to imitation, they fail in original composition and can never attain to the ideal of art."[52] This denial of artistic creativity depended upon and reinforced the perception of the deaf as "savages," for James Hunt, founder of the Anthropological Society in England, attributed the same failure to Africans: "The Negro had had the benefit of all the ancient civilizations, but there was not a single instance of any pure Negro being eminent in science, literature or art. . . . What civilization they had was imitated, and they had never invented an alphabet nor reasoned out a theological system."[53] Needless to say, part of the construction of that civilization was the belief that no "savage" could ever master it, and any evidence to the contrary was, as de Gérando showed, simply dismissed as an anomaly. The deaf artist had to be both normal and original at the same time, an impossible conundrum, analogous to that in which women artists found themselves, deemed capable copyists but no more: "The female sex possesses a remarkable talent for translation, adaptation, interpretation. In the domain of imitation, she is inimitable. . . . Her natural inclinations lead her less toward invention than toward imitation. Where receptivity dominates, originality is weak."[54] For the deaf, for the African, and for women, originality was a category from which they were excluded by definition, reinforcing the long-lasting and complex analogies between these three groups, which had originated in the myth of the harem. The clichéd analogy of white women and "primitive" men in racial science of the period could be supported by their joint use of the pathological language of signs in the harem.[55]

Deaf intellectuals paradoxically celebrated their capacity for imitation in their theorization of sign language, which they called *mimicry* (*mimique* in French). The imitation was not, however, of the word but of the idea itself. Deaf sign language was equated with speech as a secondary representation of the original idea held in the mind or perceived in exterior reality. Roch-Ambroise Bébian, an educator of the deaf from Guadeloupe, was the first hearing person to teach the deaf by using what he saw as the natural language of mimicry. He learned French Sign Language from his pupil and friend Ferdinand Berthier. The interaction between them was such that the works of both writers can be seen as two parts of a whole, the theory of mimicry. Bébian's identification with his deaf pupils led to his expulsion from the institute in 1822. In 1817 Bébian published an essay comparing sign language to "natural" language, which challenged the etymological obsessions of the new science of linguistics.[56] Bébian saw gesture as an original, natural language to which voice was an accessory: "The gesture, without any auxiliary, expresses both the idea and the relations of ideas."[57] Mimicry was thus the answer to the nineteenth-century scholarly quest, described by Foucault, for "the exact reflection, the perfect double, the unmisted mirror of a nonverbal knowledge."[58] Berthier saw mimicry as that doubled image in which: "thought is reflected whole as if in a mirror, complete with its most delicate contours. It materializes there, so to speak."[59] Mimicry refused to be pathologized as speech's Other, but used the arguments of phonocentrism to justify sign language. This construction strikingly confirms Derrida's observation that "phonologism does not brook any objections as long as one conserves the colloquial concepts of speech and writing which form the solid fabric of its argumentation."[60] Mimicry was both a part of, and an accommodation with, linguistic theories that, although prioritizing speech, allowed for an original moment of meaning that had dispersed into numerous different languages, analogous to the polygenetic formation of the human races. The mimickers came to a successful accommodation with their hearing contemporaries' theories of language, race, and representation, which enabled the deaf community in Paris to flourish between 1830 and 1890 in a fashion that has yet to be equaled.

Mimicry was nonetheless not a perfect double. Deaf people were often described as being "separated by a murky veil from their fellows [*semblables*]."[61] In Homi Bhaba's important analysis of the ambivalence of the colonial mentality, they were "almost the same but not quite."[62] Bhaba's formula, derived from the colonial situation, is clearly an apt one for the so-called savage deaf, one of the many human races then believed to exist. Mimicry was the supplement to a language that has full presence in speech. But the logic of the supplement cuts both ways, indicating both plenitude and lack in the original that it supplements, like the supplements to a dictionary. This destabilizing characteristic of the supplement gives it a subversive and dangerous force, undercutting Berthier's argument that mimicry could and should exist in a stable relationship to speech, being "that primitive language of which the infant makes use instinctively before and after the appearance of its nascent reason." In this sense, it was something prior to speech, whose fullness was not disturbed by this addition. But he continued: "It slips, at a more advanced age, into daily conversation, unknown to the speakers and becomes, without their noticing it, the obligatory auxiliary for people who shine at the bar, at the political hustings, on the throne, as on the stage, whether tragic, comic or even comic opera. What is an exactly reproduced ballet, if it isn't above all an excellent lesson in mimicry?" Here, the supplementarity of mimicry revealed the lack in spoken language, that is, its inability to present the full and complete representation of the idea, prevocal and fully known. Speech in its most crucial, as well as its most everyday, moments lacks, in the logic of Berthier's argument, the means to represent itself without being completed by the addition of gesture. Mimicry's relationship to speech was thus crucial for the production of meaning, and any attempt to pathologize or eliminate sign language would thus have profound consequences for speech itself.

The Hand, the Touch, and the Gesture

In tracing how the pathologization of deafness was achieved, despite the resistance of the deaf, I want to focus first on the hand itself, then on its gestures and signs, before pulling back to see the deaf in full frame. Sensory perception had been a central concern of French philosophy since Condillac's *Traité des Sensations.* Nineteenth-century writers continued this speculative tradition, in investigating the sense of touch. But the explosion of professional literature on deviance and its correlative, sexuality, found the hand itself to be the object of suspicion, investigation, and discipline. In these works concerning the care of the deaf, blind, and insane or the regulation of sexuality, a distinction emerges between the transforming, material hand and the invisible, spiritual sense of touch. Condillac's single category of touch was now understood as comprising two distinct processes. On the one hand, touch was a process of judgment, which had neither an effect on the touched object nor any visible presence when not in use. This kind of touch includes the distinction between hot and cold, sharp and blunt, wet and dry, etc. In this sense, touch was not suspect or subject to moral condemnation. On the other hand, touch was an activity that left a mark on the object touched. In order to preserve this distinction, the active touch was often described simply as the hand. It was, however, no easy matter to separate touch and the hand. Under absolutism, the distinction emanated downwards from the body of the king, who was alone able to combine the material hand and the spiritual touch by virtue of his anointment as God's secular representative on earth. This gesture was the royal touch for the scrofula, in which French monarchs healed those afflicted with this skin disease.[63] In the early nineteenth century this practice was recalled in Gros's famous depiction of *Napoleon in the Plague House at Jaffa* (Paris: Musée du Louvre, 1804), showing the emperor healing with his touch. Without such absolute authority, incarnated in the person of Louis XIV or Napoleon, the confusion between touch and the hand became a central dilemma for artists, anthropologists, and criminologists.

This distinction emerges clearly in comparing the history of the deaf to that of the blind, who were initially housed together by the state during the French Revolution. The same language was used to describe the deaf and the blind, as here by one administrator of the school for the blind in 1817: "The moral world does not exist for this child of nature; most of our ideas are without reality for him: he lives as if he was alone; he relates everything to himself."[64] The initial breakthrough in the education of the blind was the invention of a raised typeface by Valentin Haüy, condensed by Louis Braille (1809–52) into the code of dots with which we are familiar. Although both systems depended on touch, they were not criticized for it: "but this writing has however the inconvenience of being arbitrary."[65] As discussions of the old chestnut regarding the preferability of blindness and deafness continued, they were decisively resolved (by those who could see and hear) in favor of blindness, for the loss of hearing entailed the loss of voice and hence of thought. When the blind read Braille, they converted the dots into the "pure" medium of sound, which more than compensated for its nonalphabetic character, whereas the deaf used sign, and thought without sound. By late century, official French government manuals on the care of the "abnormal" advised that, although Braille was "an intermediary system between the manuscript and the printed text," in sign language, "all spiritual ideas will be unhappily materialized."[66] Thomas Arnold, who founded a small oralist school in Northampton in 1868 believed that the blind: "mentally, morally and spiritually [are] in a more advantageous condition than the deaf." If the blind could create "a mental language of vibrations and motions" from touch, the deaf were restricted to "a language of mimic gestures, . . . which is destitute of all that phonetic language provides of antecedent progress in thought and knowledge."[67] Indeed, as Arnold reminded his readers, the oralist method of deaf education, which focused on speech training, used touch as its primary method: the deaf child felt the vibrations in the speaking person's voicebox and attempted to imitate them. Despite the curious vampiric embrace of throats that this method entailed, it was supported by moralists over sign language throughout the century. Now, nothing essential had changed in the nature of sign language and Braille in the intervening fifty years, but, whereas Braille was considered arbitrary in 1840 and mimicry had won a certain acceptance, by 1890 it was Braille that had become acceptable, and the deaf were once again considered precivilized. In Arnold's widely accepted viewpoint, the decisive factor in this change was the ability of the blind to hear. Sight, he argued, was "much inferior in providing us with available mental images and an organ of expression." The deviance and pathology of deafness was now simply self-evident in the inability to hear and in the manual nature of sign language, requiring no further proof to ensure its moral condemnation.

If touch was not the culprit in this medical and moral revision, gesture, on the contrary, left behind it the hand. Once a sentence in Braille has been read, no trace of the reading touch remains. But when a deaf person has completed a sign, the hand remains as a palpable, material entity. The hand, capable of all manner of activities, was very much at the center of moral and cultural debate in the nineteenth century, evoking discourses of sexuality, crime, and class. The much discussed sin of female masturbation was euphemistically referred to as "manualization," making it clear that the hand was to blame. Although eighteenth-century artists had used sexual and criminal gestures in their work, they had not considered the hands themselves responsible. However, the science of phrenology, founded by Franz-Joseph Gall, considered that the forms and shapes of the body were symptoms that revealed important inner truths. The hand came to be the focus of one popular offshoot of these ideas, known as chiromancy, from which palmistry is derived. For chiromancers, such as Adolphe Desbarrolles, the hand was held to be every bit as legible as was the face for Lavater, and his study of the particularities of the hand and of gesture ran through many editions.[68] Phrenology was rejected as a science by the 1850s, but the belief it inspired in the legibility of the body was retained. For example, in the confusion surrounding the birth of criminology that arose in Third Republic France, the existence of criminal types of hands and gestures was seriously discussed. In Cesare Lombroso's theory of the atavistic criminal, whose deviance was a consequence of biological deformity and racial descent from "Negroes" and "Mongloids," thieves were "frequently remarkable for the mobility of their features and their hands."[69] The similarity of this description to that of a signing deaf person was reinforced by the "finding" that: "the hearing of criminals is relatively obtuse and they are prone to disease of the ear." One Dr. Gradeningo reported that 72.5 percent of "instinctive criminals" had defective hearing.[70]

The opposing side in the criminology debate, which stressed free will and social conditions as the causes for crime, nonetheless continued to investigate the hand in support of very different conclusions. At the Third International Congress of Criminal Anthropology, held in Brussels in 1892, Lombroso's theories were decisively rejected. Nonetheless, delegates still approved a paper that held that: "The human organism forms a very complicated ensemble whose different parts influence one another. A change in one of its organs sometimes entails alterations in very distant parts."[71] The Congress thus denied that the criminal could be identified by atavistic deformities, but proposed that the pathology of criminal activity nonetheless entailed observable physical mutations. The observation of the hand had thus quickly become an established part of police practice. In 1879, Alphonse Bertillon began to create his system of photographing and measuring the criminal, in which the hand had a prominent place, for it was held to be one of those body parts that no criminal could disguise.[72] As this practice spread, English prisoners were photographed in 1880 holding their hands palm down in sight of the camera,[73] a marked change from earlier practice, which produced simple portrait photographs of convicts.[74] This means of identifying individuals by tell-tale details rather than the whole was endorsed by the art critic Edmond Duranty in his 1876 essay on "The New Painting," which we have come to know as impressionism:

> What we need is the particular note of the modern individual, in his clothing, in the midst of his social habits, at home or in the street. . . . By means of a back, we want a temperament, an age, a social condition to be revealed; through a pair of hands, we should be able to express a magistrate or a tradesman; by a gesture, a whole series of feelings.[75]

The artist, critic, and detective made their distinctions in the same fashion, isolating a detail and concentrating upon its salient features, which could not be disguised. Deviance, whether in the harem, the picture, or the prison had to be *visible.* This penetrative scientific gaze was felt by all concerned to be the height of modernism, united by a common belief in the forces of progress.[76]

The identification of criminals by the hands was refined by Francis Galton into the use of fingerprints as a means of distinguishing the massed ranks of the colonized. For Galton, fingerprints were "the most important of all anthropological data," even though he was unaware of the traces left by the fingers that we now call by that name. Instead, he saw the patterns on the hand as fulfilling "the need [that] . . . is shown to be greatly felt in many of our dependencies; where the features of the natives are distinguished with difficulty; where there is but little variety of surnames; where there are strong motives for prevarication . . . and a proverbial prevalence of unveracity."[77] It was now not enough to control territories at home or abroad by the police or by colonization. The microtechniques of discipline sought to colonize the body and even the hand. In concord with the long-standing classification of the deaf as "savages," administrators of the Institute for the Deaf quickly came to argue, in the words of one Inspector Arnold, that: "the language of signs is *the language of thieves,* because swindlers have the habit of talking amongst themselves in low voices often with the use of both hands and physiognomy."[78] In this pronouncement, Gall's physiognomy rubs shoulders with Lombroso's criminology and the anthropological focus on language and human activity. While the name of the science had changed from phrenonology to anthropology, the same physical characteristics remained under suspicion. The casual connection between race, crime, sexuality, and sign language that had been conveyed in the myth of the harem now had a presumed causal, scientific force.

Deafness and Degeneracy

Despite such energetic categorization, as well as the researches of eminent scientists such as Paul Broca, a visible sign of deafness still could not be found.

But Broca's location of speech capacity in the third frontal convolution of the brain seemed to give physiological force to Itard's equation of deafness with mental disorder. In order to save their mental health, it was now biologically essential for the deaf to learn to speak. It was the government of the Third Republic in France that installed such oralist education, overturning the Revolutionary commitment of 1791 to sign language education for the deaf.[79] Arguing that sign language was outdated, the Interior Ministry set about installing the oral method in all deaf institutions. The ground was prepared by a series of articles on the deaf in the influential *Revue des Deux Mondes* by Maxime du Camp, an Orientalist, who published an account of his visit to the harem in Istanbul, and later accompanied Flaubert to Egypt and North Africa.[80] Du Camp transferred his anthropological curiosity to Paris, writing a multivolume study of the city, followed by an equally long tome on charity in the capital. In his assessment of the deaf, he argued that it was possible to take either a pessimistic or an optimistic view. The optimistic view held that the achievements of deaf writers and artists demonstrated that "the evil which affects [the deaf] is local and does not at all affect the faculties of the brain." But du Camp clearly adhered to the pessimistic view, which categorized those born deaf as "defective. . . . [They are] betrayed by the ill-shaped head, the tapering forehead and chin, the prominent ears, and the nervous twitchings of the face which many cannot restrain; these are a sort of indication that the animal nature predominates."[81] Social policy concerning the deaf and other "marginal" groups in the Third Republic was all too often determined by such phrenological musings dressed up as scientific observation.

In 1879, the Interior Ministry ordered the Institute for Deaf Women in Bordeaux to commence "pure" oral instruction. A year later, an international conference for deaf educators was held in Milan, the center of Lombroso's operations since 1876, to resolve the question of methods. But, as nearly all the delegates were French and Italian oralists, the result was never in doubt, from the interminable opening address of the chair, Giulio Tarra: "Gesture is not the true language of man. . . . Gesture, instead of addressing the mind, addresses the imagination and the senses. Thus, for us, it is an absolute necessity to prohibit that language and to replace it with living speech, the only instrument of human thought."[82] The proceedings consisted of endless repetitions of this theme, as if the failure to find a visible pathology of deafness had to be countered by a numbing torrent of words, to make the mute deaf body speak its deviance. Finally, in what can only be described as Derrida's Nightmare, the Congress closed to shouts of "Vive la parole!" led by French Academician Adolphe Frank. The Milan Congress banned sign language from all areas of deaf education, a ruling that held good until 1976 in France. *Vive la parole!* was thus a watchword throughout the High Modern period.[83]

Sign language culture did not, of course, disappear at the wave of this Italian wand. Deaf culture was widespread in France and the United States, and, to take one example, more deaf people were becoming artists than ever before. At the Salon of 1886, thirteen deaf artists exhibited, including the sculptor Félix Martin who had won the *Légion d'Honneur*, France's highest civilian honor, for his statue of the *Abbé de l'Epée* (Paris: Institut National des Jeunes Sourds) in 1878. Deaf activists campaigned against the Milan congress resolutions in such newspapers as *La Defense des Sourds-Muets*, with radical politicians, and in the deaf community.

But such cultural resistance found itself unable to compete with the framing of the "primitive" passions of the deaf, as attested by the harem, within a frame of medico-pyschology. The deviance of the deaf was now made visible, not on the body, but on the graph and diagram. In 1883, Alexander Graham Bell presented a paper to the National Academy of Sciences in Washington, D.C., entitled "Upon the Formation of a Deaf Variety of the Human Race." Bell took data, which he himself admitted were incomplete, subjected them to a "worst-case scenario," and was able to produce a graph that demonstrated that: "[t]he indications are that the congenital deaf-mutes of the country are increasing at a greater rate than the population at large; and the deaf-mute children of deaf-mutes at a greater rate than the congenital deaf-mute population."[84] Mapped on the graph, the deaf children of the deaf seemed to be dramatically outbreeding the hearing. Bell attributed the problem to the deleterious effects of sign language, which "causes the intermarriage of deaf-mutes and the propagation of their physical defect." No evidence was advanced to support this theory, given credence solely by the long-standing belief in the sexual deviancy of deafness. Bell further raised the specter that a signing deaf nation might be established in the West of America, noting ominously that "24 deaf mutes, with their families, have already arrived [in Manitoba] and have settled upon the land. More are expected next year."[85] The use of sign language would lead to more marriages, more deaf people, and eventually a "deaf variety of the human race" might be established in the American West. Bell's work was later cited by eugenicists, but his short paper was found salient and convincing at the time, because it could be projected onto the "silent screen" of deafness, whose emergence I have described. Bell's arguments were accepted by the Royal Commission established in Britain in 1889 to report on the question of deaf education. Advising oral education, the Commission also advocated the segregation of the sexes, as "the passions of the deaf are undoubtedly strong," for which they advanced no evidence.[86]

Perhaps the members of the Royal Commission had visited the 1887 Salon exhibition in Paris, where P.-L. Bouchard (1853–1937), a now forgotten History painter, had exhibited his major canvas *Les Muets du Serail* [The Deaf in the Harem].[87] The catalogue announced that: "In the harem, the mutes, subalterns to the eunuchs, were skilled at tightening the fatal bowstring, and when the Sultan had pronounced a sentence, they executed it immediately and without noise."[88] The painting shows four "mutes" entering through a door at the right of the canvas, carrying the famous bowstrings as identification. But they were also depicted as black, merging the categories of race and deafness into one multiply deviant body. Furthermore, the muscular black "mutes" are advancing on a group of scantily clad women at the left, who vainly clamber up onto a bed to try to escape the executioners. The intended victim seems to be a bare-breasted woman lying unconscious on the bed, giving the image an eroticism that is more than a little tinged with sadism. In a pictorial pun, one woman attempts to flee by drawing back a curtain placed in the exact center of the image, otherwise known as the vanishing point of the perspective, only to be blocked by a male figure, perhaps the sultan himself or his head eunuch. Bouchard's play with modes of spectatorship was a marked feature of Orientalist depictions of the harem, but in the wake of Bell's paper and the Milan Congress, this picture was not simply an idle fantasy. It enjoyed a "spectacular success with the public," and, as one critic remarked: "The dramatic subject, relieved by a somewhat gaudy virtuosity, chills women's skin with an irresistible shiver and provokes elsewhere intense fantasy and reflexion."[89] This chilling interaction among deafness, race, and sexuality evoked by Bouchard motivated the oralist campaign of the 1880s and was soon backed up with a new host of anthropometric and psychological studies, seeking to make the deviance Bouchard had so clearly envisaged into a scientific "fact." Meanwhile, as the harem tradition declined, until its eventual abolition in 1908, the first school for the deaf in Istanbul was opened in 1891 by the Frenchman F. G. de Grati, teaching the oral program.[90]

The modern period was marked by a propensity to make difference visible. In seeking to understand and cure deafness, doctors, linguists, and psychologists have had to confront the invisibility of the complaint. Definitions of deafness were and remain a highly contested area. Prior to the nineteenth century, it is reasonable to assert that deafness, in the modern sense, did not exist. There were deaf people, certainly, but their affliction was regarded as an incurable mystery sent by God. For the materialist philosophers of the Enlightenment, deaf people were an interesting case study of the importance of sensory input to the formation of intelligence and rationality. Philosophers such as Diderot and Condillac were intrigued to know if the world could be understood by those with only four senses. It so, what kind of world did deaf people know? But the appearance of categories of the normal and the pathological marked a radical discontinuity in Western attitudes to the body. A medical condition known as deafness was defined and a cure sought. The sign language of the deaf was a visible symptom of deafness, and researchers sought an equally visible cause for the problem. The primacy attached to speech throughout the modern era has ensured that deafness, and its concomitant difficulties in speaking, has always been defined as pathological. Deaf sign language is a system of visual signs that fulfills all the functions of speech. William Stokoe, the pioneer linguist of sign language in this century, has put it succinctly: "Because American Sign Language is the medium of communication used by a community of people . . . anything expressible in another language can be expressed in it."[91] This seemingly unexceptionable statement was in fact part of Stokoe's effort to undo one hundred fifty years of opposition to sign language. At the Milan Congress of Deaf Educators (1880), sign language was outlawed in favor of speech in all aspects of deaf education and has only recently returned to schools for the deaf.

This history I have offered in this chapter cannot be concluded, for it is not yet over. The very definition of deafness, which is at the heart of both the cultural interpretation of deafness and social policy toward the deaf, remains an area of contestation. The Americans with Disabilities Act (1990) included the deaf as one of the categories of the disabled, as most hearing people might expect. But Carol Padden and Tom Humphries have argued that, for the deaf themselves, " 'disabled' is not a primary term of self-identification, indeed it is one that requires a disclaimer."[92] For Padden and Humphries, the deaf are not a medically disadvantaged group, but a cultural minority, distinguished primarily by their use of Sign Language and a shared cultural tradition. They distinguish between the condition of deafness and being Deaf, which they indicate by the use of the upper case.[93] The former medical condition does not necessarily bring with it membership in the latter cultural group. Indeed, the sign HARD-OF-HEARING is used by the Deaf to refer to people with hearing loss who are not part of the Deaf community. For advocates of Sign Language, it is clear that the method known as Total Communication, much favored by hearing educators, retains the structure and syntax of spoken English and is consequently difficult for the profoundly deaf to learn. More importantly, the argument goes, imposing a language that is in effect foreign upon deaf children denies them the right to their own Deaf culture and language. This debate, which has now reached the national media, is symptomatic of the crisis over minority rights in a multicultural society after the Cold War. It has come to center around a device known as cochlear implants, surgically implanted devices that seek to mimic the auditory nerve.[94] Doctors continue to argue that they cannot stand by and allow deaf people to remain deaf. But advocates in the Deaf community, more mindful of the conflicted history of the Deaf, remain hostile to such "medicalized" methods, demanding, in one writer's words, "the right to live deaf," rather than the traditional right to difference.[95] In many universities across the United States, bitter disputes are in progress as to whether the study of American Sign Language can fulfill foreign language requirements. The arguments revolve around the traditional objections to sign language, which would have been familiar to Epée and Berthier, namely that it lacks a written format, that it is not a "real" language, and even that it is not a foreign language. The manner in which such conflicts are handled will determine whether the postmodern era can use its hindsight to advantage, or whether Hegel continues to be right, and the only thing we learn from history is that we learn nothing from history.

Notes

1. Corneille Le Bruyn (1702), quoted by Marcia Pointon, "Killing Bodies," in John Barell (ed.), *Painting and the Politics of Culture: New Essays on British Art, 1700–1850* (Oxford and New York: Oxford University Press, 1992), p. 63n. 58.

2. Ali Behad, "The Eroticised Orient: Images of the Harem in Montesquieu and His Precursors," *Stanford French Review* 13, no. 2–3 (Fall 1989), p. 110.

3. On phonocentrism, see Jacques Derrida, *Of Grammatology*, trans. Gayatri Chakravorty Spivak (Baltimore: Johns Hopkins University Press, 1976), which is a very practical guide to the hostility to deafness in the nineteenth century. Many issues touched upon in this article are dealt with at greater length in my book, *Silent Poetry: Deafness, Sign and Visual Representation 1750–1920* (Princeton University Press, forthcoming).

4. Michel Foucault, *Discipline and Punish: The Birth of the Prison*, trans. Alan Sheridan (Harmondsworth: Penguin, 1977), pp. 200, 205.

5. Jules Janin, Gérard de Nerval, Théophile Gautier et al., *Deburau* (Paris: Daubussson and Kugelmann, 1856), pp. 11–12.

6. Judith Wechsler, *A Human Comedy: Physiognomy and Caricature in Nineteenth-Century Paris* (Chicago: University of Chicago Press, 1982), p. 53.

7. See N. M. Penzer, *The Harem: An Account of the Institution as it Existed in the Palace of the Turkish Sultans with a History of the Grand Seraglio from Its Foundation to the Present Time* (London: George G. Harrap, 1936), pp. 29–31.

8. Paul Rycaut, *The Present State of the Ottoman Empire*, ed. Harry Schwartz (1668); reprint (New York: Arno Press and the New York Times, 1971).

9. Joseph Pitton de Tournefort, *Relation d'un voyage du Levant, faite par ordre du Roy.* 2 vols. (Paris: Imprimérie Royale, 1717), 2: 287.

10. Tournefort, *Relation d'un voyage du Levant*, 2: 305.

11. Jean Racine, *The Complete Plays*, 2 vols., trans. Samuel Solomon (New York: Random House, 1967), 2: 58.

12. Tournefort, *Relation d'un voyage du Levant*, 2: 304.

13. Behad, "The Eroticised Orient," p. 112.

14. Tournefort, *Relation d'un voyage du Levant*, p. 288.

15. Jean-Jacques Rousseau, *On the Origin of Language,* trans. John H. Moran and Alexander Gode (Chicago: University of Chicago Press, 1966), p. 9.

16. In the *Cratylus,* Socrates uses the example of deaf sign language to argue that names are an imitation, just as painting is an art of imitation; *The Dialogues of Plato. Translated in English with Analyses and Introductions,* by B. A. Jowett, 2d ed. rev. and corr. (Oxford: Clarendon Press, 1875), sections 422–26.

17. Abbé Charles-Michel de l'Epée, *La véritable manière d'instruire les sourds et muets* (1784; reprint, Paris: Fayard, 1984), p. 9.

18. Robert Hawker, *The History of the Asylum for the Deaf and Dumb* (London: Williams and Smith, 1805), p. 5.

19. Abbé Charles-Michel de l'Epée, *Institution des sourds et muets* (Paris: Nyon l'aine, 1776), pp. 181–82.

20. Georges Canguilhem, *The Normal and the Pathological,* introduction by Michel Foucault, trans. Carolyn R. Fawcett (New York: Zone, 1991), p. 246. Reprint. Originally published as *On the Normal and the Pathological* (Dordrecht: Reidel, 1978).

21. Thomas H. Gallaudet, *A Sermon on the Duty and Advantages of Affording Instruction to the Deaf and Dumb* (Concord, N.H.: Hill, 1824), p. 8.

22. See Ludmilia Jordanova, *Sexual Visions: Images of Gender in Science and Medicine between the Eighteenth and Twentieth Centuries* (Madison: University of Wisconsin Press, 1989), p. 95.

23. Canguilhem, *The Normal and the Pathological,* p. 40.

24. All materials cited from the file "Lefebvre" in Archives Nationales, Paris, série F 15/2587.

25. Philippe Pinel, *Traité médico-philosophique sur l'aliénation mentale ou la manie* (Paris: Richard, Caille et Ravier, an IX [1801]), pp. 165–69.

26. See Harlan Lane, *The Mask of Benevolence: Disabling the Deaf Community* (New York: Knopf, 1992), pp. 212–13 for details of Itard's procedures. Horrific as these "treatments" may seem today, it is only reasonable to note that they were no more or less barbaric than other nineteenth-century medical procedures in the era before anesthetic. It is our knowledge that they could never have had a beneficial effect that makes them seem so disturbing.

27. [Berthier et al.], *Les sourds-muets au XIXe siècle* (Paris: Institut National des Sourds Muets, 1846), p. 10.

28. Johanne Christoph Hoffbauer, *Médécine légale relative aux aliénés et aux sourds-muets,* trans. A.-M. Chambeyron (Paris: J.-B. Baillière, 1827), with notes by Itard and Esquirol, p. 181.

29. Ibid., pp. 210–11.

30. Such attitudes have not disappeared. A 1985 psychiatric report concluded that "[p]rofound deafness that occurs prior to the acquisition of verbal language is socially and psychiatrically devastating." Symptoms ascribed include "suspiciousness, paranoid symptomatology, impulsiveness, aggressiveness." *Plus ça change* . . . Quotations from Harlan Lane, *The Mask of Benevolence,* p. 35.

31. Harlan Lane, *When the Mind Hears: A History of the Deaf* (New York: Random House, 1984), p. 134.

32. Quoted by Christian Cuxac, *Le langage des sourds* (Paris: Payot, 1983), p. 121.

33. Robert's Dictionary suggests 1833 as the first usage, but it was in use at the Institute for the Deaf prior to this date.

34. *Journal des sourds-muets,* no. 8 (5 April 1895): 114.

35. Jacques Lacan, *The Four Fundamental Concepts of Psychoanalysis,* ed. Jacques-Alain Miller, trans. Alan Sheridan (New York: Norton, 1981), p. 107.

36. Jacques Derrida, *The Truth in Painting,* trans. Geoff Bennington and Ian McLeod (Chicago: University of Chicago Press, 1987), p. 68.

37. This theme could be pursued at length in a rereading of Maurice Merleau-Ponty's *Le Visible et l'Invisible; suivi de notes de travail* (Paris: Gallimard, 1964) with Paul de Man's "The Rhetoric of Blindness," in *Blindness and Insight: Essays in the Rhetoric of Contemporary Criticism,*

2d rev. ed. (Minneapolis: University of Minnesota Press, 1983), and Jacques Derrida's essay accompanying his exhibition catalogue, *Mémoires d'aveugle: L'autoportrait et autres ruines* (Paris: Réunion des musées nationaux, 1990).

38. Malek Alloulah, *The Colonial Harem*, trans. Myrna Godzich and Wlad Godzich (Minneapolis: University of Minnesota Press, 1986), p. 18.

39. Quoted in Timothy Mitchell, *Colonizing Egypt* (Cambridge: Cambridge University Press, 1988), p. 112.

40. See Claude Blanckaert, "On the Origins of French Ethnology: William Edwards and the Doctrine of Race," in George W. Stocking, Jr. (ed.), *Bones, Bodies, Behavior: Essays on Biological Anthropology* (Madison: University of Wisconsin Press, 1988), 18–55; Robert Nye, *Crime, Madness, and Politics in Modern France: The Medical Concept of National Decline* (Princeton: Princeton University Press, 1984); and George W. Stocking, Jr., *Victorian Anthropology* (New York: Free Press, 1987).

41. Hippolyte Taine, *The History of English Literature*, trans. H. Van Laun (London: Chatto and Windus, 1871), pp. 15–16.

42. André Regnard, *Contribution à l'histoire de l'enseianment des sourds-muets* (Paris: L. Larose 1902), p. 51. See Cuxac, *Le Langage des sourds*, pp. 144–47, for commentary on Regnard.

43. Naomi S. Baron, *Speech, Writing, and Sign: A Functional View of Linguistic Representation* (Bloomington: Indiana University Press, 1981), p. 207.

44. Taine, *History of English Literature*, p. 9.

45. Quoted by Nina Athanassoglou-Kallmyer, *French Images of the Greek War of Independence (1821–1830): Art and Politics under the Restoration* (New Haven: Yale University Press, 1989), p. 92. This book provides a useful, if unashamedly pro-Hellenistic, survey of attitudes to the war.

46. Lee Johnson, *The Paintings of Eugène Delacroix: A Critical Catalogue*, 6 vols. (Oxford: Clarendon, 1981), 1: 114.

47. Ignatius Mouradgea d'Ohsson, *Tableau général de l'empire othoman* (1820) and Alphonse Lamartine *Voyage en orient*, quoted by Théophile Denis, "Les Muets du Sérail," in *Études variées concernant les sourds-muets* (Paris: Imprimerie de la Revue française de l'enseignement des sourds-muets, 1895), pp. 138, 135.

48. I have pursued this line of inquiry in my *The Body in Visual Culture from Enlightenment to Postmodernism* (Routledge, forthcoming).

49. Quoted by Timothy Mitchell, *Colonising Egypt*, p. 29.

50. Edward W. Said, *Orientalism* (New York: Pantheon, 1978), p. 43.

51. Adolphus Slade, *Records of Travels in Turkey, Greece, etc., and of a Cruise in the Black Sea with the Capitan Pasha in the Years 1829, 1830 and 1831*. 2 vols. (Philadelphia and Baltimore: Carey and Hart, 1833), 1: 239.

52. Joseph Marie de Gérando, *De l'education des sourds-muets de naissance*. 2 vols. (Paris: Méquignon, 1827), 2: 594.

53. James Hunt, "On the Negro's Place in Nature," in *Journal of the Anthropological Society of London*, vol. 2 (1864), p. xvi. See also Henry Louis Gates, Jr., "Authority, (White) Power, and the (Black) Critic; or, It's All Greek to Me," in Ralph Cohen (ed.), *The Future of Literary Theory* (New York: Routledge, 1989), p. 326.

54. Quoted by Tamar Garb in "*L'art féminin*: The Formation of a Critical Category in Late-Nineteeth-Century France," *Art History* 12 (March 1989): 62n. 41. Although Garb's example is from 1907, the attitude has a much longer history.

55. Nancy Leys Stepan, "Race and Gender: The Role of Analogy in Science," *Isis* 77 (1987): 261–77.

56. Roch-Ambroise Auguste Bébian, *Essai sur les sourds-muets et sur le langage naturel* (Paris: J. G. Dentu, 1817), pp. 78, 87–89.

57. Ibid., p. 106.

58. Michel Foucault, *The Order of Things: An Archaeology of the Human Sciences* (New York: Random House, 1970), p. 296.

59. Ferdinand Berthier, *Les sourds-muets avant et depuis l'Abbé de l'Epée.* (Paris: J. Ledoyen, 1840), p. 53.

60. Derrida, *Of Grammatology*, p. 56.

61. Adolphe Franck, quoted in Ch. Le Pere, *Statue de l'Abbé de L'Epée, oeuvre de Félix Martin* (Paris: Institution Nationale des Sourds-Muets, 1879), p. 46.

62. Homi Bhaba, "Of Mimicry and Man: The Ambivalence of Colonial Discourse," in *October: The First Decade, 1976–1986*, ed. Annette Michelson et al. (Cambridge: MIT, 1987), p. 318. See also Robert Young, *White Mythologies: Writing History and the West* (New York: Routledge, 1990).

63. See Marc Bloch, *The Royal Touch*, trans. J. E. Anderson (New York, 1973).

64. Sébastien Guillé, quoted by William R. Paulson, *Enlightenment, Romanticism, and the Blind in France* (Princeton: Princeton University Press, 1987), p. 95.

65. Abbé C. Carton, *Le sourd-muet et l'aveugle* 2 vols. (Bruges: Vandescasteele-Werbrouck, 1840), p. 1.

66. E. Hamon de Fougeray, *Manuel pratique des méthodes d'enseignement spéciales aux enfants anormaux.* (Paris: Alcan, 1986), pp. 131, 19.

67. Thomas Arnold, *The Languages of the senses, with special reference to the education of the deaf, blind, deaf and blind* (Margate, England: printed by Keble's Gazette, 1894), pp. 9–15.

68. Adolphe Desbarrolles, *Chiromancie nouvelle. Les mystères de la main revelés et expliqués* (Paris: E. Dentu, 1859), p. 419.

69. Havelock Ellis, *The Criminal* (New York: Scribner & Welford, 1890), p. 83.

70. Ibid., p. 118.

71. G. Jelgersma, "Les caractères physiques intellectuals et moraux reconnus chez le criminel né sont d'origine pathologique," *Actes du Troisième Congrès International d'Anthropologie Criminelle* (Bruxelles: F. Hayez, 1893), p. 32–33.

72. See Donald E. English, *Political Uses of Photography in the Third French Republic, 1871–1914* (Ann Arbor: UMI Research Press, 1984), pp. 75–79. See also Allan Sekula's influential essay describing the connection between anthropometry and photography, "The Body and the Archive," *October* 39 (1986).

73. Reproduced in John Szarkowski (ed.), *Photography until Now* (New York: Museum of Modern Art, 1989), p. 89.

74. See, for example, the portraits of Birmingham prisoners from 1860–62 reproduced in John Tagg, *The Burden of Representation: Essays on Photographies and Histories* (Amherst: University of Massachusetts Press, 1988), p. 58.

75. Edmond Duranty, "The New Painting: Concerning the Group of Artists Exhibiting at the Durand Ruel Galleries," in Linda Nochlin (ed.), *Impressionism and Post-Impressionism, 1874–1904: Sources and Documents* (Englewood Cliffs, N.J.: Prentice-Hall, 1966), p. 5.

76. On the artist as a detective, see Carlo Ginzburg's "Clues: Roots of an Evidential Paradigm," in *Clues, Myths, and the Historical Method*, trans. John and Anne C. Tedeschi (Baltimore: Johns Hopkins University Press, 1989).

77. Francis Galton, *Finger Prints* (London: Macmillan, 1892), pp. 2, 14.

78. Quoted in *Journal des Sourds-Muets* 16 (5 August, 1985): 245.

79. For a discussion of the influence of biological theory in Third Republic France, see Robert A. Nye, *Crime, Madness, and Politics in Modern France* (Princeton, N.J.: Princeton University Press, 1984).

80. He took a series of photographs of the ancient monuments of Egypt—and of Flaubert in Arab dress—that have won him a place in the canon of photographic pioneers. He himself disparaged this achievement: "To learn photography is no great thing; but to transport the equipment on the backs of mules, camels and men was a difficult problem." Quoted in Abigail Solomon-Godeau, *Photography at the Dock: Essays on Photographic History, Institutions, and Practices* (Minneapolis: University of Minnesota Press, 1991), p. 160.

81. Maxime du Camp, from *Revue des Deux Mondes* (15 April 1873), translated as "The Paris Institution," in *The American Annals of the Deaf* 21, no. 1 (January 1877): 12.

82. Quoted by Lane (*When the Mind Hears,* p. 391), who provides a detailed history of the Milan Congress (pp. 377–95).

83. One of the features of postmodernism is the reemergence of political, cultural, and intellectual currents, which modernism believed it had disposed of forever. It is no accident that, even as the scientific measures of the deviance of sign language have become discredited, what A. de Candolle, the French translator of Bell, called "the antipathies and distrust caused by the silence and uncouth gestures of the deaf-mutes who use signs," persists. (Quoted in *American Annals of the Deaf* 21, no. 2 [April 1886]: 148.)

84. Alexander Graham Bell, "Upon the Formation of a Deaf Variety of the Human Race," *Memoirs of the National Academy of Sciences,* vol. 2 (1883), p. 216.

85. Ibid., p. 221.

86. The Commission led to the abolition of sign language education, which had remained widespread in Scotland, Wales, Ireland, and England outside London.

87. Bénezit gives the location of this work as Munich, but no museum there now owns it, and I presume it has been lost.

88. Société des artistes français, *Catalogue illustré du salon* (Paris: L. Baschet, 1887), p. 236.

89. Denis, "Les Muets du Serail," p. 134.

90. Volta Bureau (U.S.), *International Reports of Schools for the Deaf.* (Washington, D.C.: Gibson, 1898), p. 27.

91. William Stokoe, quoted by Naomi S. Baron, *Speech, Writing, and Sign: A Functional View of Linguistic Representation* (Bloomington: Indiana University Press, 1981), p. 212.

92. Carol Padden and Tom Humphries, *Deaf In America: Voices from a Culture* (Cambridge, Mass.: Harvard University Press, 1988), p. 44.

93. The capital *D* in *Deaf* signifies the distinction between the medical condition and the culture. It is also a mark of identification for those who consider themselves to be members of the Deaf community. As this paper is largely concerned with periods prior to the emergence of this strategy, I have used the lowercase throughout.

94. For the cochlear controversy, see the article "Pride in a Soundless World: Deaf Oppose a Hearing Aid," *New York Times,* 16 May 1993, A1, and resulting letters from very different points of view 27 May 1993, A26. The case against the implants is made by Lane, *The Mask of Benevolence.*

95. Cuxac, *Le Langage des sourds,* pp. 181–85.

3

Colonizing and Transforming the Criminal Tribesman

The Salvation Army in British India

Rachel J. Tolen

INDIA HAS LONG been linked in the Western world with two particularly marked social phenomena. The first is its caste system, which has been understood to be based on a special understanding of the body, its substance, its inlets, and its boundaries (Daniel 1984; Marriott 1976; Marriott and Inden 1977).[1] India has also retained a powerful hold on the Western imagination as the opulent "Jewel in the Crown," the prized colony of the British Empire, a land of adventure and conquest. Yet scholars have only rarely acknowledged that these two powerful concerns, body and empire, bear upon each other. This article examines a particular instance of the way in which Indian bodies and bodies of Indians were recruited into the discourse and polity of colonial India.

In anthropology, the concern with how the social order is impressed upon the body can be traced to Marcel Mauss (1973[1936]). Mauss recognized that the "techniques of the body" are not simply a matter of nature, but also, universally, a social and *historical* matter (Mauss 1973[1936]:83). Later theorists have addressed a similar problem, but their work has signaled a shift in focus from the body's significance in the maintenance of tradition and structures of solidarity to its pivotal role in the social order's structures of domination (Bourdieu 1977; Foucault 1979), as well as in resistance to domination (Comaroff 1985).

Recent studies have suggested that the techniques of the body under British rule had powerful transformative effects within Indian society. Bernard S. Cohn's essay "Cloth, Clothes and Colonialism" has initiated the study of habiliment under British rule, examining the play of contrasts in dress between ruler and ruled and the way in which rebellion was clothed in the twentieth century (Cohn 1989). As many writers have noted, Gandhi's body became the site of a battle for *swaraj*, or self-rule, expressed in a struggle over the body's vestiture, gender, carnality, production, and consumption (Appadurai 1978; Gandhi 1927; Nandy 1983; Rothermund 1963). Carol A. Breckenridge has drawn attention to the ways in which India's representation as the "Jewel in the Crown" at the 1851 Crystal Palace exhibition in London highlighted the sumptuary politics of the adornment of the body (Breckenridge 1989). Christopher Pinney has examined the confluence of anthropometric techniques and official theories of caste in a normalizing grid for classifying peoples in British India (Pinney 1989). And Ranajit Guha has shown that while bodily conduct and gestures of subservience were salient features of relations of domination, the "reversal" of such signs constituted one of the "elementary aspects of peasant insurgency" in colonial India (Guha 1983). Suggesting that the body did not emerge untouched by colonial rule, these studies indicate a shift away from transhistorical models of "Hindu" or "South Asian" systems of bodily practice (exemplified by the work of Marriott and Inden), toward analyses that (with Mauss) situate the body in historical processes, processes that have generated a great deal of heterogeneity in "South Asian" thought and practice.

Bernard S. Cohn's much cited, but only recently published, essay entitled "The Census, Social Structure and Objectification in South Asia" (1987) maintained early on that an understanding of caste required an understanding of colonial history. Under colonial rule, hundreds of "castes" and "tribes" in India came to be legally defined as "criminal," and recent studies have begun to take a critical look at caste and criminality in colonial India. David Arnold has examined the twin concepts of the "martial races" and "criminal tribes" and their relation to British perceptions of Indian society (Arnold 1979, 1984), while Sandria Freitag has underscored the importance of space and place in shaping relations between criminal tribes and the colonial state (1989). They and other contributors to the volume *Crime and Criminality in British India* have provided useful benchmarks for the present analysis (Arnold 1985; Freitag 1985; Yang 1985). While these studies have drawn attention to the relation of British conceptualizations of caste and criminality to structures of domination, my analysis situates that relation in bodily practice. I focus on the techniques of representation and reform that were used to produce a particular kind of body, one subjected yet productive. Central to the processes of colonizing the body of the criminal tribesman was the objectification of the body's activity as labor.

The judicial instrument for transforming indigenous communities into "criminal tribes" was the Criminal Tribes Act, first passed in 1871 but applied at that time only to the Northwest Provinces, Oudh and Punjab. The original act provided for the registration, surveillance, and control of criminal castes and tribes. In 1911 the act was extended to include the Bombay and Madras presidencies and to provide for greater police and judicial control over criminal castes. These judicial measures set the scene for the body of the criminal tribesman to become a subject of colonial discourse and reform, during a period extending from the era of "High Empire" and utilitarian reform at the close of the nineteenth century to the end of colonial rule in the middle of the twentieth.

While this analysis deals with the public discourse on criminal caste in the colonial state of India as a whole, it also includes a specific regional concentration: the middle section of the article focuses on representations of criminality in the administrative literature produced by the Madras Presidency. The final sections of the article again transcend regional boundaries, in an analysis of practices of reform enacted by government and Salvation Army functionaries throughout the subcontinent, an analysis connecting these practices to the growing dominion of the empire.

The Constitution of the Notion of a Criminal Caste

The constitution of the notion of a "criminal caste" was a gradual process, involving changes not only in the way in which particular Indian communities were represented but also in the way in which history itself was represented. In 1904, the *Madras District Gazetteer* for the Bellary District of southern India described the predicament of two Indian communities thus:

> The railways, again, have robbed some of the people of their only employment. Before the days of trains the wandering Korachas and Lambadis lived by trading with the west coast, driving down there once or twice a year large herds of pack-cattle laden with cotton, piece-goods, etc., and returning with salt, areca, cocoanut and so forth. This occupation is now gone and these two castes, driven to less reputable means of livelihood, are responsible for much of the crime of the district. (Francis 1904:184)

Detailed paragraph-by-paragraph revisions of the *Gazetteers* were issued periodically. In 1930 the government appended the following revision:

> The excuse that the original occupations of the Lambadis and Korachas have disappeared, has become threadbare with age. It may be safely asserted that these people always had criminal leavings [*sic*] even if their tours to the West Coast were ostensibly for purposes of trade. (Government of Madras 1930:97)

The transformation of hundreds of communities like those of the Lambadis and Korachas into criminal castes or tribes marked the confluence of several strains of social thought. Two major bodies of knowledge were instrumental in constituting the notion of a criminal caste: (1) ideas about the nature of Indian society derived from colonial anthropology; and (2) ideas drawn from the contemporary discourse on crime and class in Britain. I will consider each of these spheres of thought independently and then examine their convergence in a theory of criminality in India.

Colonial rule was realized not only through the consolidation of overt "political" power but also through the consolidation of knowledge about the ruled. Such massive projects as the census, the *Gazetteers,* and a series of *Castes and Tribes* volumes written about the different regions of India were undertaken with the goal of amassing a body of knowledge about the various peoples of India, their customs, and their manners, in order to aid in their efficient administration (see, for example, Thurston 1909). In the *Castes and Tribes* volumes, sections of Indian society were described with unconditional authority regarding their physical appearance, traditional occupations, religious practices, personal habits, and behavior. All such traits were objectified (Cohn 1987), associated with the names of "castes," defined with taxonomic precision, and arranged in alphabetical order. For British administrators, "caste was a 'thing' " (Cohn 1968:15), and its reification became one of the dominant modes of representation of Indian society. Throughout the administrative literature is woven the assumption that great uniformity of behavior can be expected from all persons of a particular caste. Thus, Edgar Thurston, the author of *Castes and Tribes of Southern India,* could write of the Kuravars (or "Koravas"): "Their conduct is regulated by certain well-defined rules" (Thurston 1909:IV, 497).[2]

British ideas about the nature of Indian society were not the only source of the notion of a criminal caste, for the British had an extensive body of beliefs about crime and class based on conceptualizations of their own society (Yang 1985). In the nineteenth century, many social theorists, legislators, philanthropists, and middle-class reformers became engaged in a discourse on crime and its causes, a discourse that cannot be easily extricated from the general discourse on class in Britain (Jones 1982). By the mid-nineteenth century, the idea of the "dangerous classes"—who were composed of the unemployed, vagrants, the poor, criminals, drunkards, and prostitutes—was firmly ensconced in Victorian thought, and a common discourse identified their physical characteristics, habits, and locale. Various causes were proposed to explain the criminality of the dangerous classes: strong drink, ignorance, poor upbringing, indigence, character defects, and hereditary predisposition. The theory that certain people had an inborn propensity for crime implied that nothing, other than overt control, could prevent them from acting on such propensities. Yet theories of hereditary criminality did not rule out attempts at reform, and proponents of reform often invoked biological as well as social explanations of crime (Mayhew and Binny 1968[1862]).

It is clear that the contemporary discourse on crime in Britain influenced British officials' perceptions of, and responses to, Indian society. "Class" and "caste" converged in the discourse on crime in India. In the first major work to catalogue the criminal sections of southern India, *Notes on Criminal Classes of the Madras Presidency,* Frederick S. Mullaly used the terms "caste," "class," and "tribe" interchangeably (Mullaly 1892). Some of the later *Gazetteers* also used "criminal caste," "criminal class," and "criminal tribe" interchangeably. However, over time usage generally shifted to "criminal tribe" or "criminal caste," as the British notion of the "criminal classes" was adapted to the colonial administrative view of Indian society.

A number of theories about the nature of Indian society converged in the notion of a criminal caste. The behavior of Indians was guided by caste, caste defined occupation, and caste was easily, if vaguely, related to racial and hereditary theories of criminality. If certain Indians committed crime, it was because their castes prescribed such behavior. During the period when the judicial apparatus for the control of criminal castes was being set in place, James Fitzjames Stephen testified before the legislature that

> traders go by castes in India: a family of carpenters now will be a family of carpenters a century or five centuries hence, if they last so long. . . . If we only keep this in mind when we speak of "professional criminals," we shall then realize what the term really does mean. It means a tribe whose ancestors were criminals from time immemorial, who are themselves destined by the usages of caste to commit crime, and whose descendants will be offenders against the law, until the whole tribe is exterminated or accounted for. (Governor-General of India 1871:419–20)

The formulaic phrase "from time immemorial" would be repeated throughout the later literature on criminal castes (for two examples, see Thurston 1909:VII, 503 and Haikerwal 1934:144). Indian society was seen as static, archaic, and primitive, in contrast to the progressive societies of the West, and criminality was itself a sign of India's primitive condition.

I have noted that "criminal caste" and "criminal tribe" were used somewhat interchangeably. But the use of the term "tribe" served a distinct representational function. It evoked both an evolutionary stage and certain values and images. There was a long-standing debate on how to differentiate "tribal" groupings from "caste" groupings in India, a debate that persisted in anthropology up to the 1960s (Bailey 1961). "Tribe" was situated on a lower rung than "caste" on an evolutionary scale. But the peculiar use of the term "tribe" to evoke a set of images is clear from the logic used to explain the causes of crime: criminal *tribes* committed crime because it was dictated by their *caste* to do so. "Caste," rather than "tribalism," was the distinctive causal feature of this breed of criminality. But the term "tribe" could evoke qualities of savagery, wildness, and otherness in a way that "caste" could not. The term "criminal tribe" was often favored because of the signs it was able to produce in British consciousness.

While colonial rule often proceeded according to preservationist policies that had the effect of fixing indigenous forms, it sometimes compelled the remolding of the society of the colonized. In such instances, "tradition" often appeared to be an obstacle to this process of transformation. Indian behavior was guided by tradition and by caste, and each represented an obstacle to the remolding of Indian society in general and the reform of the criminal tribes in particular. Crime was one manifestation of the debased nature of "tradition" in Indian society. It was the result not "of personal demoralization, but of group traditions" (Cressey 1936:18). Therefore, reform was to be directed toward the "traditional attitudes and conduct of the group as a whole" (Cressey 1936:18). Some communities may indeed have come to espouse "crime" as a "caste tradition," in response to asymmetries of power in the indigenous social order or in resistance to (or even open defiance of) domination in the newly emerging social and economic order instituted by the colonial state.[3] The British struggle to gain taxonomic control over social relations in Indian society represented an effort at remodeling, rather than a playing out of pure fantasy. As no British writer failed to observe, one of the meanings of the word "Kallar" (the name of one prominent criminal caste) was thief.[4] But theft could only make one a "criminal" in a particular type of social order. The procedures used to classify persons as "criminal tribesmen" carried new categorical emphases with new associations.

The Kallars were one of a number of prominent groups in the Madras Presidency that had been involved in the indigenous *kaval* system. *Kaval* was a system whereby a *kavalkaran,* or watchman, was collectively paid by a village for protection of its property. Some of the communities involved in *kaval* had been part of an organized military and intelligence force whose ties with indigenous systems of rule in southern India had caused them to come into conflict with British forces during the early days of the subcontinent's subjugation by the East India Company. By all British accounts, the *kaval* system had become one of "protection" payments demanded by organized criminal gangs under the threat of arson or other means, if necessary. But the overriding concern of the British was not that *kaval* was a system of exploitation of poor villagers, but that it represented an alternative and contending focus of power in the region. In the words of one administrator, the Kallars (the chief caste against which anti-*kaval* measures were brought) represented a "thorn in the flesh of the authorities" (Francis 1906a:89).

Certain of these prominent groups, most notably the Kallars and the Maravars, have received considerable scholarly attention (Blackburn 1978; Dirks 1987; Dumont 1986). In focusing on particular groups and regions, however, many scholars have failed to come to adequate grips with the multiple meanings of the concept of the "criminal caste" itself and its multiple uses under colonial rule.[5] Stuart Blackburn goes so far as to write that, in Madurai, the Criminal Tribes Act was "in essence a Kallar Control Act" (Blackburn 1978:48). But hundreds of communities were brought under the Criminal Tribes Act. In 1931, 237 tribes were being treated under the act in the Madras Presidency alone (Government of Madras 1933:117). The concept of the criminal caste, as I have shown, drew on a number of different discourses, divorced from application to any single community. I have therefore chosen an alternative approach, one that centers on the multiple uses of the concept of the criminal caste, a concept derived from *many places* (Appadurai 1988). In order to focus this perspective, I will now turn to an analysis of representations of criminality in the administrative literature produced by the Madras Presidency.

Representations of Criminality and the Morphology of the Criminal

By 1911, when the Criminal Tribes Act was extended to the Madras Presidency, an established body of knowledge about criminal castes, including theories about their constitution, had already been developed. Following the 1892 publication of Mullaly's *Notes on Criminal Classes,* further descriptions appeared in the *Gazetteers,* Thurston's *Castes and Tribes of Southern India* (Thurston 1909), and various other publications. These accounts presented their subjects in remarkably uniform and conventional terms. Thurston's and Mullaly's accounts remained authoritative and were cited heavily in later descriptions. Often, the native assistants used in the collection of information had been drawn from the ranks of the prestigious and the powerful, and too often the portrayals of "criminal castes" appeared to rely on hearsay and innuendo, undoubtedly filtered through the particular prejudices of the chosen informants. Vanniyans, for instance, were said to "have rather a bad name for crime" (Hemingway 1907b:111).

Many of the writers on criminal castes relied heavily on information provided by the police, who concerned themselves with identifying the castes of people who were suspected of criminality, and some of these writers, such as Mullaly, had official careers in the police force. But the task of collecting information about the castes of "criminals" naturally met with a number of obstacles, and attempts to extract information from those actually apprehended in offenses encountered some resistance. As Mullaly acknowledged: "[I]t has been a difficult task to induce the people to gratify curiosity when questioned as to their caste customs; this, however, may possibly be accounted for by their being themselves ignorant of their own folk-lore" (Mullaly 1892:v).

Nor was the attempt at codification of the "criminal castes" always a simple matter, for there was often a good deal of confusion about the names of "castes" and their distinction from the names of "gangs" or other groups. One "criminal caste," the Donga Dasaris, was found to be not a "true caste at all," for its members did not exhibit any of the "ordinary signs of caste organization," such as caste *panchayats* or endogamy (Francis 1906b:251). Mullaly cited "Alagiri" as the general name of a broad class of criminals who worshiped Kalla Alagar, a south Indian deity. Many criminal castes were supposed to worship Kalla Alagar, who was associated with theft, and thieves were said to devote a portion of their loot to him. Mullaly (1892) identified some of the "classes" of "Alagiries" as "Capemaries," "Donga Dasaries," "Gudu Dasaries," and "Padayachies." Yet the term "Alagiri" was also "applied by the police to persons suspected by them whose castes are not positively known and who give a vacillating account of themselves, and more particularly to persons found in crowds at fairs and festivals who cannot give a full and satisfactory account of themselves" (Mullaly 1892:1). As colonial rule was realized through a consolidation of knowledge about the colonized, those who were unknown to the law were conceived to be alien to it and outside the sphere of legality in British India.

Yet the administrative literature did not simply codify criminality, it fashioned images of it. It drew the attention of the administrative mind to a shadowy but ever-present figure. In the discourse on criminal caste, criminality was divorced from any particular act of crime, for it was something that ran much deeper in Indian society. Criminality was a potentiality that might be manifested in certain persons and practices at any time or place. It was a pervasive, underlying tone that suffused the fabric of Indian social life. Criminality was an inward *quality*, a potentiality, that certain persons were predisposed to manifest.[6]

This aspect of criminality may be seen by examining a number of the ways in which criminality was represented. Much of the discourse on criminal caste emphasized the pervasive and concealed criminality of large sections of the Indian population. In the literature there are many descriptions of persons who appear to be ordinary, law-abiding subjects but who conceal a hardened and dangerous criminality. In *Notes on Criminal Classes,* Mullaly described the Kepumaris (or "Capemaries") thus:

> All the large fairs and festivals throughout the presidency are visited by these people in parties of four or five, usually accompanied by women and

> boys. The men adopt the disguise of respectable traders; well dressed, wearing caste marks and other outward signs of respectability, they gain admission to temples, caste chatrams and places resorted to by travellers and pilgrims; they acquaint themselves with the names of persons of good social standing and ascertain some facts connected with their private life; thus they are able to pass themselves off as acquaintances, and are taken into the confidence of their victims. (Mullaly 1892:3–4)

In later descriptions of other criminal castes, apparently harmless sections of Indian society were portrayed as a hidden but ever-present menace.[7] One section of the Kepumaris, for example, was said to "dress decently, imitate the ways and speak the languages of different castes" (Government of Madras 1931:120). Kepumaris were also said to "avert suspicion by their respectable appearance and pleasant manners" (Francis 1906b:252).

The theory that a concealed criminality pervaded Indian society was further manifest in the concept of the "ostensible occupation." Criminal castes were said to engage in legitimate pursuits, but only in order to divert attention from their "real" occupation, crime. Of Kallars it was written that their "usual ostensible occupation is cultivation, but as a matter of fact they live largely by crime and blackmail" (Hemingway 1907b:107). A description of the Kuravars (or "Koravas") further illustrates this feature of criminality:

> To all outward appearance they lead respectable lives. To guard against the security sections of the Penal Code, they purchase a little, not very valuable, land, and lease it out for a small fee for others to cultivate. When asked by the Police how they earn a livelihood, they can point to their land and cattle, and pose as agriculturalists. (Richards 1918:93)

An inversion of the theme of concealed criminality was also reflected in the portrayal of criminal castes. Rather than being hidden, that is, criminality was an open manifestation of defiance. Some communities, it was suggested, lacked any sense of shame, even going so far as to legitimate their criminal activities through "traditions" and religious practices. Kallars were said to be not "ashamed of the fact" of their criminality (Francis 1906a:89). One Kallar apparently "defended his clan by urging that every other class stole—the official by taking bribes, the vakil [lawyer] by fostering animosities and so pocketing fees, the merchant by watering the arrack [a type of liquor] and sanding the sugar. . . . [T]he Kalla[r]s differed from these only in the directness of their methods" (Francis 1906a:89). Such may in fact have been the attitude of many displaced by the new colonial order. Accounts of criminal castes also frequently described the religious practices of criminals who devoted a portion of their loot to various deities (Hatch 1928:139–53). Such "traditions" legitimizing theft appeared to the British to be a ritualized negation of just relations of property ownership. Gangs of thieves may indeed have participated in such activities. But such representations also emphasized the British observers' view that Indian behavior was guided by "tradition" and that the chief locus of tradition in Indian society was caste.

The discourse on criminal castes in the administrative literature endeavored to define both criminality and the criminal. In general, the descriptions of criminal castes dealt extensively with the personal characteristics and habits of particular communities represented as *tokens* of their criminality. These representations aimed at defining a morphology of the criminal. Yanadis, for instance, were characterized as exhibiting "Mongolian" features. The men were "broad about the cheek bones . . . with a pointed chin, a slight moustache, no whiskers, and a scanty straggling beard over the forepart of the chin," while the women, though "tolerably well featured," had a "wild, timid, shrinking look" (Mullaly 1892:22). Thurston drew attention to the hairless appearance of the Kuravar ("Korava"), as he felt it conflicted with the "typical criminal of one's imagination"; "even the innocent looking individuals," he warned, "are criminal by nature" (Thurston 1909:VII, 471).

Modes of dress, deportment, disposition, habits, and mannerisms were important tokens as well. Jogi women could be known by their appearance, "decked in gaudy attire and laden with brass bangles and bead necklaces," at fairs and festivals, where they "profess[ed] to tell fortunes" (Mullaly 1892:19). In many cases, personal hygiene was central to the critique of the habits of criminal castes. Kuravar (or "Koravar") men were described as "dirty, unkempt looking objects" who wore "their hair long and usually tied in a knot on the top of the heads, and indulge[d] in little finery" (Mullaly 1892:61). Consuming inferior types of food was also a token of criminality. Mullaly claimed that the Voddas (or "Wudders") ate "every description of animal food" and were "much addicted, like most of the lower classes, to spirituous liquors" (Mullaly 1892:78); the Pariahs were "a dark-skinned race, eating every species of food, hardworking, thriving, yet intensely ignorant and debased" (Mullaly 1892:118). Consumers of hard liquor and eaters of rats, frogs, lizards, and jackals, criminal castes partook of forms of subsistence that not only offended the standards of British administrators, but also were undoubtedly considered defiling by the elites from whom the administrators obtained much of their information.

The nomadic habits of certain communities were particularly associated with criminality. Many criminal castes appear to have been nomadic, and the association of these nomadic communities with crime was influenced by the association of vagrancy with crime in Britain (Yang 1985). One of the *Gazetteers* described the Dommaras as a "wandering caste," noting that they subsisted "partly by breeding pigs and making date-leaf mats" and that their women, whose "frail morals [were] somewhat of a by-word," went about begging (Francis 1906b:252). "By nature" nomadic, Yanadis made a "scanty living by catching jackals, hares, rats and tortoises, by gathering honey, and by finding the caches of grain stored up by field-mice" (Hemingway 1907a:198).

Criminal castes in India, like vagrants in Britain, represented an obstacle to the institution of the British ethic of work-discipline. Yanadis, who were "very partial to sour and fermented rice-water," were "stopped by the Forest officers from drinking" it, as it made them "lazy, and unfit for work" (Thurston 1909:VII, 423). The marginality of these communities, their disinclination to "self-improvement," their lack of work-discipline, their poverty, their poor standards of hygiene, their nomadic tendencies—all were tokens signifying their criminality, and criminality represented resistance to wage labor and changing systems of production as well as to the maintenance of public order. Such traits were implicated not only in the definition of criminal castes in India, but also in the definition of "criminal classes," and especially vagrants, in Britain.

This section has focused on the representation of the Indian "criminal tribesman" in administrative literature written primarily for colonial officers. But this representation was not confined to documents prepared for administrators. The figure of the Indian criminal tribesman also found a niche in a distinctive genre of literature geared to entertaining and edifying a British audience at home. In *The Land Pirates of India: An Account of the Kuravers a Remarkable Tribe of Hereditary Criminals Their Extraordinary Skill as Thieves Cattle-lifters & Highwaymen &c and Their Manners & Customs* (Hatch 1928), W. J. Hatch revealed the Kuravar to be an incorrigible rogue. In the series "Seeley's Books of Travel," criminality was presented as the distinctive interesting feature of Indian society. Here the Kuravar found a place among *Arabs in Tent & Town, An Unknown People in an Unknown Land, Savage Life in the Black Sudan,* and *The Menace of Colour.* Many of the authors of the "Seeley's Books of Travel" were retired colonial officers whose literary mode of reminiscence allowed them to "tour" their pasts in the same way that collecting objects for display allowed them to represent their colonial careers to others (Breckenridge 1989:210). Hatch himself served with the London Missionary Society, but his work typified the genre of "collectible" ethnographic knowledge that colonial officers (and others who had "non-official" colonial careers) shared with popular audiences in Britain and the rest of the English-speaking world through the "Seeley's Books of Travel."

It is noteworthy that criminality was the "essence" selected to represent India in this survey of a world of others. But, as Arjun Appadurai has noted, "ideas that claim to represent the 'essences' of particular places reflect the temporary *localization* of ideas from *many* places" (Appadurai 1988:46). Notions of Indian criminality had complex genealogies, as this article shows. The practices of representation of the criminal essence of India provided a symbolic infrastructure for the confinement of the native in a place (Appadurai 1988). With this symbolic infrastructure went a material infrastructure that sought to reform the bodies of criminal tribespeople. Implicated in reform was submission to capital relations in the colonial economy, such that in the 1930s it could be written that: "There has been a marked movement amongst these tribes towards reformation and several members of them have now settled down to honest labour" (Ayyar 1933:197). I will now turn to signs and practices manifested in the system of reform of criminal tribes.

Systems of Control and Reform

The Criminal Tribes Act provided for a variety of measures aimed at the discipline and control of criminal castes. District magistrates were charged with keeping a register of the members of criminal castes in their designated districts. Upon registration, a trace of the body of each member of a criminal tribe was recorded in a fingerprint. This body was then transcribed into a space, as each member's residence was recorded. According to the law, criminal tribespeople were to notify the authorities of any changes in residence. The legislation also provided for police surveillance, for a pass system requiring each registered person to obtain a pass in order to leave his or her village or surrounding area, and for the relocation of whole communities to reformatory settlements.

The systems of reform of criminal castes, like those adopted for the reform of the "dangerous classes," encoded social relations in spatial geography (Freitag 1989). During the nineteenth century authorities undertook a variety of measures to combat vagrancy in Britain—most notably, labor colonies and a pass system. Under the pass system, persons apprehended for vagrancy were issued a ticket to pass on to the next district, where they were required to report to the police (Jones 1982:191). In Victorian England, class relations were encoded in spatial relations on the land: the homes of the dangerous classes were located in specific sections of the urban areas. But, embodied in the figure of the vagabond, the dangerous classes spread throughout the cities and countryside. The labor colonies in Britain served to contain the spread of the dangerous classes by enclosing the vagrant in a realm of concentrated labor.

Some of the criminal tribes reformatory settlements in the Madras Presidency were controlled directly by the government, under the supervision of the Labor Department (Government of Madras 1929:127). The stated objective of these reformatory settlements was "to reform members of so called criminal tribes by providing them with the means, under supervision, of earning an honest livelihood" (Government of Madras 1939:61). The presidency governments farmed out many of the reformatories to various missionary organizations (such as the London Missionary Society, the American Baptist Mission, and the Salvation Army) that were eager for the opportunity to proselytize to a literally captive audience. These organizations received financial support from the presidency governments to help cover the administrative costs of running the reformatories. There were many different types of settlements, with varying practices of control, and although many organizations sponsored settlements, the Salvation Army was perhaps the most influential. An examination of the Salvation Army criminal tribes reformatories throughout the Indian subcontinent reveals not only the character of the reform of criminal tribes, but also the relation between "colonizing and transforming" the criminal tribesman and the enterprise of imperial rule.

The Salvation Army in India

The Salvation Army began as a missionary organization on the east side of London in the mid-nineteenth century. Founded as the "East London Christian Revival Society" by William Booth, a follower of Wesley, the mission was dedicated to the salvation and reform of those most marginal to middle-class Victorian values. William Booth's injunction to his followers was "Go for souls and go for the worst" (Salvation Army 1922:59). The "worst" were generally drawn from the dangerous classes: William Booth's converts were former drunks, thieves, prostitutes, and "other notorious characters" (Neal 1961:8). The mission approached the problem of converting the dissolute by adapting its meeting form to reach those estranged from the conventions of regular churches. Taking its message out into the streets of the urban slums, it attempted to attract the heathen to open-air meetings.

The Salvation Army focused on the eradication of "sinful habits," and abstinence from the consumption of alcohol was a requisite for all members (Salvation Army 1940:33). Personal salvation, it was believed, required obedience, discipline, and the shunning of "[w]orldly amusements and unworthy associations" (Salvation Army 1940:33). As Catherine Booth wrote concerning the "normal experience" of holiness: "While there is a spark of insubordination or rebellion or dictation, you will never get it. Truly submissive and obedient souls only enter *this* Kingdom" (quoted in Blackwell 1949:15). To achieve its goals of disciplinary reform, the Salvation Army organized itself along the lines of a military order. The change of name to "The Salvation Army" in 1878 had an "immediate effect," for, since it was "an important step on the way to clothing fully the ideal of Christian soldiership," it set off "a whole train of logical consequences" (Sandall 1948:18). After the change in name, its members sought to appropriate the signs and practices of a military order. They soon began wearing secondhand British Army uniforms of various styles and periods. Later, custom-made uniforms were designed to invest the bodies of members with the signs of a military order in a more standardized manner. "General" Booth founded an order of "Articles of War" that all Salvation Army members were to sign, and a military hierarchy was replicated in a ranking system of lieutenants, commissioners, captains, and cadets.

India was the Salvation Army's first mission territory outside England. Salvation Army mission activities in India were begun by Frederick de Lautour Tucker (Commissioner Booth-Tucker), who had originally been with the Indian Civil Service but had resigned his post after becoming acquainted with the Salvation Army on a trip to England. There he had met and married Emma Booth, the daughter of the Salvation Army's founder (see Figure 3.1). The 1882 "invasion" of India by Booth-Tucker and his Salvation Army followers was met by government disapprobation (Booth-Tucker ca. 1930:12). British officials viewed the Salvation Army with disfavor, perhaps in part because of the working-class background of its members but also in part because of the relatively unorthodox methods it used in winning converts. Those in government saw the street processions and open-air meetings as a noisy breach of public order with the potential to result in embarrassment for the Crown, if not in outright rioting. Salvation Army missionaries were frequently arrested at public meetings and processions during their first few years of missionization. But these disputes were settled so that the Salvation Army could pursue its attempts to convert Indian subjects, and eventually the Salvation Army and government agencies embarked on collaborative projects, such as the reform of criminal tribes.

Booth-Tucker advocated adapting Salvation Army practices to customs familiar to the peoples of India. As the legend goes, clothed "in turban and dhoti, with a sack for his bedding and a small tin box for his papers, he set out barefoot" (Begbie ca. 1913:13). This novel presentation of self may also have annoyed the sensibilities of government officials, who sought to preserve a distance between Indian and British dress and customs (Cohn 1989). As "Fakir Singh," Booth-Tucker carried a begging bowl and played the role of a *sadhu,* or Hindu mendicant (Carpenter 1957:89).

The original uniforms prescribed for Salvation Army officers in India were a long white jacket, scarf, ordinary trousers, European boots, and turban with "Muktifauj" (Salvation Army) badge for men, a white sari for women. Booth-Tucker devised the use of "caste marks" for Army officers (bands of red, yellow, and blue painted on the brow), which he felt would enhance their attempts to

Figure 3.1 Frederick and Emma Booth-Tucker. Reproduced from Frederick Booth-Tucker's *Muktifauj, or Forty Years with the Salvation Army in India and Ceylon* (ca. 1930).

accommodate Indian custom. The uniforms were later changed to emphasize military associations: red military-style jackets were added to the ensembles of both men and women (Booth-Tucker ca. 1930:93). Other military signs were modified according to what was perceived to be Indian "custom": instructions were given for the fabrication of flags displaying "appropriate mottoes" and Salvationists mounted on horseback ("sowars") (Booth-Tucker ca. 1930:103).

The Salvation Army's project was conceived as a military mission driven by "the spirit of holy aggression" (Salvation Army 1923:27). The Army's conquest of India was envisioned in spatial terms, and the country was mapped out into various territories "for purposes of Army administration" (Salvation Army 1922:60). One of the organization's primary goals was the creation of institutional spaces; although open-air meetings were to remain important foci of missionization, there was a concern to establish spiritual spaces in the form of Salvation Army halls in every village. The Salvation Army set about an iconic replication of the structures and practices of institutional types conceived in Britain. Leper colonies, hospitals, homes for women, boys' industrial homes, beggars' homes, vagrants' colonies, dispensaries, orphans' homes, village schools, and boarding schools were just some of the institutions established in order to transform the nature of life on the subcontinent.

In part, institutional spaces were deemed important because they made it possible to use one of the Army's primary instruments of conversion, the "mercy seat." The mercy seat or "penitent-form" was a bench at the front of the meeting hall, at which "persons anxious about their spiritual condition [were] invited to seek Salvation" by kneeling and praying (Salvation Army 1940:176). The mercy seat seems to have required a spiritual space set aside for its use, since the annual yearbooks published by the Salvation Army repeatedly emphasized the need to establish permanent buildings in Indian villages, and those buildings were considered "essential to the effective use of the penitent-form" (Salvation Army 1928:78). In the use of the penitent-form, the deportment of the body was seen as an outward manifestation of inward transformation of character, and the viewing of this prostrate body was the climax of worship, a moment when every eye "would be riveted on the mercy seat" (Booth-Tucker ca. 1930:241).

The Salvation Army in India tried to maintain strict standards of discipline and order during public assemblages. The excitement with which General Bramwell Booth (son of William Booth) described the spectacle upon his arrival in the city of Trivandrum reveals the profound degree to which discipline was seen as an indication of spiritual development:

> First meeting at 4 o'clock. . . . The scene a remarkable one indeed. People seated for the most part on the ground, proper formation preserved, aisles marked out, men and women quite separate, thousands of Salvationists carrying small flags of our three colours, which they raise every now and again to emphasize some point of song, or show special approval of something said. (cited in Booth-Tucker ca. 1930:152)

Disorder at public meetings elicited immediate action: "In the Prayer Meeting, the singing in so vast a gathering [was] difficult to control. We counted fifty drums scattered about, most of them the centre of a singing group. But we managed to unite them by playing Daniel's cornet through the magnavox—a great hit!" (cited in Booth-Tucker ca. 1930:154). By means of practices such as these, the Salvation Army sought to organize a perfectly orchestrated social body in India within a divine military order.

The Salvation Army Reformatories and Transformations of Habitus

The Salvation Army first became formally involved in the reform of criminal tribes in 1908, when Commissioner Booth-Tucker met with officials of the government of the United Provinces and agreed to take over the administration of a reformatory established for criminal tribes (Booth-Tucker ca. 1930:206–207). Following their achievements in the United Provinces, the Salvation Army took over the administration of reformatories in other provinces. The various provincial governments continued to subsidize the reformatories, but the Salvation Army was responsible for their everyday management.

The practices of reform in criminal tribes settlements were intended to transform individual characters and attributes. This was intended to be an overall transformation, one that did not stop with an eradication of the propensity for committing individual acts of crime or with an eradication of the outward tokens of an inward state of criminality. It aimed at something deeper, at making a profound assault on the "habitus," a term Pierre Bourdieu derives from the work of Marcel Mauss (1973[1936]).[8] Bourdieu lays emphasis on the durability of the habitus, but, drawing on Erving Goffman's analyses of "total institutions," he does acknowledge that dominant structures tend to fashion practices for re-forming the habitus:

> If all societies and, significantly, all the "totalitarian institutions" . . . that seek to produce a new man through a process of "deculturation" and "reculturation" set such store on the seemingly most insignificant details of *dress, bearing,* physical and verbal *manners,* the reason is that, treating the body as a memory, they entrust to it in abbreviated and practical, i.e. mnemonic, form the fundamental principles of the arbitrary content of the culture. (Bourdieu 1977:94)

As criminality was manifested not simply in individual acts of crime but also in dispositions, gestures, modes of deportment, habitual behaviors, and types of bodily adornment, these were all the focus of methods aimed at deconstructing persons and reconstituting them as disciplined individuals. "By means of the firm yet kindly discipline of Christlike love," those in the settlements were taught "industry, cleanliness, honesty, and obedience," and their children were "trained to become worthy citizens of India" (Salvation Army 1923:60).

When these measures of reform did not succeed, the tendency of criminal tribespeople to cling to a life of crime was referred to as an indication of their "Habitualism" (Booth-Tucker ca. 1930:222). Both within the reformatory and beyond its confines, the disciplinary practices of the Salvation Army were focused on reaching and transforming the habitus, in keeping with William Booth's motto "Get into their skins!" (Booth-Tucker ca. 1930:163).[9]

The Salvationists invested a great deal of effort in trying to change modes of ornamentation of the body. For some criminal tribespeople, this meant converting to an entirely different mode of adornment. The Lambadi woman was made to "give up her heavy unwashable dress and take to the ordinary sari" (Aiyappan 1948:165). In the Salvation Army reformatories, strict standards of grooming and hygiene were enforced. In one reformatory, the "women's hair was dishevelled, their clothing ragged, and their general appearance untidy and dirty" until it was announced that no woman could obtain a pass to leave the reformatory to go to market unless she "presented a neat appearance," for, the officers felt, the residents could not be allowed to go about "like so many vagrants" (Booth-Tucker ca. 1930:234). Conflicts over standards of hygiene broke out between the reformatory authorities and the inmates, and the residents of one institution were ordered to wash their own clothing:

> "Wash our clothes;" said one Tribe, when we suggested that their garments might be a little better for making acquaintance with some water. "Do you take us for Dhobis [launderers]?" "But could not your wives do it?" "Certainly not! It would spoil the taste of our food." But this very same tribe now comes to our meetings, well dressed, clean and tidy. (Booth-Tucker 1916:44)

General Bramwell Booth, commenting on his visit to a reformatory settlement, made a note in his diary of the apparent transformation of the inmates evident in the change in their deportment: "The simply wonderful difference in appearance and demeanour of the settlers lately arrived, with those already here. . . . All quite captivating" (cited in Booth-Tucker ca. 1930:154).

There were other practices suited to inculcating a disciplined habitus in criminal tribespeople. One of these was the regulation of time. E. P. Thompson has traced the relationship of changes in conceptions of time to the growth of industrial capitalism in Britain through the restructuring of work habits (1967). Capital relations generated a system of production that compelled an increased synchronization of bodies and acts. In capitalist systems of production, time becomes a commodity: the worker sells his or her labor-time for a wage. Labor-time, as a commodity, becomes alienated from the volition of the person; thus, "work" becomes distinguished from "life" and one's "own time" from the employer's (Comaroff 1985; Thompson 1967). Time should not be "wasted" but rather made useful. The commodification of time requires that the rhythms of labor become increasingly regular and regulated—in short, disciplined according to units of time.

Salvation Army officers from Britain came to India with a rigid internal sense of time-discipline that guided them in all their actions on the subcontinent. In the reformatory settlements the regulation of time was one of the central semiotic media for inculcating a disciplined habitus: "At daybreak, before commencing work, all the villagers assemble in two centres for prayer and Bible reading, together with a brief exhortation regarding the duties of the day," noted Booth-Tucker (ca. 1930:234). There were daily roll calls to establish presences and absences of inmates. The pass system stipulated by the Criminal Tribes Act required that criminal tribespeople adhere to a time schedule in reporting to the police if moving from one jurisdiction to another. In other cases, persons registered under the Criminal Tribes Act might be required to report to the police station twice each night, at 11:00 p.m. and 3:00 a.m. In part, such practices were intended to insure that members of criminal tribes did not rove about at night committing crime. But such practices also served to reinforce the power of the clock to determine the behavior of the person, to inculcate the habit of "being on time," and to make manifest the *value* of time.

In the reformatories, members of criminal tribes were inducted into commoditized relations. On the one hand, this entailed an induction into the processes of commodity production and the conventions of selling one's bodily activity as labor for a wage. This initiation took place in factories, workshops, and industrial schools under the supervision of Salvation Army officers. External markets were established for the sale of leather-work, silk products, carpets, mats, and other commodities produced in the reformatories. According to Booth-Tucker, the Salvation Army sought to "organise the work in such a way" that it would "produce only those goods for which there was a sure market, and for which there would be a reasonable profit" (Booth-Tucker 1916:54). The Salvation Army assumed an instrumental role in the agency of capital, for it was maintained that the inmates were "totally unable to market their own produce" and that, in any case, to allow them to do so "would serve as an opportunity and inducement for crime" (Booth-Tucker 1916:54). By so alienating the producers, as well as the process of production, from the exchange of the products of labor, the reformatory established a distinctive feature of capitalist commodity production. Also noteworthy was the Salvation Army officers' conviction that remunerated labor had a more positive effect than did unremunerated labor (Booth-Tucker ca. 1930:218), perhaps because it taught one the value of exchanging one's labor for a wage.

On the other hand, through the practices of the system of reform, criminal tribespeople came to know commodities themselves as bearers of value. The officers of the Salvation Army knew the power of purchase well: as one writer observed, it was best to distribute Salvation Army literature "by purchase, which is far more effective than by free gift, seeing that every man values and takes notice of what he purchases more than [of] what is thrust gratuitously into his hands" (Salvation Army 1931:26). Writing in the 1960s, one police inspector reflected on the successes of the reformatory system, as signified by the incorporation of criminal tribespeople into commoditized relations. The "far reaching changes in the economic and social life" that reform had brought about among some of the criminal tribespeople were exemplified in the fact that many of them now had "their own houses and . . . decent furniture, sewing machines, cycles, etc." (Ramanujam 1966:420). The possession of commodities, as tokens of the commodity form of relation, came to signify the transformation of criminal tribespeople into "free" and law-abiding citizens.

In the Salvation Army reformatories, the inmates learned the value of one particular object above all others, the Bible. According to Booth-Tucker, inmates frequently asked to be given this "wonder Book," which was "not to be placed on the shelf and produced at intervals for a few brief moments," but was to be "read and re-read and pore[d] over" (Booth-Tucker ca. 1930:235). Of all the objects in the Salvation Army reformatory, the Bible was perhaps the most fetishized, because it was seen to confer upon its (British) owners both the power and the knowledge of command.

But the power of the written word took many forms in the system of reform. The power of the written word to codify, define, and control was reflected in the registration and pass system; few of the people upon whom this system was imposed could read at all (unless they had spent some time in the reformatory schools), and so it manifested the power of the written word with great cogency. The pass system required that one carry messages about oneself on one's person, messages meant for others that the bearer could not decipher.

It is apparent that some criminal tribespeople came to affirm the power of the written word. Just as writing could define and transform a person into a "criminal tribesman," it could also transform a criminal tribesman into a "free citizen." In the Salvation Army reformatories, an individual could eventually earn a certificate of good conduct and so obtain a "free pass" from the government. This lesson apparently had great cogency, and the certificate of good conduct came to be evidence to the inmates themselves of their own transformation. In one reformatory, several young women composed a song about this process of transformation, proclaiming, "We are Nekmashes, you must know! If you doubt it, you can see our Certificates!" (Booth-Tucker ca. 1930:212).[10]

But of all the practices for disciplining the body, labor itself was the most important. According to the theory of the Salvation Army, the "problem" of caste was essentially one of labor organization. Caste was said to be "a sort of gigantic *hereditary trades-unionism*" (Booth-Tucker ca. 1930:9). Since caste was the distinctive agent in causing the criminal tribesman to commit crime, and the problem of caste was essentially a labor problem, labor would be instrumental in the reform of the criminal tribesman. The reformatory system instilled in criminal tribespeople the moral value of discipline labor: "In healthy surroundings and under the direction of experienced and sympathetic Officers thousands of these erstwhile outlaws are being reclaimed and taught the way of righteousness, and, as well, the dignity of labour" (Salvation Army 1925:44). But labor had to be performed according to a precise economical calculation. In order to achieve the maximum disciplinary effect, it had to be neither too taxing nor too light: "every prisoner was weighed at regular intervals to see if he were losing flesh, and should such be the case, a lighter task was assigned" (Booth-Tucker ca. 1930:225). Commissioner Booth-Tucker termed the process of transforming the criminal through industry "Criminocurology" (Salvation Army 1915:80).

One of the central objectives of the system of reform was to transform criminal tribespeople into productive and subjected bodies. Although labor itself was the most prominent practice involved in that transformation, all of the disciplinary practices in the reformatory could be seen as having their effects for productive relations. As Michel Foucault has noted:

> it is largely as a force of production that the body is invested with relations of power and domination; but, on the other hand, its constitution as labour power is possible only if it is caught up in a system of subjection. . . . [T]he body becomes a useful force only if it is both a productive body and a subjected body. (Foucault 1979:26)

What the criminal tribes reformatories endeavored to produce were subjected and productive bodies.

Commissioner Booth-Tucker has left us a visual representation of what this transformation entailed (see Figure 3.2). Like other pictorial representations of the "before and after" type, this one depicts a transformation of bodies. The page is divided into three parts: at the top is a group of persons who, having "given evidence of transformation in character, have been released from Government surveillance" (Booth-Tucker ca. 1930:228–29); at the bottom is a group of unreformed criminal tribeswomen; and in the center is a mediating image, that of a group of Salvationists assembled "for the purpose of publicly destroying idols surrendered by converts"—that is, for an organized ritual of destruction of representations of Indian "tradition" (Booth-Tucker ca. 1930:228–29). In

Figure 3.2 Top and bottom: Reformed and unreformed criminal tribespeople. Center: Salvationists destroying idols. Reproduced from Frederick Booth-Tucker's *Muktifauj, or Forty Years with the Salvation Army in India and Ceylon* (ca. 1930).

the bottom photo, the women stand in a huddled mass; dressed in various "native" costumes and bedecked in exotic jewelry, they exhibit a variety of expressions and placements of hands. The top photo, in contrast, depicts reformed individuals dressed in saris and turbans (reconstructions of Indian "tradition"). They stand erect, arms at their sides, all vestiges of superfluous gesture and movement excised: bodies subjected, but ripe with productive potential.

Conclusion

What were the significant effects of the discourse on criminal caste, the Criminal Tribes Act, and the criminal caste reformatories? From one perspective, it might be concluded that the effects were of little significance, since only a small part of the population of the subcontinent was brought under the Criminal Tribes Act or confined to a criminal tribes reformatory. The Criminal Tribes Act was repealed by the legislature of independent India in 1952. The act of repeal provided for the "denotification" of criminal tribes and their release from reformatories.

Yet the notion of a criminal caste had been internalized by the society of the colonized. A detective manual written by a police inspector nineteen years after India had become an independent nation-state catalogued the habits, dress, and identifying characteristics of various criminal castes in order to aid police in detecting crime. It gave police officers the following advice on how to detect "Lambadi crime":

> [The Lambadis] are fond of smoking hookahs at home and chilams or cleverly rolled leaf pipes (chuttas) away from home. The discovery of these or of pansupari bags ornamented with cowri shells or mirrors at the scene of a crime points to Lambadis as the culprits. . . . When indications point to Lambadis as the culprits, an immediate muster of the members of the suspected tanda [settlement] should be taken, and the absence of any member should be carefully enquired into. (Ramanujam 1966:424–25)

Thus, the tokens indexing the various manifestations of criminality in British India remained part of the official methodology of "police science."

The endeavor to reform the criminal tribes had a further significance, one that extended to the core of imperial rule. The missionary activities of the Salvation Army were linked to the spatiotemporal extension of the British Empire.[11] An article in the 1931 *Salvation Army Year Book* addressed the issue of "How the Army's World-Wide Uniformity of Action Is Maintained" (Salvation Army 1931:17). The iconic replication of institutional spaces furthered the spatiotemporal extension of this "world-wide uniformity." The "underlying principles" of the Salvation Army's methods were applied uniformly throughout the world because "the fundamental needs of the human heart are so alike the world over" (Salvation Army 1931:17). The Salvation Army envisioned the process of reforming criminal tribes as one that would result in "the boundaries of the 'Social City' being extended to take in the most troublesome of these people"

(Salvation Army 1924:19), and so this process was joined to the wider processes of spatiotemporal extension.

The spatiotemporal extension of the boundaries of the Salvation Army's "Social City" was expressly linked with the spatiotemporal expansion of the empire. In an article entitled "The Political Significance of the Salvation Army," the linkage to empire was explicitly acknowledged: "Few people . . . realize that the work which The Salvation Army does is of measureless importance to England as an Empire builder; and still fewer look below the surface, and see its work both as a uniter and a divider" (Barnett 1922:21). This linkage to imperial rule extended to the transformation of habitus among the criminal tribes in the Salvation Army reformatories, for restoring "order among these lawless people" was considered a "priceless contribution to the political and economic stability of the Empire" (Barnett 1922:21).

In his studies of the growth of modern institutional structures that encode the values of confinement, exclusion, and discipline, Foucault has traced the transformation of the principle of confinement into the institutions of the prison, the asylum, and the clinic. In *Madness and Civilization* he comments on the transition from the confinement and exclusion of lepers to other practices of confinement and exclusion:

> What doubtless remained longer than leprosy, and would persist when the lazar houses had been empty for years, were the values and images attached to the figure of the leper as well as the meaning of his exclusion, the social importance of that insistent and fearful figure which was not driven off without first being inscribed within a sacred circle. (Foucault 1965:17)

It could perhaps be said that Indian civilization had developed its own form of exclusion in the form of untouchability, and it seems apparent that many of the nomadic criminal tribes were ostracized by the communities through which they passed on their excursions. Since much of the information gathered by British officials was filtered through indigenous elites, some of the scorn for these communities may have derived in part from others in Indian society. But what was new in the colonial social order was the value placed on confinement. The criminal caste reformatories were a model for the institutional world that the colonized were to inhabit.

Yet the criminal caste reformatory was part of the more far-reaching dream of imperial rule. For within this confined, enclosed, and highly controlled circle of barbed wire it was possible to establish the relation between the colonizer and the colonized in concentrated form to maximum effect. The Salvation Army was acknowledged to act as an agent in the establishment of this relation within the reformatory, as the title of an article in the 1925 *Year Book* suggests: "The Army Tackles an Oriental Problem: Colonizing and Transforming Hereditary Criminals in India" (Salvation Army 1925:44). The colonization of the body inscribed within a space was a marked feature of the overarching mission of colonial rule. The criminal tribes reformatory was linked to the spatiotemporal expansion of an empire that sought to extend into the habitus of its subjects, an empire that sought in its vastness to annihilate the everyday boundaries of space-time, making manifest the dream of an empire on which the sun would never set.

Notes

Acknowledgments. During the various phases of production of this chapter I have benefited from advice, commentary, and criticism provided by the following people: Frank F. Conlon, E. Valentine Daniel, Kris Hardin, Charles F. Keyes, Pratyoush R. Onta, Samuel Sudanandha, and Rebecca Tolen. I am especially grateful to Arjun Appadurai and Carol A. Breckenridge for their insight, guidance, and patience throughout several revisions of this essay. Special thanks to my family for their unwavering support, even when my academic preoccupations have seemed exceedingly obscure. I am grateful also for the help and friendship of the late Michael Bazinet, who reproduced the photographs for the original publication of the article in *American Ethnologist.* Finally, thanks are due to Jennifer Terry for her help in getting the photographs reproduced for publication in the present form.

1. There are, of course, alternative views of the caste system. The most prominent of these is the one proposed by Louis Dumont, who regards the caste system as a manifestation of the fundamental value of hierarchy in Indian thought (1970).

2. British officials often used fairly idiosyncratic transliterations of indigenous terms, and, as this article argues, there was a good deal of confusion about the distinctions between various groups. Where there is variation among primary sources in transliteration of names, or where an author has used a particularly unusual variant, I have substituted a more common variant of the caste name in the text. In these instances, the variant used by a particular author in the original text appears in parentheses.

3. In his studies of crime in the Madras Presidency, David Arnold observes that the crime rate generally rose and fell in relation to periodic famines (Arnold 1979, 1985). During periods of famine, patron-client relations between the wealthy and the poor changed, and many of the "ordinary" poor turned to illegal activity to sustain themselves. Yet many continued in these activities after such crises had passed, becoming "professional thieves."

4. I cannot fully attest to the etymology of this word. Samuel Sudanandha has suggested "spy" as an alternative gloss for the term "Kallar" (personal communication, 1988).

5. *Crime and Criminality in British Indian* (1985), a volume edited by Anand Yang, represents an exception to this trend and has made a substantial move in the direction of looking at relations between colonial modes of domination and criminal tribes throughout the subconti-

nent. Another interesting comparative case is the discourse on martial races and British recruitment of "Gurkhas" from Nepal into the Indian military (Onta 1990).

6. In this sense, criminality was what Charles S. Peirce has called a "qualisign." A qualisign cannot be known or perceived through a direct process of signification, but only through tokens, or signifying manifestations, of a quality (Daniel 1984:31). As a qualisign, criminality could not be represented directly; it had to be represented by means of other signifiers.

7. Many of these descriptions of criminals masquerading as ordinary citizens echo an earlier discourse on the "thugs," who were said to practice cult-like murders (Gordon 1969).

8. Bourdieu defines habitus as "a system of dispositions" (Bourdieu 1977:214). According to him, the word "disposition" most closely approximates the meaning of habitus, since it encompasses a "way of being," a "habitual state," and a "predisposition, tendency, propensity, or inclination" (Bourdieu 1977:214).

9. Many of the practices of missionary reform in India paralleled practices in other colonized areas. Jean and John Comaroff's studies of the Methodist mission in southern Africa have brought out the significance of certain methods of reform that were also used by the Salvation Army in India (Comaroff 1985; Comaroff and Comaroff 1986).

10. A Nekmash is one who earns his or her living legally, in prescribed ways.

11. Nancy Munn uses the concept of "spatiotemporal extension" in reference to "the capacity to develop spatiotemporal relations that go beyond the self, or that expand dimensions of the spatiotemporal control of an actor" (Munn 1986:11). In the case that I am outlining, it is not the control of a single actor but rather structures of domination that are extended through spatiotemporal processes.

References

Aiyappan, A.
1948 Report on the Socio-Economic Conditions of the Aboriginal Tribes of the Province of Madras. Madras: Government Press.

Appadurai, Arjun
1978 Understanding Gandhi. *In* Childhood and Selfhood: Essays on Tradition, Religion and Modernity in the Psychology of Erik H. Erikson. P. H. Homans, ed. pp. 113–43. Lewisburg, PA: Bucknell University Press.
1988 Putting Hierarchy in Its Place. Cultural Anthropology 3(1):36–49.

Arnold, David
1979 Dacoity and Rural Crime in Madras, 1860–1940. Journal of Peasant Studies 6(2):140–67.
1984 Criminal Tribes and Martial Races: Crime and Social Control in Colonial India. Paper presented at a seminar entitled "Comparative Commonwealth Social History: Crime, Deviance and Social Control," 4 December, University of London Institute of Commonwealth Studies, London.
1985 Crime and Control in Madras, 1858–1947. *In* Crime and Criminality

in British India. A. A. Yang, ed. pp. 62–88. Tucson: University of Arizona Press.

Ayyar, K. N. Krishnaswami
1933 Madras District Gazetteers: Statistical Appendix and Supplement to the Revised District Manual (1898) for Coimbatore District. Madras: Government Press.

Bailey, F. G.
1961 "Tribe" and "Caste" in India. Contributions to Indian Sociology 5:7–19.

Barnett, S. A.
1922 The Political Significance of the Salvation Army. *In* The Salvation Army Year Book: 1922. pp. 21–22. London: Salvation Army Book Department.

Begbie, Harold
ca. 1913 The Light of India. London: Hodder and Stoughton.

Blackburn, Stuart H.
1978 The Kallars: A Tamil "Criminal Tribe" Reconsidered. South Asia (n.s.) 1:38–51.

Blackwell, Benjamin H.
1949 Doctrinal Foundations. *In* The Salvation Army Year Book: 1949. pp. 14–18. London: Salvationist Publishing and Supplies, Ltd.

Booth-Tucker, Frederick
1916 Criminocurology; or, the Indian Crim and What to Do with Him: Being a Review of the Work of the Salvation Army among the Prisoners, Habituals and Criminal Tribes of India. 4th ed. Simla, India: Liddel's Printing Works.
ca. 1930 Muktifauj, or Forty Years with the Salvation Army in India and Ceylon. London: Marshall Brothers, Ltd.

Bourdieu, Pierre
1977 Outline of a Theory of Practice. Richard Nice, trans. Cambridge: Cambridge University Press.

Breckenridge, Carol A.
1989 The Aesthetics and Politics of Colonial Collecting: India at World Fairs. Comparative Studies in Society and History 31(2):195–216.

Carpenter, Minnie Lindsay
1957 William Booth: Founder of the Salvation Army. London: Wyvern Books.

Cohn, Bernard S.
1968 Notes on the History of the Study of Indian Society. *In* Structure and Change in Indian Society. M. Singer and B. S. Cohn, eds. pp. 3–28. Chicago: Aldine Publishing Company.
1987 The Census, Social Structure and Objectification in South Asia. *In* An Anthropologist among the Historians and Other Essays. pp. 224–54. Delhi: Oxford University Press.
1989 Cloth, Clothes and Colonialism: India in the Nineteenth Century. *In*

Cloth and the Human Experience. A. B. Weiner and J. Schneider, eds. pp. 303–54. Washington, DC: Smithsonian Institution Press.

Comaroff, Jean
1985 Body of Power, Spirit of Resistance. Chicago: University of Chicago Press.

Comaroff, Jean, and John Comaroff
1986 Christianity and Colonialism in South Africa. American Ethnologist 13:1–22.

Cressey, Paul Frederick
1936 Reforming the Criminal Tribes of India. Sociology and Social Research 21(1):18–25.

Daniel, E. Valentine
1984 Fluid Signs: Being a Person the Tamil Way. Berkeley: University of California Press.

Dirks, Nicholas B.
1987 The Hollow Crown: Ethnohistory of an Indian Kingdom. Cambridge: Cambridge University Press.

Dumont, Louis
1970 Homo Hierarchicus. Chicago: University of Chicago Press.
1986 A South Indian Subcaste: Social Organization and Religion of the Pramalai Kallar. M. Moffatt, L. Morton, and A. Morton, trans. Delhi: Oxford University Press.

Foucault, Michel
1965 Madness and Civilization: A History of Insanity in the Age of Reason. New York: Mentor Books.
1979 Discipline and Punish: The Birth of the Prison. New York: Vintage Books.

Francis, W.
1904 Madras District Gazetteers: Bellary. Madras: Government Press.
1906a Madras District Gazetteers: Madura. Madras: Government Press.
1906b Madras District Gazetteers: South Arcot. Madras: Government Press.

Freitag, Sandria B.
1985 Collective Crime and Authority in North India. *In* Crime and Criminality in British India. A. A. Yang, ed. pp. 140–163. Tucson: University of Arizona Press.
1989 Community and Space: The Case of the "Criminal Tribes." Paper presented at a conference entitled "Culture, Consciousness and the State," 23–27 July, Isle of Thorns, United Kingdom.

Gandhi, Mohandas K.
1927 An Autobiography; or, The Story of My Experiments with Truth. Ahmadabad, India: Navajivan Trust.

Gordon, Stewart N.
1969 Scarf and Sword: Thugs, Marauders and State-Formation in 18th Century Malwa. Indian Economic and Social History Review 4(4):403–29.

Government of Madras
1929 Report on the Administration of the Madras Presidency of the Year 1927–28. Madras: Government Press.
1930 Madras District Gazetteers: Statistical Appendix for Bellary District. Madras: Government Press.
1931 Madras District Gazetteers: Statistical Appendix for Trichinopoly District. Madras: Government Press.
1933 Report on the Administration of the Madras Presidency of the Year 1931–32. Madras: Government Press.
1939 Madras Administration: 1937–1938. Madras: Government Press.

Governor-General of India
1871 Abstract of the Proceedings of the Council of the Governor-General of India, Assembled for the Purpose of Making Laws and Regulations, 1870. Vol. 9. Calcutta: Office of the Superintendent of Government Printing.

Guha, Ranajit
1983 Elementary Aspects of Peasant Insurgency in Colonial India. Delhi: Oxford University Press.

Haikerwal, Bejoy Shanker
1934 Economic and Social Aspects of Crime in India. London: George Allen & Unwin, Ltd.

Hatch, W. J.
1928 The Land Pirates of India: An Account of the Kuravers a Remarkable Tribe of Hereditary Criminals Their Extraordinary Skill as Thieves Cattle-lifters & Highwaymen &c and Their Manners & Customs. London: Seeley, Service, & Co. Limited.

Hemingway, F. R.
1907a Madras District Gazetteers: Godavari. Madras: Government Press.
1907b Madras District Gazetteers: Trichinopoly. Madras: Government Press.

Jones, David
1982 Crime, Protest, Community and Police in Nineteenth-Century Britain. London: Routledge & Kegan Paul.

Marriott, McKim
1976 Hindu Transactions: Diversity without Dualism. *In* Transaction and Meaning: Directions in the Anthropology of Exchange and Symbolic Behavior. B. Kapferer, ed. pp. 109–142. Philadelphia: Institute for the Study of Human Issues.

Marriott, McKim, and Ronald B. Inden
1977 Toward an Ethnosociology of South Asian Caste Systems. *In* The New Wind: Changing Identities in South Asia. K. David, ed. pp. 227–38. The Hague: Mouton.

Mauss, Marcel
1973[1936] Techniques of the Body. B. Brewster, trans. Economy and Society 2(1):70–88.

Mayhew, Henry, and John Binny
1968[1862] The Criminal Prisons of London and Scenes of Prison Life. London: Frank Cass & Co., Ltd.

Mullaly, Frederick S.
1892 Notes on Criminal Classes of the Madras Presidency. Madras: Government Press.

Munn, Nancy D.
1986 The Fame of Gawa: A Symbolic Study of Value Transformation in a Massim (Papua New Guinea) Society. Cambridge: Cambridge University Press.

Nandy, Ashis
1983 The Intimate Enemy: Loss and Recovery of Self under Colonialism. Delhi: Oxford University Press.

Neal, Harry Edward
1961 The Hallelujah Army. Philadelphia: Chilton Company.

Onta, Pratyoush R.
1990 Obsessions of Empire: Trade with Nepal and the Colonization of the Gurkhas. MS, files of the author.

Pinney, Christopher
1989 Representations of India: Normalization and the "Other." Pacific Viewpoint 29(2):144–162.

Ramanujam, T.
1966 Prevention & Detection of Crimes. Madras: Madras Book Agency.

Richards, F. J.
1918 Madras District Gazetteers: Salem. Vol. 1, Part 2. Madras: Government Press.

Rothermund, Indira
1963 The Philosophy of Restraint: Gandhi's Strategy and Indian Politics. Bombay: Popular Prakashan.

Salvation Army
1915 The Salvation Army Year Book for 1915. Theodore Kitching, ed. London: Salvation Army Book Department.
1922 The Salvation Army Year Book: 1922. London: Salvation Army Book Department.
1923 The Salvation Army Year Book: 1923. London: Salvationist Publishing and Supplies, Ltd.
1924 The Salvation Army Year Book: 1924. London: Salvationist Publishing and Supplies, Ltd.
1925 The Salvation Army Year Book: 1925. London: Salvationist Publishing and Supplies, Ltd.
1928 The Salvation Army Year Book: 1928. London: Salvationist Publishing and Supplies, Ltd.
1931 The Salvation Army Year Book: 1931. London: Salvationist Publishing and Supplies, Ltd.

1940 The Salvation Army Year Book: 1940. London: Salvationist Publishing and Supplies, Ltd.

Sandall, Robert
1948 From Mission to Army. *In* The Salvation Army Year Book: 1948. pp. 17–18. London: Salvationist Publishing and Supplies, Ltd.

Thompson, E. P.
1967 Time, Work-Discipline and Industrial Capitalism. Past and Present 38:56–97.

Thurston, Edgar
1909 Castes and Tribes of Southern India. 7 vols. Madras: Government Press.

Yang, Anand A., ed.
1985 Crime and Criminality in British India. Tucson: University of Arizona Press.

4

This Norm Which Is Not One

Reading the Female Body in Lombroso's Anthropology

David G. Horn

Introduction: Criminal Anthropology and Social Bodies

THIS ESSAY EXPLORES the social scientific construction of the female criminal, the prostitute, and the normal woman, and the elaboration of a new anthropology of the body in late-nineteenth-century Italy. In the criminology of Cesare Lombroso, the body of the "female offender" was constituted as a particular kind of social text: an index of present and potential risks to the larger social organism. As such, it was made an object of a variety of specialized reading practices: scientific measurements and evaluations that struggled to overcome the female body's epistemological resistance and contain its dangerousness.

At first glance, the discourses and practices of criminal anthropology seemed to construct durable and reassuring boundaries around bodies that were doubly other (cf. Gilman 1988). They promised to make intelligible the special nature of female criminality and to map dangerous women in relation to criminal men, children, "savages," nonhuman primates, and, especially, "normal" women. But here, as in other modern projects of boundary drawing (Haraway 1989), borders proved permeable, and the scientific efforts to shore them up destabilized them further. On the one hand, a statistical and probabilistic construction of criminality threatened, in the case of women, to collapse the categories of the deviant and the normal.[1] On the other hand, the female body was found to resist and subvert the exegetical practices of anthropologists, to mask its truths and seduce its readers.

In the end, what emerged from Lombroso's studies was less the (hoped for) transparent pathology of the female offender than the barely legible *potential* dangerousness of the normal woman. As a result, not only were the criminal woman and the prostitute made objects of new practices of surveillance, prevention, and punishment, but the normal woman was placed at the center of a whole range of modern discourses and technologies called "social," which ranged from social medicine to social hygiene to social work. Against the background of struggles for women's suffrage (Pieroni Bortolotti 1963), anthropologists constructed women as beings who were incapable of balancing rights and duties, who lacked an innate respect for property, and who embodied a latent criminality that could be held in check only by the experience of maternity and the action of new techniques of government.

The "Social" Construction of Criminality

Lombroso's studies of criminal bodies rarely figure in the origin stories anthropologists tell about themselves; more often his science of deviance has been constructed as "deviant science."[2] And yet Italian criminal anthropology constituted itself as a modern social scientific discourse and practice, with links not only to the evolutionary theories of Darwin (whose works were first translated into Italian in 1864), but also (and perhaps more importantly) to the emergence of the science of statistics, to the "discovery" of social facts, and to the identification of the national population as an appropriate object of scientific knowledge and government. Each contributed in the mid-nineteenth century to the reconfiguration of crime as a "social problem": a patterned and predictable, if undesirable, consequence of social life, rather than the sum total of individual acts threatening the sovereignty of law or the king. Crime became a "risk" that criminal anthropologists proposed to know and manage through detailed knowledge of social laws and embodied criminality (cf. Ewald 1986). The point is not that Italian criminal anthropology "took account" of the "social context" of criminality. Rather, the criminal body and society were (re)objectified in relation to one another, as different kinds of "social bodies" (Horn 1994). Scientists simultaneously constructed Italian society as a body threatened by the disease of crime; and the body of the criminal, as a sign and "embodiment" of social dangerousness and deviations from statistical norms.

The construction of crime as a social fact and problem, and of criminology as a social science, was ironically linked to a new focus on the individuality of the offender and to new scientific practices of individuation. Lombroso's "positive school"[3] of criminology was distinguished from its predecessor and rival, the "classical school," by a shift of objects from the crime to the criminal and his or her environment (Ferri [1901] 1968:60) and by a new anthropology in which "penal man" (*homo penalis*) gave way to "criminal man" (*homo criminalis*) (Pasquino 1980:19–20; cf. Foucault 1978a). The classical school operated around the triangle of law, crime, and punishment. The problem was to adjust penalties to offenses; the individual to be punished (*homo penalis*) was interesting only insofar as his or her identity and legal responsibility (the capacity of free will) were at issue.

The new "anthropological" school socialized crime, centering instead on the two main poles of the criminal (*homo criminalis*) and society (Pasquino 1980:20). On the one hand, the anthropologists expressed a new concern with statistical regularities. They complained that the classical school had not been able to explain (and indeed had never thought to explain) why there were 3,000 murders every year in Italy, and not 300 or 300,000 (Ferri [1901] 1968:72). On the other hand, the anthropologists argued that it was necessary to take account of the social dangerousness of individual offenders (Foucault 1978a). They proposed replacing the classical school's typology of crimes, which Enrico Ferri termed a "juridical anatomy" of deeds ([1901] 1968:71), with a typology of criminals and, as we will see, an anatomy of deviant and dangerous bodies, grounded in scientific measurements.

In this way, anthropologists sought to break classical theory's link between responsibility and practices of punishment. As Lombroso's collaborator Raffaele Garofalo argued, the social dangerousness of an individual might, in fact, be greatest when his or her legal responsibility was least (Lombroso et al. 1886:197). Indeed, for the first time, one could be a criminal—that is, a danger to society—without having committed a crime (Fletcher 1891:210), something that had been literally unthinkable for the classical school.

This "social" construction of the problem of criminality also implied a new regime of governmental practices focused on the prevention, rather than the punishment, of crime. The anthropological problem was to determine "in what manner and to what degree it is necessary for the health of society to limit the rights of delinquents" (Lombroso et al. 1886:201). Remedies and penalties gave way to practices of surveillance, preventive detention, parole, and proposals for making talented criminals "serviceable to civilization" (Lombroso-Ferrero [1911] 1972:212–16). Vagabonds, Lombroso suggested, might be used to colonize wild and unhealthy regions; murderers might perform surgery or serve in the military; and swindlers might pursue police work or journalism (Lombroso [1911] 1968:447). These interventions depended in turn on estimations of the social risks posed by particular kinds of individuals; specific crimes figured merely as "indications" of dangerousness (Lombroso et al. 1886:198).

At the center of these new social sciences, punitive rationalities, and governmental practices stood the body of the criminal, which in the early writings of the anthropologists was typically male. Attention to bodies, Lombroso argued, promised to deliver criminology from the idealism and "metaphysics" of the classical school. His anthropology shared with other nineteenth-century scientific (and political) projects the assumption that only the body could ground and locate difference (Stocking 1988). In particular, a science of bodies promised to make intelligible the "dangerousness" that was the object of the new criminological discourse, but which threatened to remain invisible. Here Lombroso drew on the sciences of physiognomy and phrenology, which had purported to find signs of interior intellectual and moral states on the body's surfaces, particularly at the level of the head and face (see Sekula 1986:11–12).[4] However, Lombroso rejected what he termed the "qualitative and deterministic" readings of the phrenologists (Lombroso et al. 1886:5) in favor of anthropometry: the precise measurement of the dimensions and relations of parts of the body, a practice that had been joined to social statistics by Adolphe Quételet.[5] For Lombroso, anthropometry appeared (at least initially) to be "an ark of salvation" (Lombroso and Ferrero 1895:1), able to fix bodily difference as deviation from statistical norms.

Anthropometry promised both to make potential dangers known and manageable[6] and to specify the social, historical, and evolutionary place of the criminal body. Lombroso argued that the criminal was linked by his abnormal anatomy and physiology to the insane person and the epileptic, as well as to those other "others" who were constituted as the objects of the social sciences: the ape, the child, woman, prehistoric man, and the contemporary savage. For Lombroso, the criminal was, in his body and conduct, an "atavism"—a reemergence of the historical and evolutionary pasts in the present. The pederast was tied by his problematic desires to Periclean Athens (Lombroso-Ferrero [1911] 1972:xiv) as surely as the Italian murderer was tied by his violent, bloodthirsty acts to the Australian aborigine (Niceforo 1910:112).[7] The slang of criminals was said to recall primitive languages (Niceforo 1910:116), the artwork of criminals to recall Australian handicrafts, and the handwriting of criminals to recall pictography and hieroglyphics (Lombroso-Ferrero [1911] 1972:134–35).[8] Finally, behaviors such as tattooing and "sexual excesses," as well as morphological features such as the slope of the forehead and the shape of the ears, linked criminals to lower, less evolved forms of animal life, as well as to children and savages. Among the latter, criminal behavior was identified as "natural," indeed "normal" (Lombroso-Ferrero [1911] 1972:134).[9]

Reading the Body: Physiognomy and Statistics

Although criminal anthropology privileged the body, especially the head and face, as signs of social dangerousness, knowledge of deviant bodies was not presumed to be exclusive to anthropologists. Artists, writers, the "lower classes" (*il volgo*), and even children, according to Lombroso, were aware of and could reproduce in paintings and poems the contours of criminal physiognomy (Lombroso et al. 1886:11; Lombroso-Ferrero [1911] 1972:48–51). Folk taxonomies were, in Lombroso's view, rooted in "natural instincts": honest men and (especially) women were innately repulsed by the ugliness of the criminal type, and the conclusions of "instinctive observers" had found expression in proverbs, folk songs, and jokes. Lay observational practices could, in fact, be put to the test: Lombroso's daughter recounts that her father "once placed before forty children, twenty portraits of thieves and twenty representing great men, and 80% recognized in the first the portraits of bad and deceitful people" ([1911] 1972:50). Even Lombroso's mother, we are told, knew a potential murderer when she saw one (Lombroso 1896:310).

If these folk typologies (which were themselves made objects of anthropological analysis)[10] served to reinforce the findings of the criminologists, they might also have risked calling into question the privileged position of the anthropological observers, or (as some critics charged) the scientific status of their theories (Lombroso et al. 1886:11). However, for Lombroso and his contemporaries, the anthropologist was distinguished from the artist, the observant folk, and the child (and indeed from the nonanthropological folklorist and the phrenologist) by his specialized techniques for reading the body, by a corporeal literacy that made possible both an exegesis and a diagnosis. As Lombroso put it, what the anthropologist did was not "guesswork, or a prophecy, but a reading" of the body's forms and gestures (Lombroso et al. 1886:8). Lombroso, who spent a good deal of time collecting "prison palimpsests" (the overlaid inscriptions made by prisoners on cell walls, drinking vessels, uniforms, and the sands of the exercise yard) (Lombroso 1888), identified the body as a "palimpsest in reverse" (Lombroso et al. 1886:8), a text that when read correctly yielded up its submerged truths: the signs of degeneration and atavism. He advised experts in criminal law not to waste time reading books in libraries, but to read the "living documents" contained in prisons (Lombroso-Ferrero [1911] 1972:302).

However, it is important to note that, although the criminal's physiognomy was seen to make visible and legible (in a relatively unambiguous, unmediated fashion) the degeneration of the *race* (Lombroso et al. 1886:6),[11] it was a sign of the *individual's* social dangerousness only in a statistical, probabilistic sense (Horn 1993). According to Lombroso, fewer than 40 percent of convicted male criminals had any physical anomalies (Lombroso et al. 1886:12), and still fewer bore the combination of factors (the "criminal type") that was considered a reliable predictor of dangerous conduct. Indeed, the infinite possible combinations of measurable signs of atavism seemed to frustrate the effort to construct typologies and taxonomies, and Lombroso struggled in his writings to find the appropriate metaphor to capture the relation between physical deviance and dangerousness. At times, Lombroso referred to anomalies as notes in a musical chord: the isolated anomaly had to be taken together with other physical and moral notes for the criminal type to take shape (Lombroso et al. 1886:38; Lombroso-Ferrero [1911] 1972:48–49). This musical metaphor was in turn linked to a modern notion of visibility. Lombroso compared patterns of atavism to impressionist paintings: "examined from up close, they seem shapeless, colored blotches, while at a distance they prove to be wonderful" (Lombroso et al. 1886:34). A wide (statistical) perspective was necessary for any image of the criminal to emerge.

In sum, while the criminal may have had a fixed appearance in popular culture, there were no *sure* anatomical guides to social dangerousness—that is, to that feature that was placed at the center of anthropological discussions of criminality. In the end, the criminal anthropologists made everyone potentially (if not equally) dangerous—just as, to use an analogy favored by Enrico Ferri, Pasteur's studies of disease had made everyone a possible source of microbial contagion ([1901] 1968:99). In both cases, science pointed to the need for globalized practices of prevention and social hygiene. And in both cases, a dramatic reimagining of society was linked to the expanding power of the experts of the social.[12] This expansion is perhaps nowhere more evident than in the social knowledges and technologies focused on the bodies of women.

Female Criminality and the Exhaustion of the Body

The studies of female offenders came rather late in Lombroso's project to know the criminal body, a project begun in the 1850s. *La donna delinquente, la prostituta e la donna normale* was first published in 1893, amid expanding debates on the scientific status of Lombroso's work.[13] In the late nineteenth century, the practice of anthropometry seemed in fact to be experiencing a crisis. During what Elizabeth Fee has called the "baroque period of craniology" (1979:426), there was an unbounded multiplication of cranial angles and indices, of other anatomical measurements, of mechanical measuring devices, and indeed of the parts of the face and body that could be distinguished as candidates for measurement. However, the significance of the new measurements and correlations was no longer always obvious to those who collected them. Lombroso had earlier regarded anthropometry as "the backbone . . . of the new human statue of which he was at the time attempting the creation" (Lombroso and Ferrero 1895:1). Now the body threatened to become exhausted as a source of scientific meanings.

This epistemological (and narrative) crisis was reflected in the text of *La donna delinquente* and its subsequent translations, which presented pages of numbers and statistical correlations without any developed analysis or interpretation.[14] Some correlations might have seemed meaningful in a transparent way, or were made to appear so in the text: for example, the fact that female poisoners (who were said to depend on practices of concealment and distraction) turned out to have the highest cranial capacities, while those women guilty of simple assault or assisting in rape had the lowest (Lombroso and Ferrero 1895:24). But what, after all, was the criminal anthropologist (or the reader) to make of the fact that the length of the face was greatest among women who had been convicted of wounding others and least among those guilty of arson (1895:24)? Though in principle the undigested numbers of anthropometry lent scientific authority to the text, they risked revealing themselves as gratuitous quantification.

As in later editions of his studies of male criminals, Lombroso (now joined by his son-in-law Guglielmo Ferrero) felt obliged to look elsewhere, to turn to anomalies (particularly facial and genital deformations and asymmetries), as well as to other qualitative materials and sources (anecdotes, paintings, folk sayings), in order to construct a portrait of the female offender.[15] The result was an unwieldy and heterogeneous text, one that moved from discussions of crime among female animals (the alcoholism of ants and the adultery of pigeons) to anthropometry, to folklore, to vivid accounts of the deeds of prostitutes and monstrous mothers.

Of particular importance, however, was the lengthy discussion of the "normal" woman that opened the book and that in many ways provided its framework.[16] The normal woman, a figure that had no real counterpart in Lombroso's studies of male criminality,[17] was ostensibly constructed as a background against which the female offender might become distinct, visible, and legible. As we will see, this aim was largely frustrated; on the contrary, the portrait of the normal woman contributed substantially to locating *all* women within the domain of the social expert.

The Normal and the Pathological

The detailed discussion of the normal woman, based almost exclusively on the writings and research of others, worked to resolve two apparent contradictions: that female criminals seemed to be too few in number and that, though doubly other, they displayed few anatomical signs of their social dangerousness.[18] These results were unexpected only because Lombroso's anthropology had explicitly linked "woman" and "criminal" (together with savage, primate, and child) as atavistic, a project that the portrait of the normal woman continued. Indeed, much of the discussion was given over to detailing the continuities between European women and the females of other, less advanced species.

In the formulations of Lombroso, which at times resembled the teleologies of Aristotle, woman was described as a "big child" (Lombroso and Ferrero 1915:98) and as a "man arrested in his intellectual and physical development" (1915:4). The signs of woman's failure to evolve or mature were varied, ranging from an underdeveloped moral sense to a predisposition to cruelty to a "physiological incapacity" for truth telling. That women habitually lied was, the authors reported, confirmed by proverbs, and rooted both in atavistic biology ("savages always lie") and in the social need to hide menstruation from men and sex from children (1915:95–98).

Women's bodies were, moreover, marked by traces of youth and the past: prehensile feet, left-handedness, dullness of the senses, an inability to experience pleasure,[19] and a resistance to pain that could be measured in the laboratory. With the aid of the "Lombroso Algometer," Lombroso and his colleagues were able to quantify the sensitivity of normal and criminal women to an electrical current induced on the hands, tongue, throat, breasts, and clitoris.[20] The "greatest deadness" was found in the hands of peasant women and in the clitorises of prostitutes. Twenty-eight percent of prostitutes were found to be completely insensible to pain, though sensitivity increased if the prostitutes had borne children (Lombroso and Ferrero 1895:138–39). (Lombroso and Ferrero dismissed the popular notion that women were *more* sensitive to pain, explaining that the fact that they reacted more loudly was simply a sign of their "greater irritability" [1915:52].) These characteristics were seen to link women not only to the indigenous peoples of Africa, who would be subjected by other researchers to similar evaluations of pain tolerance, but also to children and criminals (1896:42). The irascibility, vengeance, jealousy, and vanity common to all four meant none was able to achieve "an equilibrium between rights and duties, egoism and altruism" (1915:112)—that crucial component of *political* participation in a modern, civilized state.[21]

What, then, was one to make of the anthropological observations that women committed fewer crimes than did men, and that the heads and bodies of female criminals presented fewer signs of degeneration than did those of males (Lombroso and Ferrero 1895:27)? The cranial anomalies of male criminals were, by Lombroso's count, three to four times more frequent, and the "complete criminal type," characterized by four or more atavistic characteristics, could be found in only 18 percent of female criminals (as compared with 31 percent of males) (1895:104; 1915:208). The frequency of anomalies among normal women was also low. Lombroso recounted a day spent observing 560 women on a boulevard—only nine, he reported, presented the "complete degenerate type" (Lombroso and Ferrero 1915:29).

Ironically, it was the preponderance of the norm among women that was marshaled to provide a solution: it emerged as a sign of the "diminished diversity" of the female relative to the male, another sign of her failure to evolve fully and of her resemblance to females of lower species. Citing Darwin and Milne Edwards on the "organic monotony" of females, a condition of the stable reproduction of species, Lombroso and Ferrero constructed both women and the female body as essentially conservative. Women's views of social order were weighed down by the "immobility of the ovum"; their habits of dress remained relatively constant throughout history; and their anatomies showed few variations (Lombroso and Ferrero 1895:108–109; 1915:210). Only the hymen and labia, the authors noted, were subject to great variability (1915:31).[22] Even the tattoos of female criminals, Lombroso complained, were "monotonous and uniform," evidence of woman's "smaller ability and fancy" and the lower degree of differentiation in the female intellect (1895:122).

In the preface to later editions of *La donna delinquente,* Lombroso and Ferrero were thus able to turn an apparent contradiction into a "vindication of the observational method." Though at first the evidence from anthropometry had seemed to call their assumptions and predictions into question, the two authors "remained faithful to the method" and followed the facts with "blind confidence" (Lombroso and Ferrero 1915:1). The relative rarity of signs of degeneration on the female body, which had seemed potentially a sign of woman's superiority, could now be linked to a lesser degree of variability, itself a sign of inferiority and weakness (1915:2).

Thus, ironically, at the precise moment that woman was identified as "normal" and "normalizing," as embodying and conserving the norms of the species, she was marked as other, if not pathological, and as opposed to history and civilization. Moreover, as we will see in a moment, this construction of the female body as closely bound to the norm presented additional reading difficulties for the anthropologist and the social worker: how might the deviant (and dangerous) female body be distinguished from the normal body?

The remaining problem—apparently low rates of "female criminality"—was solved by reconstructing this category in order to include prostitution. Among women, argued Lombroso and Ferrero, criminality had almost always taken the form of prostitution, from prehistory to the nineteenth century: "The primitive woman," they wrote, "was rarely a murderess; but she was always a prostitute" (Lombroso and Ferrero 1895:111). This move, of course, required the authors to define prostitution rather broadly—to include, for example, all forms of marriage in "savage" societies (Lombroso and Ferrero 1896:212–14) and the conduct of wealthy as well as poor women in modern Italy. Lombroso and Ferrero observed that even the Mediterranean rituals of hospitality (which have continued to attract the attention of ethnographers of the region) had their roots in savage and medieval practices of prostitution (1896:214–15, 238).

If, in Lombroso's view, other female criminals rarely showed signs of degeneracy, this was far from the case among prostitutes, who were the very images of the evolutionary and historical pasts. Their "precocity," minor degree of differentiation from males, and resemblance to Hottentot women were all identified as signs of atavism, and 51 percent of prostitutes were found to have more than five anomalies (Lombroso and Ferrero 1895:113–14, 32). There was, in fact, a "crescendo of the peculiarities as we rise from the moral women, who are most free from anomalies, to prostitutes, who are free from none" (1895:107). (If this seemed to credit criminal women with more deviance than was their due, Lombroso was quick to observe that "it is incontestable that female offenders seem almost normal when compared to the male criminal, with his wealth of anomalous features" [1895:107].)

Lombroso illustrated the (limited) degeneracy of the prostitute with a series of photographs of Russian and French women, collected by Pauline Tarnowsky: uniform portraits whose composition and backgrounds already marked their subjects as objects of official concern.[23] Lombroso was unable to use pictures of Italian women because of a prohibition in his own country against measuring, studying, or photographing criminals once they had been convicted.[24] As might be imagined, he deplored these limits on the power of social experts. He complained that the treatment of prisoners as "sacred" contrasted with the treatment of mere suspects, who could be discredited and held up to publicity, as well as with the treatment of other women who were made objects of science: "Consumptive patients, pregnant women, may be manipulated, even to their hurt, by students for the good of science, but criminals—Heaven forfend!" (Lombroso and Ferrero 1895:89).

The reliance on photographs, both here and in the discussion of male criminals, returns us to the importance of the visual and reading practices in Lombroso's anthropology. As Allan Sekula has noted, photographs were central to a variety of attempts to manage social dangers, from the nominalist "criminalistics" of Alphonse Bertillon, seeking knowledge and mastery of individual criminals, to the realist criminology of Francis Galton, seeking knowledge and mastery of the "criminal type" (Sekula 1986:18; cf. Green 1986). In Lombroso's writings, although the photographs ostensibly were included simply as verifications of arguments made on the basis of anthropometry, they also struggled to make concrete and visible the abstract notion of statistical dangerousness.[25]

Lombroso's method was rather different from the "pictorial statistics" elaborated by Francis Galton (Green 1986:18). The goal was not to construct a composite image of "the criminal" and then to extract typical features, but to identify anomalies and deviations from commonsensical norms that might be used as aids in reading bodies.

In Lombroso's text, unlike in Galton's work, photographs seemed to resist "taming," the transformation of the idiosyncratic into the typical (Sekula 1986:17). Though he ignored the "messy contingency of the photograph" (Sekula 1986:17), Lombroso had to acknowledge again the rarity of the female criminal "type," offering up the metaphor of a "physiognomic kinship" among female offenders to describe the patterned distributions of visible anomalies (Lombroso and Ferrero 1915:198). Precisely because the "type" was not common, it was necessary to multiply rather than collapse or superimpose the pictures of the heads and faces of the dangerous, and to look and read with an attention both to family resemblances and to idiosyncratic difference.

At times, the gaze of the anthropologist resembled that of the connoisseur or voyeur, delighting in surfaces and making frank aesthetic appraisals that the (presumed male) reader was invited to check (*controllare*) and ratify (1915:195; cf. Lombroso and Ferrero 1896:300).[26] Although in some sense the pictures were constructed to "speak for themselves" (Tagg 1988:78) and were presumed to speak the truth of female criminality, the reader was addressed repeatedly and directly in the text. He was urged to observe particular numbered photographs and to note a "vile, repulsive air", "oblique glances", heavy jaws, rudimentary breasts, hair that was black and thick, and lips that were coarse, full, cracked, or sensual (Lombroso and Ferrero 1895: 88–102). In short, the reader was invited to read for himself the criminal texts.[27]

At other times, surface readings would not do.[28] Here, the anthropological gaze began to resemble the penetrating vision of the Baconian natural scientist, aiming to uncover nature's secrets (Merchant 1980:164–90)—especially when these were hidden by women's makeup and clothing.[29] Indeed, in Lombroso's view, the difficulties of reading the body and of "positively affirming" anthropological conclusions were multiplied in the case of prostitutes. After all, Lombroso wrote, prostitutes were unlike other female offenders because they were generally free from the anomalies that produced "ugliness," a result of the laws of sexual selection and of the marketplace (Lombroso and Ferrero 1895:85). "Let a female delinquent be young and we can overlook her degenerate type," Lombroso and Ferrero wrote, "and even regard her as beautiful; the sexual instinct misleading us here as it does in making us attribute to women more of sensitiveness and passion than they really possess" (1895:97). Still, the outwardly beautiful prostitute concealed underlying signs of her degeneracy: internal anomalies (overlapping teeth and a divided palate), atavistic genitals, and facial signs that were covered by the mask of makeup or by the "freshness" and "plumpness" of youth (1895:101–102). The ugly truth was revealed only by the penetrating gaze of the scientist, able to see beyond seductive surfaces to detect anomalies, or (more surely) by the passage of time:

> [W]hen youth vanishes, the jaws, the cheek-bones, hidden by adipose tissue, emerge, salient angles stand out, and the face grows virile, uglier than a man's; wrinkles deepen into the likeness of scars, and the countenance, once attractive, exhibits the full degenerate type which early grace had concealed. (1895:102)

Here, as elsewhere, the affirmation of the expert reading practices of the social scientist was accompanied by a recognition of the difficulty, if not the impossibility, of any reliable readings of the deceptive female body. There was, in fact, a layering of uncertainties: the erosion of boundaries and taxonomic categories effected by a probabilistic understanding of criminal dangerousness, and the absence of adequate markers caused by the tendency of woman to adhere to the norm, were here joined by bodily acts of epistemological resistance. Together, these worked to blur the very boundaries between normal and deviant that criminal anthropology sought to draw, and to multiply the possible objects of social scientific concern and technical intervention.

Ostensibly, Lombroso and Ferrero constructed a stable, triangular scheme, announced by the book's title.[30] At one apex stood the prostitute, the true expression of female degeneration (Lombroso and Ferrero 1896:596). At another apex was the born criminal, who in Lombroso's view was an entirely exceptional being. She was "doubly abnormal" because she did not follow the "natural" form of retrogression in women—that is, prostitution. Born female criminals were the true "monsters" of their sex, marked anatomically by virile characteristics that makeup could not disguise and behaviorally by a rejection of maternity that typically took the form of infanticide.[31] Gender and sexual inversions were at once the signs and logical consequences of female criminality, since, as we have seen, neither the category "woman" nor the category "female" allowed for deviation from norms.

At the final apex stood the normal woman, the most unstable figure because she was physically indistinguishable from the "occasional criminal" and was indeed linked to her by a "fund of immorality" latent in all women, the principal sign of which was an innate lack of respect for property (Lombroso and Ferrero 1895:216). The normal woman, in a sense, embodied potential criminality. "In ordinary cases," wrote Lombroso and Ferrero, "the child-like defects of the semi-criminal are neutralized by piety, maternity, want of passion, sexual coldness, weakness, and undeveloped intelligence" (Lombroso and Ferrero 1895:151).

In sum, woman was constructed as both normal in her pathology and pathological in her normality. This construction not only removed all women from the domain of rights, duties, and politics, but also inscribed them in the domain of the social. It made women suitable objects of ongoing surveillance and corrective interventions that, in an effort to restrict "opportunities" for criminality, blurred the lines between penal practices and social work.

Penality and Social Technologies

The social technical interventions and punitive measures Lombroso proposed for female criminals were linked to his tripartite typology.[32] He argued, for example, for the social usefulness of prostitution, a "safety valve" (Lombroso and Ferrero 1915:3) that "serves to relieve the passions of men" (Lombroso-Ferrero [1911] 1972:181), and advised state regulation rather than penalization. He recommended imprisonment (as in the case of men) only for the monstrous born criminal who posed an ongoing social danger. By contrast, the "occasional criminal," who was presumed to be passive, determined by her environment, and open to the suggestions of others, was to be made the object of educational interventions and penalties appropriate to her sex: "in view . . . of the important part played by dress, ornaments, etc. in the feminine world, penalties inflicted on vanity—the cutting off of the hair, the obligation to wear a certain costume, etc., might with advantage be substituted for imprisonment" (Lombroso-Ferrero [1911] 1972:181).[33] In addition, Lombroso advocated special tribunals for women, adjusted to "the milder nature of feminine criminality, the usefulness of women in the home, and the serious injury inflicted on the family and the society in general by the segregation of the wife and mother (if only for a short period)" (Lombroso-Ferrero [1911] 1972:181). Finally, Lombroso proposed a politics of prevention centered on the regulation of marriage and maternity. He recommended the liberalization of divorce and the prohibition of marriage before the age of eighteen as ways of reducing violent crimes against husbands (1915:453–54). And, in a move that joined woman's autonomy to her potential dangerousness, Lombroso argued that "the child bearing function" could act as "a moral antidote" to female criminality (1895:154).[34]

However, to view Lombroso's anthropology simply as calling women back to their (maternal) natures is to overlook the construction of the female body as a *social* body—no longer enclosed in the private spaces of the home, but accessible in the social domain to the investigations and intervention of social scientists and social technicians (social workers, hygienists, home efficiency experts, nutritionists, and architects). In Italy, as elsewhere in Europe, the ongoing exclusion of woman from the political domain was accompanied by her location in the domain of the social, both as a special object of social science and social technologies, and as a subject responsible for reproduction, for the welfare of the family, and, by extension, for the health of the larger social body (Horn 1994; Riley 1988).[35]

In this connection, it is important to remember that the positive school of criminal anthropology elaborated not only a new view of the criminal, but also a new view of society. The object of this and other sciences of the social (sociology, social medicine, social hygiene) was no longer the society envisaged by liberalism: a collection of autonomous individuals, each equipped with free will, and responsible for his or her own actions. Rather, it was a social body, with its own laws, regularities, and pathologies, which had to be known by new sciences and managed according to new rationalities of government. In this sense, criminal anthropology was linked to the emergence of what Foucault called "governmentality": modern forms of power and knowledge concerned with the management of risk and the promotion of the health and welfare of the biological population (Foucault 1978b).

Writing genealogies of our modernities may, in fact, be frustrated by efforts to contain or marginalize the discourses and practices of "discredited" sciences. Only the blurring or transgressing of comfortable boundaries—between science and pseudoscience, the social and the biological, heredity and milieu—makes it possible to situate criminal anthropology in relation to reformulations of penal practices, to other projects to know bodies in anthropology, to the elaboration of social technologies focused on women, and to the emergence of new reading practices (including the "decoding" of the human genome) that seek to make human difference intelligible. And it may be precisely through its *failure* to isolate the dangerous female body that Lombroso's anthropology establishes its links to the present, a present in which social life (including reproduction) is frequently imagined in actuarial terms, and in which the body continues to be pored over for signs of risks to be managed.

Notes

Acknowledgments. An earlier version of this essay was presented at the 1990 Meetings of the American Ethnological Society. I am grateful to Anne Fausto-Sterling, Eugene Holland, Antonia Mortimer, Victoria Powers, Nicole Rafter, Jennifer Terry, Jacqueline Urla, and the students in my seminar on Science and Difference for their comments on various drafts.

1. Hacking identifies a fundamental tension (and source of power) in the modern idea of the normal, which can be traced back to Auguste Comte: "the normal as existing average,

and the normal as figure of perfection to which we may progress" (1990:168). Also see Canguilhem ([1978] 1989:151–79).

2. Of course, *anthropology* did not mean the same thing in nineteenth-century Italy that it means today. In Italy, as in Germany and France, the unmodified noun typically referred to what we would now call physical or biological anthropology (cf. Stocking 1988:9). There was in practice, however, a considerable permeability of boundaries, at least until World War II; the training of most early anthropologists (and of many folklorists) was in medicine, while the entry on *etnologia* in the *Enciclopedia italiana* was written by a geographer (Puccini and Squillacciotti 1979:202). However, it is worth noting that both Lombroso and his colleague Alfredo Niceforo imagined themselves at the center of a holistic "science of man," a "general anthropology" embracing physical anthropology, ethnology, ethnography, and folklore (Niceforo 1910:3; Puccini and Squillacciotti 1979:204).

3. Italian criminal anthropology variously described itself as the "Italian School," the "Anthropological School," the "Positive School," the "Modern School," the "Scientific School," and the "New School" (Lombroso-Ferrero [1911] 1972:vi). Its members shared no particular disciplinary training, but rather a manner of problematizing criminality and deviance.

4. While the focus on the head was in some ways "commonsensical" and continuous with popular practices of reading faces, Sekula suggests it also worked to "legitimate on organic grounds the dominion of intellectual over manual labor" (1986:12). As we will see, the anthropology of criminal women also privileged the genitals as loci of deviance. See also Fausto-Sterling (this volume).

5. On the links between anthropometry and photography, see Sekula (1986:19–23); on anthropometry's relation to racist evolutionary thought, see Gould (1981:73–122).

6. On the uses of anthropometry to manage populations in other domains, see Blanckaert (1988).

7. Lombroso claimed to have discovered atavism when, while examining the skull of the brigand Vilella during a postmortem, he detected a formation typically found in rodents:

> This was not merely an idea, but a revelation. At the sight of that skull, I seemed to see all of a sudden, lighted up as a vast plain under a flaming sky, the problem of the nature of the criminal—an atavistic being who reproduces in his person the ferocious instincts of primitive humanity and the inferior animals. Thus were explained anatomically the enormous jaws, high cheek-bones, prominent superciliary arches, solitary lines in the palms, extreme size of the orbits, handle-shaped or sessile ears found in criminals, savages, and apes, insensibility to pain, extremely acute sight, tattooing, excessive idleness, love of orgies, and the irresistible craving for evil for its own sake, the desire not only to extinguish life in the victim, but to mutilate the corpse, tear its flesh, and drink its blood. (Lombroso-Ferrero [1911] 1972:xxiv–xxv)

The notion of atavism was rejected by Raffaele Garofalo and Gabriel Tarde, who emphasized the determinant power of the social environment (Fletcher 1891:209; Nye 1976).

8. For Lombroso, the fundamental nature of the criminal did not change, despite changing historical and cultural definitions of "crime." However, Lombroso argued that some forms of crime, as well as female criminality, tended to increase with the increase in the level of civilization (Lombroso-Ferrero [1911] 1972:151). See also Horn (1993).

9. Here, "civilization" was marked by the emergence of "criminality" as a feature of a special segment of the population and as deviations from norms that were no longer congruent with the "natural." Paradoxically, deviance became a sign of progress and (as we will see in the case of female offenders) adherence to norms an indication of a failure to evolve. Thus, Lombroso could argue that the most serious crimes in "savage" societies were violations of *custom* (1896:69–75).

One anthropologist went beyond this relativizing of "the normal" in an effort to subvert its usual meaning. Paul Albrecht, in a paper read at the 1885 Congress of Criminal Anthropology, defined the modern criminal as "normal"—that is, like all other animals—and the honest, law-abiding citizen as an abnormal being: "Abnormal or honest man kills or punishes normal or criminal man because the latter refuses to allow himself to be abnormalized" (cited in Fletcher 1891:235).

10. Alfredo Niceforo traced the roots of his "anthropology of the poor" to the early-nineteenth-century work of Cadet de Gassincourt, which, basing itself in popular beliefs, had sought to explain the misanthropy of bakers, the cruelty of butchers, the debauchery of hosiers, and "the seditious spirit" of masons and typographers (Niceforo 1910:9).

11. The degeneration of the race was due, in Lombroso's view, to the action of alcoholic beverages and "inheritance." Its effects included sterility, madness, and crime, and it manifested itself in anomalies of the ear, skull, and genitals (Lombroso et al. 1886:6).

12. As Bruno Latour notes, Pasteur's microbiology "metamorphosed" the very composition of the "social context" and endowed Pasteur with "one of the most striking fresh sources of power ever": "Who can imagine being the representative of a crowd of invisible, dangerous forces able to strike anywhere and to make a shambles of the present state of society, forces of which he is by definition the only credible interpreter and which only he can control?" (1983:158). On Alexandre Lacassagne's use of Pasteur to construct a critique of the Italian school of criminology, see Sekula (1986:37).

13. It is worth noting that *La donna delinquente* was the first book by Lombroso to be partially translated into English—and after only two years (Lombroso and Ferrero 1895). The earlier study of male criminals, *L'uomo delinquente* (1876), was not made available (and in a greatly abbreviated form) until 1911. However, in his introduction to *The Female Offender,* W. Douglas Morrison made no mention of the sex of the book's object; he simply described the volume as "an example of the method in which [Lombroso's] inquiries are conducted" (1895:xv).

14. The data presented were the collective production of criminologists working throughout Europe, including Lombroso and his colleagues. Lombroso cited studies of a total of 1,033 criminal women, 685 prostitutes, and 225 "normal" women in hospitals, as well as studies of 176 crania of deceased criminal women and 30 of "normal" women (Lombroso and Ferrero 1915:164).

15. Lombroso remarked in the preface to *La donna delinquente* that his critics had, in any case, attached too much importance to anthropometrical measurements (Lombroso and Ferrero 1915:1–2).

16. The entire discussion of the "normal" woman was omitted in the abridged English translation, which also deleted virtually every reference to sexuality, fertility, and menstruation. Lombroso complained in the preface to the French translation that these editorial decisions were "absurd" (Lombroso and Ferrero 1896:xiii).

17. In *L'uomo delinquente,* occasional reference was made to "normal" and "honest" man, but neither was marked as an object of urgent scientific scrutiny (cf. Lombroso and Ferrero 1915:3).

18. These problems had been identified by Gabriel Tarde, the French crowd psychologist and a persistent critic of the atavistic elements of Italian anthropology (Tarde 1886). For a discussion of Tarde's criminology, and of the relations between the French and Italian schools, see Nye (1976).

19. Lombroso assembled a series of scientific and anecdotal sources that agreed on the "reduced sexual sensitivity" and passivity of women (Lombroso and Ferrero 1915:43–48). Opinions to the contrary, he observed, resulted from a confusion of eroticism and the "satisfaction of maternal instincts" and from the fact that woman had three "sexual centers" to man's one: "A woman studied by Moraglia," he remarked with evident wonder, "knew how to masturbate in 14 different ways" (1915:46–47).

20. Measurements of sensitivity to pain in male subjects had confined themselves to the hands and the back of the tongue (Lombroso 1896:388–90).

21. On criminal anthropology as an effort to construct a rational basis for political inclusions and exclusions, see Pick (1989) and Horn (1993).

22. The reduced variability of females could be further explained, the authors argued, by

the operation of sexual selection. In "savage" societies, "[m]an not only refused to *marry* a deformed female, but ate her, while, on the other hand, preserving for his enjoyment the handsome woman who gratified his peculiar instincts" (Lombroso and Ferrero 1895:109).

23. For a discussion of the relations between photography and modern policing practices, see Tagg (1988) and Sekula (1986).

24. The same prohibition had required Lombroso in *L'uomo delinquente* to rely on photographs from the German prison *Album,* the *National Police Gazette* (New York), and the *Illustrated Police News* (Boston) (Lombroso 1896:221). However, Lombroso observed that the difficulties were doubled in the case of female offenders and prostitutes: "We might have offended the sense of shame of these chaste virgins" (Lombroso and Ferrero 1915:195).

25. Sekula (1986:55) describes Bertillon and Galton's use of photographs as "attempting to preserve the value of an older, optical model of truth in a historical context in which abstract, statistical procedures seemed to offer the high road to social truth and social control." I am grateful to Jennifer Terry for this reference.

26. Lombroso and Ferrero observed, for example, that French criminal women were "infinitely more typical and uglier" than Russian women, evidence of the fact that "the more refined a nation is, the further do its criminals differ from the average" (Lombroso and Ferrero 1895:94).

27. Jennifer Terry makes a similar point for scientific photographs of lesbian bodies produced in the 1930s: "Although the images of these lesbians do not appear to be at all different from what one would expect heterosexual women to look like, the very composition of the photograph and the fact that it serves as an indicator of variant characteristics invite the viewer to look for and find pathology or difference" (1990:324–25).

28. Lombroso's interest in the exegesis of bodily surfaces had its origin in his service as an army doctor in 1864, when he was "struck by a characteristic that distinguished the honest soldier from his vicious comrade: the extent to which the latter was tattooed and the indecency of the designs that covered his body" ([1911] 1972:xxii). This idea, Lombroso confessed, "bore no fruit."

29. Compare Cuvier's preoccupation with the hidden or interior genitalia of the "Hottentot Venus" (Fausto-Sterling, this volume).

30. In fact, Lombroso multiplied the objects of scientific and social technical concern to include hysterical offenders, suicides, criminal lunatics, occasional criminals, epileptic delinquents, and those guilty of crimes of passion.

31. Maternity, Lombroso suggested, was the antithesis of criminality and had never been "the motive power of crime in a woman" (Lombroso and Ferrero 1895:154).

32. The discussion of "therapies" was added by Lombroso's daughter to later editions of *La donna delinquente,* on the basis of Lombroso's notes (1915:449 n).

33. For a discussion of proposals for male criminals, see Lombroso ([1911] 1968:245–451).

34. At the same time, Lombroso recommended relaxing the punishments for abortion and infanticide (Lombroso and Ferrero 1915:453–56).

35. For a discussion of feminism in nineteenth-century Italy, see Pieroni Bortolotti (1963).

References

Blanckaert, Claude

1988 "On the Origins of French Ethnology: William Edwards and the Doctrine of Race." In *Bones, Bodies, Behavior: Essays on Biological An-*

thropology. Ed. George W. Stocking, Jr., pp. 18–55. History of Anthropology 5. Madison: University of Wisconsin Press.

Canguilhem, Georges
1989 *The Normal and the Pathological.* Trans. Carolyn R. Fawcett. New York: Zone Books. Reprint of *On the Normal and the Pathological,* Dordrecht: Reidel, 1978.

Ewald, François
1986 *L'État providence.* Paris: Grasset.

Fee, Elizabeth
1979 "Nineteenth-Century Craniology: The Study of the Female Skull." *Bulletin of the History of Medicine* 53: 415–33.

Ferri, Enrico
[1901] 1968 *The Positive School of Criminology: Three Lectures* Ed. Stanley E. Grupp. Pittsburgh: University of Pittsburgh Press.

Fletcher, Robert
1891 "The New School of Criminal Anthropology." *American Anthropologist* 4(3): 201–36.

Foucault, Michel
1978a "About the Concept of the Dangerous Individual in 19th-Century Legal Psychiatry." *International Journal of Law and Psychiatry* 1: 1–18.
1978b *The History of Sexuality. Vol. 1: An Introduction.* Trans. Robert Hurley. New York: Pantheon.

Gilman, Sander L.
1988 *Disease and Representation: Images of Illness from Madness to AIDS.* Ithaca: Cornell University Press.

Gould, Stephen Jay
1981 *The Mismeasure of Man.* New York: Norton.

Green, David
1986 "Veins of Resemblance: Photography and Eugenics." In *Photography/ Politics: Two.* Ed. Patricia Holland, Jo Spence, and Simon Watney, pp. 9–21. New York: Comedia.

Hacking, Ian
1990 *The Taming of Chance.* Cambridge: Cambridge University Press.

Haraway, Donna
1989 *Primate Visions: Gender, Race, and Nature in the World of Modern Science.* New York: Routledge.

Horn, David G.
1993 "Making Criminals, Making Italians: Science, the Body, and the Construction of Identities in Nineteenth-Century Italy." Paper presented to the conference European Identity and Its Intellectual Roots, Minda de Gunzburg Center for European Studies, Harvard University.
1994 *Social Bodies: Science, Reproduction, and Italian Modernity.* Princeton: Princeton University Press.

Latour, Bruno
1983 "Give Me a Laboratory and I Will Raise the World." In *Science Observed: Perspectives on the Social Study of Science.* Ed. Karin D. Knorr-Cetina and Michael Mulkay, pp. 141–70. London: Sage.

Lombroso, Cesare
1988 *Palimsesti del carcere.* Turin: Fratelli Bocca.
1896 *L'uomo delinquente, in rapporto all'antropologia, alla giurisprudenza ed alle discipline carcerarie,* vol. 1. 5th ed. Turin: Fratelli Bocca.
[1911] 1968 *Crime, Its Causes and Remedies.* Trans. Henry P. Horton, Montclair, N.J.: Patterson Smith.

Lombroso, Cesare, and Guglielmo Ferrero
1895 *The Female Offender.* Ed. W. Douglas Morrison. London: T. Fisher Unwin.
1896 *La femme criminelle et la prostituée.* Trans. Louise Meille. Paris: Alcan.
1915 *La donna delinquente, la prostituta e la donna normale.* 3d ed. rev. and enl. Turin: Fratelli Bocca.

Lombroso, Cesare, Enrico Ferri, Raffaele Garofalo, and Giulio Fioretti
1886 *Polemica in difesa della scuola criminale positiva.* Bologna: Zanichelli.

Lombroso-Ferrero, Gina
[1911] 1972 *Criminal Man, According to the Classification of Cesare Lombroso.* Montclair, N.J.: Patterson Smith.

Merchant, Carolyn
1980 *The Death of Nature: Women, Ecology, and the Scientific Revolution.* New York: Harper and Row.

Niceforo, Alfredo
1910 *Antropologia delle classi povere.* Milan: Vallardi.

Nye, Robert
1976 "Heredity or Milieu: The Foundations of Modern European Criminological Theory." *Isis* 67 (238): 335–55.

Pasquino, Pasquale
1980 "Criminology: The Birth of a Special Savoir." Trans. Colin Gordon. *I&C* (7):17–32.

Pick, Daniel
1989 *Faces of Degeneration: A European Disorder, c. 1848–c. 1918.* Cambridge: Cambridge University Press.

Pieroni Bortolotti, Franca
1963 *Alle origini del movimento femminile in Italia, 1848–1892.* Turin: Einaudi.

Puccini, Sandra, and Massimo Squillacciotti
1979 "Per una prima ricostruzione degli studi demo-etno-antropologici italiani nel periodo tra le due guerre." In *Studi antropologici italiani e rapporti di classe: Dal positivismo al dibattito attuale,* pp. 67–93, 201–239. Milan: Franco Angeli.

Riley, Denise
1988 *"Am I That Name?" Feminism and the Category of "Women" in History.* Minneapolis: University of Minnesota Press.

Sekula, Allan
1986 "The Body and the Archive." *October* 39: 3–64.

Stocking, George W., Jr.
1988 "Bones, Bodies, and Behavior." In *Bones, Bodies, and Behavior: Essays on Biological Anthropology*. Ed. George W. Stocking, Jr., pp. 3–17. History of Anthropology 5. Madison: University of Wisconsin Press.

Tagg, John
1988 "A Means of Surveillance: The Photograph as Evidence in Law." In *The Burden of Representation: Essays on Photographies and Histories*, pp. 66–102. Amherst: University of Massachusetts Press.

Tarde, Gabriel
1886 *La Criminalité comparée*. Paris: Alcan.

Terry, Jennifer
1990 "Lesbians under the Medical Gaze: Scientists Search for Remarkable Sex Differences." *The Journal of Sex Research* 27(3): 317–39.

5

Anxious Slippages between "Us" and "Them"

A Brief History of the Scientific Search for Homosexual Bodies

Jennifer Terry

Since the late nineteenth century, the body has been central to both scientific and popular constructions of the origins of homosexuality. As homosexuality came to be associated with pathology, questions about the causes and distinguishing features of same-sex desire propelled physicians, biologists, anthropologists, and forensic scientists to identify and measure characterological features of suspect individuals, and to classify them according to an array of sexual typologies. Over the past century, a melange of scientific studies has emerged that postulates in one way or another a vital link between the body and homosexual desire. This chapter offers an abbreviated historical sketch of this century of scientific attempts to correlate corporeal attributes with homosexuality, tracing some of the more significant ways that bodies have been scrutinized for proof of innate constitutional deficiency as well as for evidence of abnormal proclivities and unusual sexual practices.

What counts as a body? What can it reveal about the causes and manifestations of perverse desire? Is it a reliable source for determining who is a "homosexual"? If homosexuality can be found on or in the body, what are its signs? What parts or territories of the body reveal it? If homosexuality is signified through the body, are its marks the *source* or the *consequence* of experiences and desires? These questions frame my inquiry into the larger historical effort to name and police homosexuality, which persists even to this day. To undertake a comprehensive analysis of this complex history is too great a task to accomplish here, so I have chosen to focus on the earlier half of this history, beginning around 1869, when the homosexual was singled out by science as a distinct type of person with unique psychical and somatic characteristics, and ending with the publication of Kinsey's reports on human sexual behavior in the period just following World War II. This framing device allows us to chart scientific curiosity and trepidation about homosexuality as it moved from being understood in terms of an innate biological condition afflicting certain individuals to being considered one of many possible forms of sexual behavior practiced by all kinds of people. This shift from a clinical understanding of the homosexual as a distinct type of being toward a statistical variance model that mapped all manner of sexual behaviors is illustrated, in part, by the increasingly contorted and contradictory ways in which homosexuality was conceptualized in relation to the body. As we shall see, in this shift the body became not only a troublesome source of evidence, but disappeared altogether as Kinsey's methods obliterated, at least momentarily, the notion of a distinct homosexual type.

By surveying the work of several prominent sexologists and focusing specifically on an American study from the 1930s that straddles the boundary between hereditary and environmental explanations for homosexuality, this chapter analyzes changes in the status of the body as a source of scientific evidence about "abnormal" sexual tendencies and desires. I focus on two distinct but overlapping constructions of bodies that circulated through scientific studies during this historically tumultuous period. The first assumed that homosexuality was the symptom of an innate and inherited constitutional predisposition, and the second assumed that the body was a surface upon which the signs of homosexuality appeared as consequences, rather than causes, of certain practices and characterological tendencies. I examine the contrast between these two ways of conceptualizing the embodiment of homosexuality, noting how scientists and physicians deployed different kinds of diagnostic techniques for determining who was inclined toward homosexuality, and what was to be done to contain their abnormal desires. I then turn to a discussion of research on human sexual variance in the late 1940s that actually made the homosexual body disappear, if only for a brief, anxiety-ridden moment. The chapter ends with an epilogue on the implications of this history for current interests in "gay biology."

Constitutional Deviance

The medico-scientific discourse from 1869 to around 1920 presented a complex and contradictory set of explanations for homosexuality, all of which understood this "contrary sexual instinct" to be somehow rooted in the body. Here I want to highlight several main ideas that structured this discourse, focusing on how psychiatric ideas from the nineteenth century set the stage for twentieth-century studies of homosexuality and the body. Before beginning, it is important to signal the difference between the way early sexologists conceptualized the body in terms of its "constitution" and the way we understand it today. We are quite accustomed today to thinking of the biological domain of the body as nominally distinct from the psyche, as well as the larger social context in which both are situated. However early sexologists did not understand these to be clearly differentiated domains. Instead, an individual's constitution encompassed biological attributes as well as moral, intellectual, and psychical qualities, all of which were seen to be deeply reflective of each other and embedded in an individual's body. In the case of homosexuality, an individual's tendency toward perverse acts was seen as evidence of innate moral inferiority as well as biological deficiency. Conversely, those who were robust and free of perverse temptations were seen to be biologically sound and morally upright.

Two key ideas shaped the way homosexuality was seen to be embodied within this constitutional framework. The first regarded it in terms of constitutional degeneracy and deemed homosexuals to be suffering from an innate pathological condition of the body linked to disorders of the brain and nervous system. The second construed homosexuality in terms of sexual inversion and imagined that homosexuals belonged to a third sex, situated between male and female. These two models, while analytically distinct, overlapped in the writings of prominent sexologists in the nineteenth and early twentieth centuries, as we shall see.

Homosexuality and Nervousness

The idea that homosexuality was a matter of constitutional degeneracy emerged at a time when European science supported a prevailing belief that certain socially disadvantaged classes of people were intellectually inferior by nature. Thus, the bodies—and particularly the brains and nervous systems—of the poor, of women, of criminals and of nonwhite peoples were assumed to be primitive, fundamentally degenerate, or neurotically diseased. The homosexual joined their ranks around 1869, in the midst of a great deal of speculation about the body's role in expressing deviant sexual desires, when German physician Karl Westphal wrote that *contrary sexual feeling* was rooted in the body's constitution.[1]

For many of the earlier sexologists, including, notably, Viennese psychiatrist Richard von Krafft-Ebing[2] and, later, British essayist Havelock Ellis, the homosexual invert was a living sign of modern degeneracy who suffered from an underlying nervous disorder that could manifest in certain kinds of physical stigmata as well as in sexually inverted personality traits.[3] This creature's tainted body was not only a necessary precondition for the expression of homosexuality; it also inclined the individual toward even more degenerate acts and moral dissipation. Homosexuality was only one of an array of signs, albeit among the strongest, of defective development. Those who expressed it inherited some degree of neuropathic taint that could manifest in a number of other biologically based deviations, including neurasthenia, eccentricity, imbecility, and even artistic brilliance.

In general, this way of conceptualizing homosexuality as a form of constitutional degeneracy generated two different frameworks. The first of these considered homosexuality as a sign of the loss of adaptive ability.[4] Privileging the nervous system, this framework posited that degeneration, of which homosexuals were both signifiers and sufferers, was caused by an exhaustion of the nervous system due to inordinate cultural constraints and stress. Since the nervous functions were seen as constitutive of the highest and most complex system, when they were broken down by stress the individual and, by implication, the culture underwent a process of simplification. This debilitation of the nervous system allowed primitive instincts to run free, manifesting in insatiable sexual appetites and promiscuity. It also led to devolution and decreased sexual differentiation that manifested in the emergence of a third, anatomically primitive class of defective individuals who were between man and woman.

The second framework linking homosexuality with constitutional degeneracy involved the Spencerian notion of overspecialization. According to this framework, as the human species became more complex—and those of European origin were seen to be the most complex—less energy was available to be spent on reproduction. Homosexuals, whose numbers were said to be increasing, represented a pathological response to the demands of modern civilization, which manifested in their presumed refusal to procreate. They signified cultural complexity taken to the point of biological sterility.[5] The alleged preponderance of homosexuality among intelligent women and artistic men of the upper classes was supporting evidence that this contrary sexual instinct was a troublesome side effect of European cultural refinement.

Homosexuals came to symbolize sterility, madness, and decadence in the late Victorian period. Psychiatric texts of the time combined conservative sexual mores with scientific opinion in the exaltation of heterosexual marriage and reproduction. By contrast, masturbation and homosexuality were condemned for contaminating and exhausting the source of noble sentiments that would otherwise develop as a part of normal sexual instincts. Krafft-Ebing believed that masturbation could induce neurasthenia, which, in tainted individuals, could deteriorate further into homosexual perversion. As compulsive nonreproductive practices, both forms of self-pollution drained the male body of its vitality and left no offspring to show for it. Ultimately, the ongoing practice of both perversions led to a point of no return, leaving the "youthful sinner" with an excessive sex drive but in a state of "psychical impotence" that made an adjustment to heterosexual relations impossible.[6]

Like Krafft-Ebing, Ellis also understood masturbation to be both a symptom of organic weakness and an aggravating factor that could lead to further perversion. Together with homosexuality, it loomed as a menacing outcome associated with sexually precocious children. Like savages and criminals, they suffered from arrested development because they were inherently tainted.[7] In Ellis's view, inverted sexual instinct grew out of a predisposition developed in an individual's early embryonic life, similar to the conditions under which other congenital defects originated, for example, idiocy, criminality, and genius. Those with such a congenital predisposition were susceptible to becoming inverts in adulthood, but some could be spared this fate if, in childhood, they were subjected to healthy routines that fostered heterosexuality and proper gender identification.

How, then, did sexually precocious children grow up to be homosexuals? Ellis explained that human sexual instinct initially was specialized in neither a homosexual nor a heterosexual direction, but could be steered down the wrong path if a constitutionally predisposed child was subjected to unhygienic circumstances, such as attending sex-segregated schools or being exposed to sexually aggressive adult inverts. Sexually precocious children were especially vulnerable to becoming homosexuals in adulthood because, as a result of expending sexual energy at a young age, their development would be arrested. If the body's development was stalled, its sexual energy remained feeble and was more likely to go either toward masturbation or toward homosexual relationships because in these situations "there is no definite act to be accomplished."[8]

One of the apparent paradoxes that sexologists had to explain was why, if homosexuals suffered from constitutional weakness, they also appeared to have such strong sexual drives. Ellis reconciled this by classifying a hardy sexual impulse as yet another sign of overall weakness and nervousness. In his view, many inverts tended to have irritable "sexual centers," which disturbed the interlocking system of the brain, nerves, reproductive organs, and genitals. Irritations of this sort manifested in promiscuity as well as in patterns of self-sacrifice and affection. Thus, the problem was neither impotence nor indifference, but an abnormally directed libido.

While sexological ideas about homosexuality and nervous degeneration tended to be based on the male body's spermatic economy, lesbians also were assumed to be constitutionally tainted. In fact, lesbianism was construed by many sexologists as merely a form of female masturbation, not even worthy of the status of counterfeit intercourse. Like other perversions, it was seen as part of an overall destructive process. But while lesbians were often subsumed in discussions of male homosexuality, occasionally they were distinguished on the basis of being innately less passionate and sensual, by virtue of being women. For example, Krafft-Ebing granted that women had strong friendships, but, like many men of his time, he believed that sex between them was neither as powerful nor as threatening to the social order as male homosexuality. And in his view, because women lacked penises they did not suffer from impotency and so were spared the temptation to take relief in homosexuality, as many neurotic men felt they must.

Writing at length about lesbianism, Ellis warned that homosexuality among women was increasing with the march of modern progress and feminism. Women's growing independence from men and marriage was as likely to foster homosexuality as was the nervous strain men experienced in the face of intensifying business competition. Ellis explained that most women tended to be heterosexual because they were generally passive. Their bodies and personalities lacked the variations common among men, making them naturally susceptible to sexual advances and normally unlikely to initiate any sexual encounters. But those with abnormal instincts had predatory tendencies, along with an array of other masculine physical characteristics. These mannish women tended to exploit more impressionable young women, seeking affection from those toward whom men felt indifference because they were cold and unattractive. The unfortunate recipients of such attention were "womanly women" who had some good qualities but were neither robust nor fit for childbearing. Like "normal" women, these women were naturally seduced by masculine people, whether these seducers be "normal" men or sexually inverted women. Were it not for being constitutionally tainted, these womanly women would be spared the temptation to succumb to an invert's advances.[9]

The Third Sex Model

The idea that the homosexual belonged to an intermediate sex dominated nineteenth- and early twentieth-century constructions of homosexuality.[10] In 1898, Karl Ulrichs, a German lawyer, elaborated on existing discussions of sexual inversion by introducing the notion of the "urning" as the congenital outcome of an undifferentiated human embryo that resulted in a female mind in a male body.[11] A decade later, Edward Carpenter would use Ulrichs's ideas to popularize his notion of the homosexual as an intermediate sex.[12] German physician Magnus Hirschfeld, whose ideas deeply influenced subsequent research on homosexuality, characterized the homosexual as a third sex.[13] These writers generally assumed that homosexuality was an innate anomaly manifesting in the tendency for constitutionally predisposed men to behave and appear feminine and thus, like women, to be inclined toward relations with men. Conversely, lesbian tendencies in women were seen to be linked to some fundamental masculine instinct that made them act like men. Homosexual desire was thus a symptom of *sexual inversion,* or the tendency to embody physical and behavioral characteristics associated with the opposite sex. Naturally, this model had great difficulty explaining cases of "masculine" or "active" homosexual men and "feminine" or "passive" lesbians.

Krafft-Ebing believed sexual inversion was evidence that homosexuals were stuck at a more primitive stage of evolutionary development than normal (heterosexual) people. To support this, he first asserted that those inclined toward homosexual perversion would have what he called *bisexual* or hermaphroditic traits, noting that hermaphroditism existed in lower life forms with which humans shared a primordial past. Then he reasoned that, since it was from these forms that we have evolved, humans still carried the potential for the reemergence of anatomical or psychological hermaphroditism. Hermaphroditism in an individual was the sign of a lagging evolutionary process because the lesser the distinction between masculine and feminine traits in any one person, the lower the individual on the evolutionary scale. Thus, individuals who displayed what were taken to be sexually ambiguous traits—whether these be anatomical or behavioral—were interpreted to be primitive and, most likely, degenerate. Conversely, progress was signified by a greater degree of sexual difference, or dimorphism, as well as procreative heterosexuality. Homosexual inverts, because they were seen as blurring the boundaries of gender—either as masculine women or as effeminate men—were regarded as "unfinished" specimens of stunted evolutionary growth, a status they shared with "savages" and certain types of criminals.[14]

The idea that homosexuals were in some sense constitutional hermaphrodites or belonged to a third sex was eventually undermined as Freud's theories of sexuality achieved greater notoriety and influence in the scientific community.[15] His ideas shifted the question of origins away from the body's sex to the individual's psyche, arguing that the defining feature of homosexuality was not a person's gender characteristics, but his or her choice of sexual object.[16] Nevertheless, as we will see, homosexuality and gender inversion were not to be so easily disentangled; the idea that those who engaged in homosexuality had psychological and somatic qualities common to the opposite sex continued to crop up in scientific thought—and certainly in popular assumptions.[17]

Constitutional Pathology or Benign Difference?

Not all physicians and scientists who believed homosexuality was a matter of constitutional predisposition assumed that it was necessarily pathological. Ellis, alternately expressing pity and disdain toward inverts, vacillated on the issue of whether inversion was related to degeneration or was simply a natural variation.[18] Hirschfeld, one of the first advocates of homosexual rights, believed that homosexuals should be tolerated as rare but natural variants, rather than being punished for what he believed to be their biologically driven desires.[19] This view placed him at odds with many of his contemporaries who associated homosexuality with pathology and morbid degeneracy. Like Hirschfeld, Freud emphasized the random and even benign biological and psychical variations among people, in contrast to Krafft-Ebing's concept of homosexuality as a sign of innate constitutional defectiveness. From his own observations, Freud believed that many who showed homosexual tendencies contributed a great deal to society and, far from being degenerate, were of high moral and intellectual standing.

Although most physicians and scientists at the turn of the century agreed that homosexuality was a condition of a biologically distinct body, some began to question biological explanations altogether. Psychoanalysts Sandor Ferenczi and A. Brill rejected the idea that homosexuality was related to any innate biological factors.[20] Freud is remembered as one of the most influential proponents of the psychogenic origins of homosexuality, which eventually supplanted most constitutional theories, but his own opinions on the matter were actually contradictory. While he refuted common assumptions that homosexuality was primarily a manifestation of hereditary degeneration, he acknowledged that biological factors might play a role in some cases of homosexuality.[21] But more than anyone else of his time, he emphasized the influence of psychogenic or environmental factors in shaping sexual orientation, appropriating the term *constitutional disposition* from his contemporaries to refer to a complicated array of tendencies and vulnerabilities that inclined an individual toward various kinds of neuroses.[22]

In most of Freud's writings, homosexuality represented a form of arrested psychosexual development resulting from a vague underlying biological component that predisposed certain individuals toward it. But he stressed that this biological component was neither clear nor sufficient for separating out homosexuals as a group with a special nature.[23] In fact, as early as 1905, Freud questioned whether one could reasonably group together the vastly different types of people who had homosexual desire on the basis of either their biology or their psychology. He argued that same-sex object choice was the result of many complex factors and manifested in a variety of ways, some more troublesome than others. In fact, he stressed that homosexual object choice was no more or less complicated than heterosexual object choice and should not be the grounds for establishing a distinct category into which all those preferring homosexuality could fit.[24] His critical stance against constructing a singular typology of the homosexual presaged subsequent scientific research that called the entire idea of a distinctly homosexual body into question.

On the whole, early sexologists believed that homosexuality was an innate constitutional condition, but not all of them saw it as pathological. Their most pronounced scientific legacy was the idea that the homosexual was an inherently different type of person, endowed with somatic and characterological features that distinguished this creature from normal people. But Freud's thinking opened up the possibility of an alternative approach to the question of the etiology and embodiment of homosexuality that would rival, if not entirely displace, the study of the body. Nonetheless, early scientific attempts to identify, measure and classify homosexuals established a set of assumptions and standard methods that were deployed again, in more sophisticated ways, during the first half of the twentieth century. As we shall see, the results were increasingly contradictory and unstable.

Between Constitutional Homosexuality and Sexual Variation

Despite the challenges to biological determinism posed by psychoanalysis, most scientists and physicians continued to perceive the body as a central piece in the puzzle of homosexuality well into the twentieth century.[25] Early debates about the defining features of homosexuality provided an impetus to probe bodies in search of reliable evidence that might distinguish homosexuals from normal people. Krafft-Ebing claimed to find typical skull dimensions, postures, gestures and mannerisms, and concluded that homosexual degeneration originated in the brain and nervous system, where he believed the damage was most pronounced.[26] Ellis noted that defects in "sexual glands," or hormone-producing organs, could lead to sexually inverted characteristics, and regarded traits such as brusqueness, aggressiveness, and timidity as matters of both heredity and hormones.[27] By the 1930s the continued quest to define homosexuality as an inherited bodily attribute was abetted by new technological devices and methodological approaches that enabled further penetration and more minute clinical surveillance of suspect bodies. These techniques were to help determine with greater precision whether homosexuality resulted from psychological conditions or constitutional factors, which had increasingly come to be synonymous with biological factors.[28] Debates on the matter from this period illustrate, on the one hand, a growing skepticism about whether homosexuals had distinct somatic features and, on the other, a tenacious desire to find homosexuality in the body.[29]

From 1935 through 1941, an ambitious study was conducted in New York City to gain more information about both the psychogenic and constitutional factors giving rise to homosexuality. Sponsored by the Committee for the Study of Sex Variants (CSSV), this research combined aspects of earlier notions about innate homosexuality with more recent psychogenic explanations.[30] Thus, as we shall see, the study reveals a transition from the nineteenth-century constitutional framework to a model of sexual variance that came to the fore with the publication of the Kinsey studies on human sexual behavior. It also offers a very rich picture of the changing status of the body as scientific proof of homosexuality; while other American research from the period combined psychological and physical examinations, the Sex Variant study is perhaps the most thorough in its attention to the body.[31] Every inch was scrutinized, including the texture of hair, the complexion of skin, the structure of the pelvis, and even the size and shape of the genitals, in order to determine the unique physical characteristics of "sex variants," the study's synonym for homosexuals. By focusing on the physical examinations, we will be able to see how the body changed as a source of evidence about deviance.

The investigation aimed to learn as much as possible about homosexuality, which was perceived to pose a growing threat to mental and urban hygiene. By amassing comprehensive knowledge on the subject, the study was to assist physicians in identifying and treating individuals who suffered from "sexual maladjustment" and to help prevent the spread of sex variance through the "general population." It took place within a larger social context of eugenic efforts to encourage hygienic reproduction and discourage "inappropriate" sexual couplings among people of the same sex or across racial boundaries. Indeed, the Committee's founder and several of its more prominent members were formally associated with eugenics institutions.[32]

Eugenic doctrine of the first half of the twentieth century placed both racial and sexual purity at the top of its agenda. Its adherents from all points along the political spectrum were especially concerned with promoting hygienic reproduction among whomever was seen as stalwart and worthy.[33] During the 1920s and 1930s, white phobia about miscegenation and racial passing paralleled a growing sex panic that inverts and perverts were everywhere, but difficult to detect visually. Hence, an apparatus for identifying and isolating them could be justified as a matter of social hygiene. In the larger social context, some of the same logic applied to homosexuals that applied to those whose relationships crossed racial and class boundaries: sexual contact of an "unnatural" sort would corrupt the population and lead to further social disorder. Panic about miscegenation was deeply tied to eugenic fears about ensuring proper rates of reproduction among the desirable (i.e., white) members of society.[34]

It is not surprising that in this context the scientific making of the homosexual type was integrally connected to campaigns for encouraging hygienic heterosexuality primarily among white people. In these campaigns, the creature called the heterosexual was fashioned as a positive and civilized counterpart to both the menacing pervert and the atavistic savage. At a time of great anxiety over drawing and maintaining differences within the human population, the sex variant, like the passing Negro, became a confusing border creature, who existed between man and woman, who traversed class and racial boundaries, and whose masquerade was treacherous. This was the general social context in which the Sex Variant study was launched.

Gynecologist Robert Latou Dickinson, of the National Maternal Health Committee, proposed the research and secured the support of IQ specialist Lewis Terman, psychiatrist Adolf Meyer, and physical anthropologist Earnest Hooton. A panoply of experts then assembled to carry out the investigation; they included psychiatrists, child psychologists, surgeons, neurologists, urban sociologists, and a former commissioner of the New York City Department of Correction. Many committee members maintained an affinity for hereditary arguments and perceived social problems as closely related to biological factors, which fueled their interest in carefully studying the body.

As founder of the CSSV, Dickinson's particular interest in studying homosexuality, and especially lesbianism, grew out of his extensive research on female sexual satisfaction, marital adjustment, and maternal health, which he began in the 1890s.[35] Just as he believed that frigidity and female sexual frustration would ruin a marriage, Dickinson was concerned that a history of lesbian relations could indicate that a woman would never be happy in her marriage to a man. His interest in detecting the signs of homosexuality thus was linked to his larger eugenic concerns about fostering marriage and reproduction among the "fit."[36] In pursuing the Sex Variant study, he hoped to be able to devise a checklist of visible characteristics that could assist physicians in identifying homosexuals and dissuading them from getting married.

Scientific Scopophilia

Forty men and forty women from various backgrounds volunteered to be examined. Among them were writers, artists, musicians, and theater performers, living bohemian lives in New York City. Most were white, but two of the men and four of the women were African American or had at least one black parent. The only thing they all had in common was a history of homosexual relations, although very few were exclusively homosexual. Because they all admitted to having such a history, researchers felt assured that any unusual characteristics they found could be attributed to homosexuality.

The search for signs of homosexuality took the subjects through an assembly line of expert examinations aimed at locating and measuring "masculinity" and "femininity" as well as the markers of homosexual sex practices. The sequence of data-gathering, then, followed a uniform pattern: first, each was given an extensive psychiatric interview, which also functioned as an ethnographic interview revealing unique cultural practices of what researchers assumed was a nascent urban demimonde. Second, standardized attitudinal tests were administered for measuring degrees of femininity and masculinity.[37] Third, each individual was subjected to a series of general physical exams; and, finally, the women were given thorough genital examinations, which were depicted in minutely detailed sketches, appearing in a graphic appendix entitled "The Gynecology of Homosexuality." In several cases, male subjects' genitals were measured and the viscosity of their semen observed, but not nearly as much attention was paid to them as to the women in this regard, and no visual representations were made of their genitals.[38]

Drawing on both psychiatric interviews and physical examinations, the study was based on correlating the subjects' narrative accounts of their lives with their body measurements, in order to get a comprehensive picture of sex variance and to single out those who were prone to it. Therefore, techniques to produce visible differences through anthropometric measurements, drawings, photographs, or x-rays constituted the scopic regime against which the psychiatric interviews of subjects would be read. According to the logic of the physical examinations, that which could be *seen* was understood to be the objective and quantifiable truth. Thus, the imperative to survey bodies was fueled by a kind of *scientific scopophilia;* the pleasures of viewing were deeply tied to both a positivist quest for the truth in physical evidence and a desire to read the body for indications of sexual practice. But the importance given to the psychiatric interviews depicting the subjects' own stories indicates that researchers believed the body alone could not provide enough information about the causes and defining characteristics of homosexuality. Nevertheless, they doggedly examined bodies for any evidence of unusual characteristics.

One-third of the subjects allowed photographs to be taken of them in the nude, as a means for supplementing other physical data and to act as diagnostic instruments for correlating body form with behavior. But since no note was made of which individual was being photographed, such a correlation was concerned with constructing a typical or generic sex variant body, rather than linking an individual subject's personality characteristics with his or her body form. This effort to homogenize the subjects was obvious in the special treatment of the photos: throughout the study, subjects' identities were protected under pseudonyms, and in the photographs their identities were concealed through a process that intentionally wiped out that most distinct characteristic, the face (Figure 5.1). This technique further rendered the photographed individual as a specimen or object, incapable of returning the spectator's gaze. Moreover, the photographs were encoded as morbid by their stark presentation at the end of the published findings, where they recall the images of diseased bodies commonly featured in medical textbooks. Thus, their very composition, as well as their presentation in the scopic regime, invites the viewer to look for and find pathology in much the same way that photographs of prostitutes analyzed by Cesare Lombroso and Pauline Tarnowsky were read for criminality and sexual deviance.[39]

Stories Genitals Tell

One of the most remarkable aspects of the Sex Variant study is that it included graphic material on female genitalia. Septuagenarian doctor Robert Dickinson believed that this was a crucial aspect of the investigation because women's genitals offered evidence not only of innate deviance, but also of deviant sexual experiences. In his earlier clinical studies, Dickinson believed genitals offered clues for detecting a woman's proclivity toward lesbianism, masturbation, frigidity, and promiscuity. Her conscious mind alone could not reveal all the important aspects leading to either her sexual satisfaction or discontent. Instead, a woman's body must be studied first, since it provided an index of emotional or psychological conditions that the modest or repressed patient could not describe. In this way, for Dickinson, the female body functioned as a master text, revealing what was otherwise hidden in the mind.

In view of his prior experience in preventive gynecology, Dickinson wanted to examine sex variant women to see what their genitals might reveal.[40] He was aided by Dr. L. Mary Moench, who conducted the gynecological examinations of "all but the most vigorous, assertive women," who refused to be examined. All traces of modesty fell by the wayside in the pursuit of accuracy and exact representations. Dickinson encouraged Moench to measure genital parts using a small ruler and her fingers. Vaginal penetration was measured in terms of the number of fingers the examiner could fit into the subject, thus making the doctor's own body a crucial part of the research. Moench then placed a small glass plate on the vulva, outlining the external genitals upon it in soft crayon, which

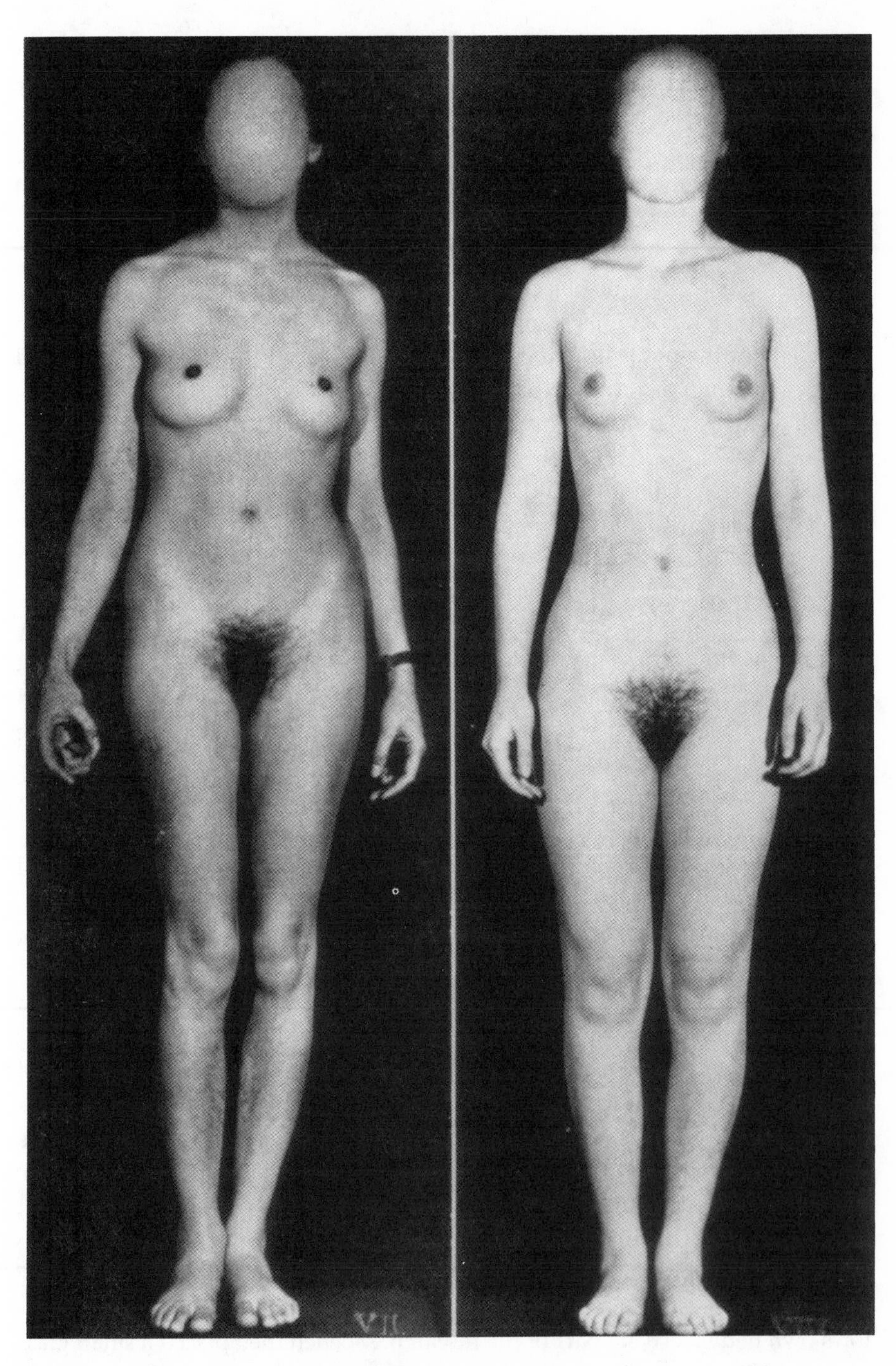

Figure 5. 1. Photograph of two female sex variants. (From George W. Henry, *Sex Variants: A Study of Homosexual Patterns* [N.Y.: Paul B. Hoeber, Inc., 1941].)

she then traced on the subject's record sheet. These line tracings were later annotated and enhanced by Dickinson, who noted remarkable or distinguishing features.

Dickinson was obsessed with accuracy and exactitude in his measurements and graphic sketches. In earlier work on human sex anatomy, he had produced laborious and highly detailed sketches of female genitals, which he drew while his patients were prone on the examining table.[41] He called his style "medical natural history" because of its graphic realism and exactly proportioned dimensions. Fancying himself both artist and scientist, he signed and dated every sketch, which led one commentator to call him a cross between Havelock Ellis and Leonardo da Vinci.[42] Dickinson's sketches illustrate how scientific practices of close observation and detailed recording bolstered the authority of the study's scopic regime. One drawing included Dickinson's careful notation of specific measurements and his recording of the duration of one subject's nipple erection at 70 seconds (Figure 5.2). Part of his obsession with detail involved the segmenting of the subject into component parts or zones that could be deciphered. This reconfiguration of the body into territories highlighting breasts, vaginas, clitorises, and labia was itself a powerful act of signification and interpretation on Dickinson's part. He and Dr. Moench assumed that markings in these highly scrutinized areas would reveal what distinguished sex variant women, and thus genitals became indices of moral character.

The gynecological exams gave rise to the wildest speculation about lesbian anatomy and experiences. Regardless of the absence of a heterosexual control group, ten typical characteristics of lesbians were established that supposedly distinguished their genitals from those of "normal women." The typical female sex variant had a larger than average vulva, longer labia majora, protruding labia minora, a large and wrinkled prepuce, a "notably erectile" clitoris, an elastic and insensitive hymen, a distensible vagina, a small uterus, and erectile nipples. The list bore a remarkable resemblance to that assembled by Havelock Ellis several decades earlier, suggesting that a standard had been set for what counted as a lesbian body.[43]

Dickinson drew a composite sketch of a "normal" woman's vulva, contrasting it to a typical sex variant vulva, placing emphasis on the presumed excesses caused by lesbian sex. Almost all terms used to describe lesbians' genitals connoted extraordinary size and hypersexuality. Pejorative adjectives, such as "wrinkled," "thickened," and "protruding," connoted excess and literally marked the subjects as pathological, while the normal unmarked female was represented in unmodified terms (Figure 5.3). Sketches isolating particular genital parts featured the unusual characteristics of particular subjects, and images were captioned with subjects' names and any significant qualities. In one sketch, the anxious doctor attempted to represent the action of the clitoris in its

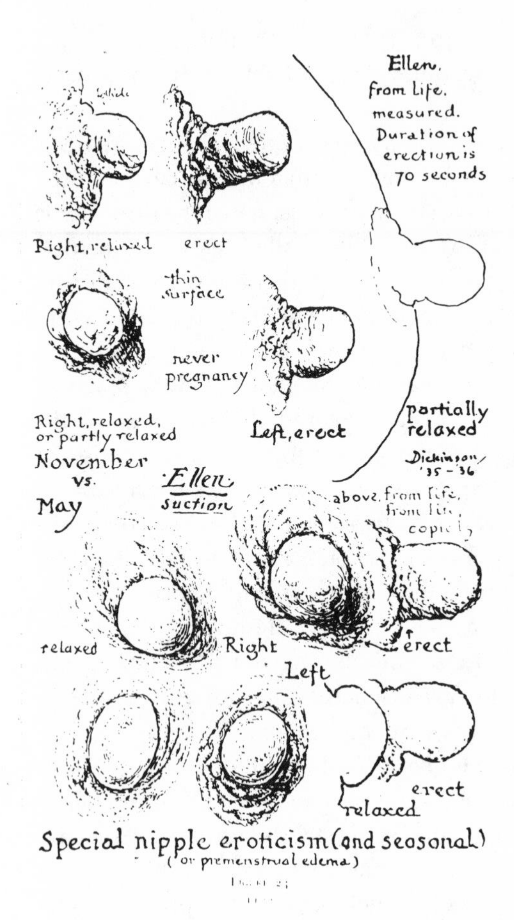

Figure 5.2. Dr. Robert Latou Dickinson's sketch of a female sex variant's breast structure and duration of nipple erection. (From George W. Henry, *Sex Variants: A Study of Homosexual Patterns* [N.Y.: Paul B. Hoeber, Inc., 1941].)

so-called excursions up and down. Elongated clitorises were noted in several of the African American lesbians, perhaps because a few of them boasted about how their lovers liked them for that attribute. Some gynecological sketches noted the race of the subject ("negress") next to what was seen to be an unusually long clitoris (Figure 5.4), recalling the lesbian counterpart to the stereotypical savage with an unusually long penis. Here, as in other representations combining racial difference and sexual deviance, we find a link in the white medical imagination between blackness and hypersexuality, this time through a clinical reading of lesbian masculinity in female genitals.

Dickinson was quite convinced that, like nymphomaniacs, sex variant women on the whole showed evidence of greater sexual excitability, especially when being examined by a female gynecologist. He suggested that an examining doctor might be able to detect a woman's perversion by watching the way

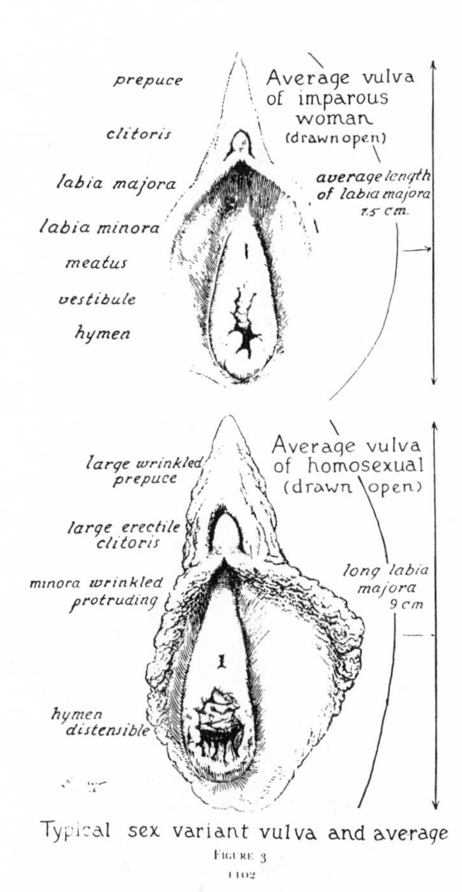

Figure 5.3. Dr. Robert Latou Dickinson's composite sketch of a "normal" woman's vulva and a typical sex variant's vulva. (From George W. Henry, *Sex Variants: A Study of Homosexual Patterns* [N.Y.: Paul B. Hoeber, Inc., 1941].)

she conducted herself during the examination. A habitual rhythmic swing of the hips, unnecessary exposure or exhibitionism, and restless behavior on the examining table could tell the doctor plenty about his patient, even before he had a chance to examine her genitals. Other more minute indications included "marked vulvar hypertrophies," clitoral erections, free mucous discharge, or general signs of vaginal congestion. He warned practitioners to be especially on the lookout for signs of arousal that could "serve to check up on the statements of the patient, being important chiefly where sexual excitability or response is flatly denied."[44]

In his final report, Dickinson asserted that, for the gynecological exam, the most clear indications of female sex variance were not constitutional or innate hereditary deficiencies, but instead those characteristics resulting from what he called "female-to-female sex play" (Figure 5.5). Thus, Dickinson clinically decoded the genital zones of the female body as evidence of a subject's behavior,

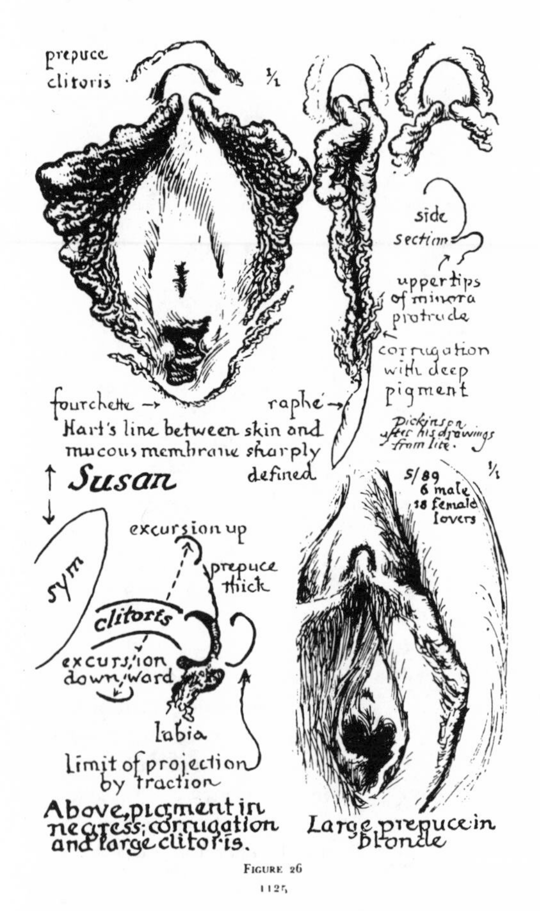

Figure 5.4. Dr. Robert Latou Dickinson's sketch of female sex variant genitalia. (From George W. Henry, *Sex Variants: A Study of Homosexual Patterns* [N.Y.: Paul B. Hoeber, Inc., 1941].)

experiences, and sexual desires, and not necessarily her genetic or congenital makeup. He paid special attention to what he called the "hypertrophy" of the prepuce and the size of the clitoris, characteristics that he asserted were primarily produced by masturbation or homosexuality. Meticulously measuring clitorises and labia, he surmised that variant "sex play" was typically focused on "digital and oral caresses." This opinion was supported by his earlier assertion that the clearest correlation between genital characteristics and experience occurred in cases where he suspected the patient masturbated or engaged in "traction, pressure, or friction."[45]

Dickinson concluded that all of the gynecological findings *could* be the result of a strong sex urge plus anything from "self-friction" to homosexual *or* heterosexual sex play. In other words, all of the typical characteristics of female sex variants could result, in fact, from any sex practices other than exclusive heterosexual intercourse in the "missionary" position. And this is precisely where

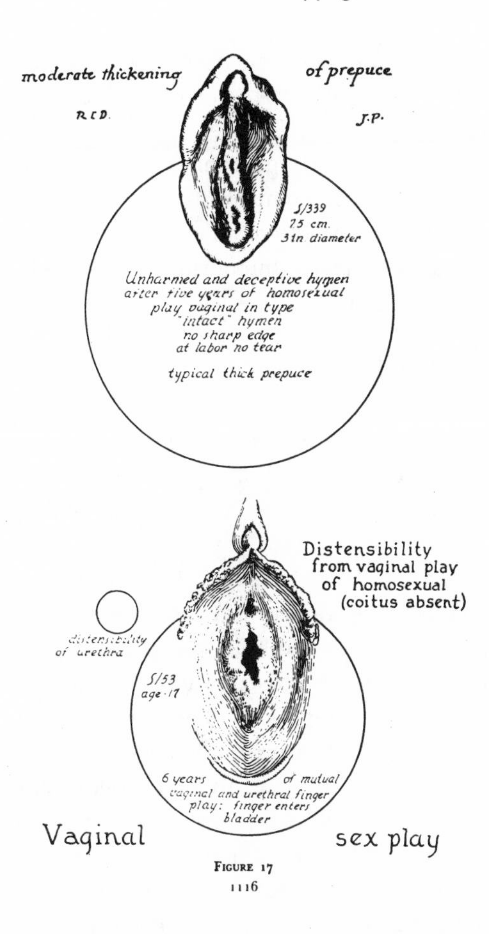

Figure 5.5. Dr. Robert Latou Dickinson's composite sketches noting the effects of vaginal sex play on female sex variants' genitalia. (From George W. Henry, *Sex Variants: A Study of Homosexual Patterns* [N.Y.: Paul B. Hoeber, Inc., 1941].)

the logic of the gynecological examinations began to unravel: elsewhere in his earlier writing, Dickinson had encouraged men to stimulate their partner's clitoris as a remedy for frigidity or sexual frustration.[46] But later in the Sex Variant study, he regarded evidence of clitoral stimulation to be a reliable sign of lesbianism. To make matters more complicated, many of the sex variant subjects Dickinson examined had extensive sexual experiences with men, making it even more difficult for him to delineate physical features that were definitively attributable to lesbian relations. Thus, what he thought were remarkable characteristics distinguishing lesbians from heterosexual women could actually have been put there by a male partner, a female partner, or the woman herself through masturbation. There was no way to confirm that certain visible characteristics resulted from specific sexual encounters, nor could a subject's sexual orientation be determined from these physical marks.

Although Dickinson did not admit it, the body had become even less reliable as evidence of lesbianism, since these marks by themselves were not absolute evidence of the crucial distinguishing factor of lesbianism—that is, the sexual object choice of a woman for another woman. However, through rendering lesbians' genitals as pathological, the doctor's findings implicitly posited a normal woman who was not only heterosexual, but also masturbated in respectable moderation and whose sexuality revolved around intercourse with her husband, primarily for the purposes of reproduction. Clearly, the normal woman was no less a scientific construct than was the lesbian.

A Body of Mixed Evidence

The Sex Variant study reiterated many aspects of the earlier constitutional framework prominent in the nineteenth century. First and foremost, it viewed homosexuality primarily in terms of sexual inversion. Thus, sex variants were thought to fall in the middle of a continuum between the two poles of masculinity and femininity, exhibiting physical and psychosexual characteristics normally found in the opposite sex. Too much masculinity in women and too much femininity in men were telling signs of variance and homosexual behavior was seen to be a product, not a cause of this anomalous distribution of gender characteristics. Researchers concluded that, as a group, both male and female subjects displayed physical characteristics of sexual inversion. For the women this meant the common deficiency of fat in the shoulders and abdomen, dense skulls, firm muscles, excess hair on the face, chest, back, and lower extremities, a tendency toward "masculine" distribution of pubic hair, a low-pitched voice, and either excessively developed or underdeveloped breasts. The women's muscles were described as small and firm, but angular, with "masculine contours."[47] The male subjects showed the inverse pattern, with higher-pitched voices and a general lack of muscle tone. In addition, aspects of general comportment, such as a tendency to swagger or to walk with a mincing gait, were regarded as innate constitutional features found among female and male sex variants, respectively. Researchers claimed that these characteristics were observed when all the subjects were taken statistically as a group, but in many cases they strained to find more than one or two of these qualities in individual subjects. Therefore, in an attempt to draw a generic or composite sex variant body, variations among subjects were homogenized in favor of a stereotypical construction of sexual inversion.

The very problem of the complex and multiple types of sex variants drove the inquiry in its ambitious undertakings to interpret the bodies and experiences of subjects in terms of the binary system of masculinity and femininity. Yet, the sexual inversion framework not only posed significant obstacles for explaining contradictory empirical evidence; it had embedded within it theoretical limitations that placed a whole range of homosexual practices outside the researchers' comprehension. This was most obvious in the tortured attempts by doctors to comprehend lesbians who did not appear to be masculine, and homosexual men who were not effeminate. Lacking an analysis of homosexual desire separate from gender, the researchers searched minds and bodies for signs of what they assumed distinguished masculine from feminine.

The nineteenth-century idea that homosexuals suffered from constitutional nervousness and arrested development lived on in the Sex Variant study's methods and conclusions. In summarizing the vast interview and examination data, psychiatrist George Henry, who authored the published report, concluded that the sex variant was indeed a distinct type and a "by-product of civilization" who was unable to adjust to modern society's high standards requiring adults to establish and maintain a home for the proper rearing of children.[48] Echoing a long-standing stereotype, Henry referred to the sex variant as immature in sexual adjustment, and noted that this was obvious in both the psychological and somatic domains. All sex variants had inherited constitutional deficiencies, troubled family relations and a lack of social opportunities for heterosexual development. By stating this conclusion, Henry clearly acknowledged that both constitutional and environmental factors contributed to homosexuality.

Aggressiveness, independence, sympathies to feminism (also referred to as "sex bitterness"), and the rejection of wifely and motherly duties were listed as attitudinal signs of masculinity common among lesbians. Promiscuity, compulsiveness, and petulance were recognized as typical qualities of sex variant men. In addition, certain character weaknesses among ancestors and family members were interpreted as factors contributing to homosexuality, and were charted in classical pedigree diagrams (Figure 5.6). Effeminate uncles, domineering grandmothers, spinster aunts, alcoholic mothers, suicidal siblings, tubercular fathers, and even artistic cousins were taken to be indicators that the germ of homosexuality could, like other manifestations of degeneracy, be inherited. Henry credited the family with a dual, if ambiguous, function of passing down hereditary traits that could contribute to sex variance and producing the environment where patterns of deviant gender behavior could be replicated by offspring who became sex variants. In sum, the family served as a setting or apparatus for producing sex variants, due to either hereditary or socioenvironmental conditions of improper gender identification. By defining the family's contribution in this dual way, the study simultaneously recapitulated nineteenth-century constitutional explanations for homosexuality, while equally stressing the significance of environmental and psychogenic factors.

At the level of the body, many other indicators of sex variance were discov-

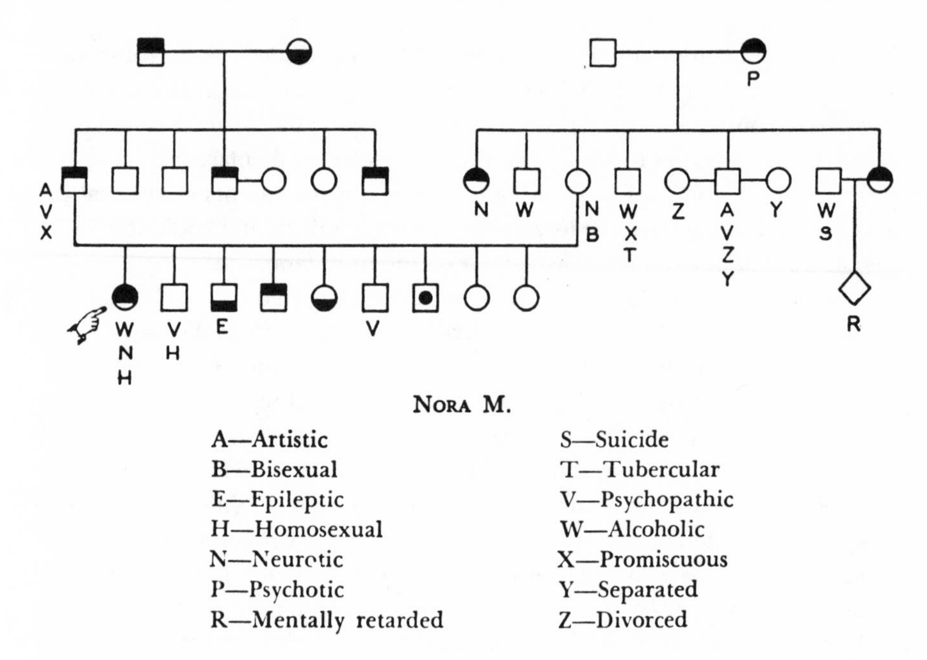

Figure 5.6 Pedigree diagram of Nora M., a female sex variant. (From George W. Henry, *Sex Variants: A Study of Homosexual Patterns* [N.Y.: Paul B. Hoeber, Inc., 1941].)

ered. While acknowledging that the physical examinations yielded no *conclusive* evidence for determining whether a person was a sex variant, Drs. Dickinson and Henry noted certain recognizable morphological patterns common to both male and female sex variants. The common presence of broad shoulders and narrow hips in both the female and male subjects was described as "an immature form of skeletal development" and presented as evidence that sex variants were at a lower stage of evolutionary development due to a lesser degree of sexual dimorphism. Noting the accompanying condition of psychosexual immaturity among subjects, the researchers combined anthropological theories of evolutionary primitivism with the psychoanalytic notion that homosexuality was a condition of arrested development resulting from an unresolved oedipal crisis. In this formulation, civilization and heterosexuality were coupled as desirable outcomes requiring some degree of expert scientific intervention in the biological realm of reproduction and the social or psychological realm of parenting.

Purported evidence of immaturity was gleaned from the gynecological examinations as well. On the one hand, Dickinson remarked that the common presence of an "infantile vulva" and the small uterus were signs of constitutional immaturity or arrested evolution in many lesbians. But while coital pain reported by many sex variant women was seen as a sign of arrested sexual development, Dickinson suggested that this could just as well be a sign of psychological immaturity as an indication of constitutional defects. Thus, a finding of infantilism in the somatic realm was again coupled with a psychoanalytic construction of lesbianism as a stage of arrested sexual development, characterized by a fixation on clitoral pleasure and not fully matured to vaginal intercourse, childbearing, and motherhood.

How Reliable Was the Body?

After summarizing the vast data from the physical examinations and noting that there were significant exceptions to general somatic patterns, Drs. Henry and Dickinson concluded that the more "pronounced deficiencies" of the sex variant were psychological rather than physical, and thus the psyche offered better information concerning who had the potential for sex variance, compared to the contradictory and complex evidence of individuals' bodies. Thus, even though Henry continued to endorse the project of looking for signs of sex variance on and in the body, he emphasized that a doctor should conduct careful interviews with each patient in order to locate and interpret the possible indications of sex variance. Bodies alone would not always reveal innate or distinct features. But, after talking with a patient, the body might be approached in a new way and interpreted to correlate certain characteristics with the patient's own narrative. While the examiner could search for physical patterns connoting innate sexual inversion or constitutional immaturity, he should be advised to look carefully for any indications of perverse practices. In other words, the body no longer merely revealed innate features, but was capable of disclosing the marks of its environment and its practices. Even then, it would not speak entirely for itself, but could be a grounding point for the narratives of the person to which it belonged.

By the end of the study, both Dickinson and Henry realized that one could not determine whether an individual was a sex variant or, at least, not by merely looking at his or her body. Over the course of the research, the goal of establishing a typical or composite homosexual body had been compromised, if not altogether thwarted, by contradictory evidence and exceptions to the study's assumptions. Furthermore, Henry asserted that a great many heterosexuals had constitutional deficiencies, so it was impossible to discern sexual orientation on this basis alone. Since human beings mated "irrationally"—that is, on the basis of desire rather than eugenic efficacy—there were many kinds of constitutional weaknesses throughout the population, and not all of them were concentrated among sex variants, Henry warned.

However, even as the notion of constitutional deficiency remained very vague, the body became anything but extraneous or obsolete in the clinical study of homosexuality. Instead, as the gynecological examinations show, it became a source of scientific knowledge about deviant *practices,* if not defective constitutions or degenerate genes. Rather than revealing distinct congenital or hereditary signs, it came to be seen in terms of surfaces or zones where experiences and behaviors left their marks. Having exhausted the possibility of determining a distinctly sex-variant skeleton, pelvis, complexion, or pattern of hair distribution, experts were urged to turn to the body instead for stories about *behavior* and *desire.*

The Variance Model and the Decline of the Deviant Type

From the outset of the Sex Variant study, its subjects were to function as specimens for determining any specifically sex variant patterns. In turn, clinical techniques were used to probe their bodies with the goal of constructing a composite or typical sex variant body. Paradoxically, this composite body would represent statistical norms or common features observed among these presumably abnormal bodies. By attempting to draw this homosexual body, researchers simultaneously sought to trace a *cordon sanitaire* around the "normal" or unmarked heterosexual body, even in the absence of a control group against which the subjects could be compared. But the study ran into serious trouble in its effort to construct a coherent and recognizable homosexual typology, especially once the typical sex variant body proved to be elusive.

Henry continued to believe it a worthy undertaking to survey the general population in order to identify statistical patterns of sexual variance, in spite of the apparent impossibility of establishing a distinct somatic and psychical type. At best, however, the typical sex variant could be little more than a statistical construct. The regime of examinations made it apparent that about the only thing sex variants had in common was what Henry called a lack of adjustment to adult heterosexual modes of sexuality. There was so much variety among these people that Henry recommended doctors study each case of sex variance for its particular form and content. But diversity among the subjects was neither the only nor even the initial force undermining the goal of establishing a distinct homosexual typology. Instead, this fictional creature's elusiveness was actually caused by the very model of sexual variance that informed the entire research project from its beginning. According to this model, the Sex Variant committee had assumed human beings were distributed across a spectrum, with masculinity concentrated at one pole and femininity at the other. Naturally, most normal men, by virtue of their masculine characteristics, were clustered around the former and most normal women were clustered around the latter. But in addressing the issue of how to categorize people, Henry declared in his conclusion that it was "scientifically inaccurate" to classify any individual—regardless of sexual preference—as fully male or fully female.[49] Instead, elements of each individual should be studied for their relative degrees of masculinity and femininity. An individual's placement on the variance spectrum depended on statistical calculations for each element, which were then regarded as indexes of that person's overall constitutional development and sexual preference. For example, a man's firm muscles and forceful enunciation were taken as signs not only of his masculinity but also his presumed heterosexuality. The same traits in a woman were signs of her masculinity, which implied she was a lesbian.

The variance model thus led experts to inspect subjects for the markers of "masculinity" and "femininity" as if these were signs of sexual orientation. But clearly, some subjects had contradictory qualities, with shoulders that might place them in a normal range for their sex, but vocal intonations or hair distribution that placed them closer to the average for the opposite sex and thus closer to homosexuality. Contradictions of this sort made it difficult to draw any clear lines between individuals or between groups of people. Clearly, the broader implication of this variance model for all men and women was that the line between the sexes could no longer be drawn sharply, nor, by logical extension, could the line between heterosexuals and homosexuals. An important consequence of this was the growing fear in the late 1930s and 1940s that people with homosexual desires may be everywhere; it was not entirely possible to determine who was who.

It is important to understand that the Sex Variant study was as much an effort to construct and maintain hygienic heterosexuality as it was to investigate homosexuality. Its conclusions implied that heterosexuality was neither a natural nor guaranteed system, but an endangered cultural institution requiring the commitment and effort of individuals and families and the ongoing intervention of scientific experts. Since their research had indicated that there may not be an innately distinct homosexual body, experts shifted their focus to thinking about sex variance as a product of one's environment and conditioning. Not surprisingly, the family became the site for engineering social hygiene, and Dr. Henry recommended a new science of parenting, in which adults would provide proper gender roles for their children by behaving in a fashion appropriate to their sex. As an instrument of social intervention, the study itself contributed to the construction and stabilization of a system for guaranteeing social order and proper reproduction of the race. In the American urban context of the 1930s, eugenicists warned that white, middle-class heterosexuality was threatened by various "unhygienic" trends—miscegenation, homosexuality, prostitution, and overpopulation among the poor. According to this vision, the crucial unit of civilized adulthood was specifically the affluent, white, companionate, heterosexual marriage, which took account of the virtues of pleasure but guided them toward patriotic reproduction, with the assistance of psychiatrists like George Henry himself.

Alfred Kinsey and the Momentary Disappearance of the Homosexual Body

In the years surrounding the Sex Variant study, the effort among scientists to describe the homosexual as a particular type of person with recognizable physical features came under severe criticism by Alfred Kinsey, a younger colleague and friend of Robert Latou Dickinson. As early as 1941, the same year the Sex Variant study was published, Kinsey presented a critique of research on androgen and estrogen levels in the urine of male homosexuals, noting a number of methodological problems, including the small size of the subject sample and inconsistencies in obtaining and measuring hormones among subjects.[50] But Kinsey saved his greatest criticism for the study's unqualified use of the categories "homosexual" and its putative opposite, "normal," to describe subjects:

> More basic than any error brought out in the analysis of the above data is the assumption that homosexuality and heterosexuality are two mutually exclusive phenomena emanating from fundamentally and, at least in some cases, inherently different types of individuals. Any classification of individuals as "homosexuals" or "normals" (=heterosexuals) carries that implication. It is the popular assumption and the current psychiatric assumption, and the basis for such attempts as have been made to find hormonal explanations for these divergences in human behavior.[51]

Kinsey was most critical of the vague and simplistic definitions given to homosexuality in scientific research and took issue with the "long-standing and widespread popular opinion that homosexual behavior depends on some inherent abnormality which, since the time of the discovery of the sex hormones, is often supposed to be glandular in origin."[52] His methodological rigor led him to reiterate that no causal relationship between anatomy and sexual history could be demonstrated without a knowledge, first, of the full range of human sexual variation, and second, of the frequency with which each kind of sexual variation occurs in the population. Indeed, these were the two premises upon which his famous studies of human sexual behavior were based.[53] "Until we know the nature of the gross behavior [of homosexuality] itself, no hormonal or other explanation is likely to fit the actuality."[54]

In his own research on human sexual behavior, which commenced in the mid-1930s, Kinsey emphasized that questions about homosexuality ought to be interpreted in the context of the "total experience" of each person, not just the selective acts of those who were self-identified homosexuals. He believed that separating out the study of homosexuality from all sexual possibilities skewed the conclusions of most scientific studies because any particular sexual behavior "always involves an adjustment in which each item is affected by all of the others in the complex."[55] Kinsey advocated mass survey methods including large numbers of participating subjects who would be asked hundreds of different questions in personal interviews. In his view, surveying randomly selected individuals was the only sound way to draw any general conclusions about sexual behavior—and homosexuality in particular. This method also did away with the distorting practice of choosing subjects on the basis of their own sexual self-identification as homosexuals. Instead, Kinsey polled all kinds of people about their sexual behavior, defining male homosexual behavior as when "one or both parties in the relation have come to ejaculation as a result of stimulation provided by another male."[56]

In preliminary research on male sexuality, Kinsey found that about thirty-five percent of the sample had been "involved in homosexual behavior" to the point of "full climax." He thus estimated that between a quarter and a third of all males in any mixed-aged group had some homosexual experience. Over the course of an individual lifetime, an estimated fifty percent of the male population would engage in homosexuality. To many who believed that homosexuals were fundamentally pathological and belonged to a separate and small group, this news was rather shocking.

Kinsey used this data to argue that scientists should study what humans actually do, rather than be concerned with dividing up the population into reified and mutually exclusive categories of heterosexuals and homosexuals:

> In brief, homosexuality is not the rare phenomenon which it is ordinarily considered to be, but a type of behavior which ultimately may involve as much as half of the whole male population. Any hormonal or other explanation of the phenomenon must take this into account. Any use of so-called normals as controls . . . should allow for the possibility that a quarter to a half of these "normals" may in actuality have had homosexual experience at some time in their lives; and . . . it must similarly be recog-

> nized that there are very few "homosexuals" who have not had at least some, and in many cases a great deal, of heterosexual experience.[57]

His research showed that some individuals had single or accidental experiences, and others confined their relations to a single male partner, while still others had contact with up to 15,000 males in the course of a lifetime. Some began their history of homosexual behavior early in life, others later; some had breaks in their homosexual activity of up to thirty-five years, and others had homosexual relations daily. Some were primarily homosexual in one period of their lives and heterosexual in another. Thus, he argued, this carefully acquired data did not warrant the separation of those who engaged in homosexual behavior into a discrete group.

Based on this great degree of variation, Kinsey refuted the popular and clinical concepts of physical stigmata that associated male homosexuality with effeminacy. "There are, in short, intergradations between all of these types, whatever the items by which they are classified."[58] Moreover, Kinsey's research showed that the heterosexual category was no more clear-cut; he found that some people engaged in both heterosexual and homosexual activity, sometimes successively but more often simultaneously in a single period of their lives. Sexual orientation for some people might even vary over the course of a few hours.

Although Kinsey, being a careful and respectful scientist himself, did not insist that biological studies of homosexuality cease altogether, his critique of categorizing people as either homosexual or heterosexual issued a damning blow to scientific projects such as the Sex Variant study. At the same time, his research, like that of the Sex Variant study, made use of a variance model, albeit focused on variability in actual sexual behavior among interviewed subjects, rather than on their degrees of masculinity and femininity. The variations he found among people led him to conclude that humans were scattered across a continuum based on their frequency of homosexual and heterosexual encounters. To chart the variations, he devised a seven-point scale for mapping sexual orientation, placing those who had no homosexual experiences at the zero degree, those who engaged in homosexual relations exclusively at the sixth degree, and those who had a mixture of homosexual and heterosexual encounters at the degrees in between.

Just as Kinsey's work obliterated the idea of a clear-cut homosexual type, it effectively erased the possibility of such a thing as a distinct homosexual body. His use of statistical survey methods, rather than the medical case history format, further undermined the typologizing method so endemic to constitutional studies of homosexuality, including those of the recently conducted Sex Variant study.

In spite of their apparent differences, both the Sex Variant study and Kinsey's research represented a shift from an earlier focus on classifying binary oppositions (male versus female and heterosexual versus homosexual) to a statistically based model of variance. The former mode of organizing difference privileged the body as a main source of truth, using techniques of physical examination in order to divide people into distinct categories based on certain features. The variance model was based on quantifying and standardizing all kinds of elements, using statistical analyses for understanding the distribution of a population across a continuum, not necessarily in clearly demarcated categories. Given the findings of the Sex Variant research, it was no longer possible to assume that the sex variant was a single and discrete type. The variance model, albeit deployed in different ways by the CSSV and by Kinsey, had punctured the *cordon sanitaire* of taxonomic difference, allowing for the possibility that aspects of variance were dispersed throughout the population.

Containing the Ubiquitous Menace

On the heels of the Sex Variant study, Kinsey's empirical method for determining variations of experience and sexual behavior gave rise to a new picture of homosexuality as a pervasive and multifaceted phenomenon. Although Kinsey argued that his data showed homosexuality to be a natural manifestation of human sexual desire, others believed that its widespread occurrence among those who seemed "normal" was cause for alarm. In the 1940s and 1950s, Kinsey's research unleashed fears that homosexual behavior was not necessarily confined to a particular population, but could be rampant and difficult to detect. The 1948 publication of his *Sexual Behavior in the Human Male* inadvertently inspired anxiety and hysteria about the invisible menace of homosexuality many believed had insinuated its way into every part of the population. The finding that about one third of all men had homosexual relations to orgasm as adults astounded the American public, especially when the mass media used these statistics to warn citizens that their priests, their bankers, their children's schoolteachers, and perhaps even their spouses or parents may be engaging in these unspeakable activities. Kinsey's subsequent volume on female sexual behavior reported similar findings of lesbianism among women, raising even greater ire among Kinsey's critics and the public who followed them. Kinsey was accused of sullying the portrait of the happy and hygienic suburban housewife by insinuating that Tupperware parties could be fronts for clandestine lesbian relationships.[59]

A great deal of hysteria among government officials was kindled unintentionally by statistical evidence that homosexuality was widespread and that homosexuals were imperceptible, even as their numbers seemed to be increasing in the population. Beginning as early as 1946, homosexuality became associated with nothing less than treason, sparking a ten-year campaign of official purges of so-called sex perverts from federal and local governments all across the United States. Within the homophobic terms of these Cold War purges, the threat of homosexuality was seen to have infected not only the family, the neighborhood, and the city streets, but also the State Department and the military, where it threatened the nation's security and strength.[60]

A growing number of American psychiatrists and psychoanalysts added to the official homophobia of the Cold War period by arguing that homosexuality was a neurotic condition that should be treated and cured.[61] They were especially eager to criticize Kinsey's methods, mainly on the grounds that he failed to take into account how subjects' psychological motivations influenced their responses to interview questions, and that he neglected to consider the role of the unconscious in shaping sexual desire and behavior.[62] Their contention was that a person's conscious admission of sexual desire was anything but the transparent truth of sexuality, and thus should not be taken at face value. Kinsey's psychoanalytic critics insisted that to understand homosexual behavior one must understand the underlying personality and neuroses that gave rise to this behavior. For the most part, they remained convinced that homosexuality was associated with a fundamentally neurotic type of personality.

Clearly, the domain of psychiatry was profoundly altered by Kinsey's shocking reports of human sexual variation. Although it was not his goal or intention, Kinsey's critique of studies linking homosexual behavior with constitutional or biological qualities actually opened up a space for the articulation of psychogenic explanations. But Kinsey himself insisted that questions about the origins of homosexuality were of little significance, and he was especially dismayed when psychogenic explanations took an increasingly homophobic tone. Indeed, much of American psychoanalytic writing from the 1940s through the 1970s shared a common goal of not merely preventing homosexuality, but of curing individuals afflicted by this deviant desire. While they otherwise lambasted Kinsey for his refusal to categorize people into heterosexual and homosexual, many homophobic psychiatrists from this period were keen to do away with the idea of a distinctly homosexual body. If homosexuality was not a matter of biology, then the chances for curing it were greater, psychoanalysts reasoned. They saw themselves as best situated to intervene therapeutically. Thus, even though by 1950 the body became increasingly problematic as a source of evidence about deviance, scientific hope to contain or abolish this sexual pathology, once advanced by early constitutional studies, came to be promoted by homophobic psychoanalysts. And even though they abandoned the idea of the homosexual body in order to institute their own therapeutic interventions, these psychoanalysts wanted to hang onto the idea that the homosexual was a psychopathological type of person.[63]

In the history of scientific research on homosexuality, Kinsey's two copious reports appeared at first to punctuate the end of biological studies through voluminous evidence that human sexual activities varied widely throughout the population. But during the postwar decade, the cultural fear and anxiety provoked by the dissolution of boundaries between homosexuality and heterosexuality overpowered the possibility that sexual variation could be accepted just because seemingly "normal" people experienced it. The perceived threat that homosexuality, like communism, was everywhere, inspired a reaction formation of widespread cultural disavowal and repulsion, headed up by government officials and psychiatric experts. Not surprisingly, great efforts were made to contain this threat through a resuscitated insistence that homosexuals were somehow a distinct group—either psychologically or biologically. What better way to allay fears about the dissolution of cherished boundaries than by refuting the variance model and thus denying the complexity of human gender and sexual relations?

Epilogue: How Big Is Your Hypothalamus?

As we have seen, scientific assumptions about the physical signs of homosexuality changed significantly from the early constitutional studies of the nineteenth century to the statistical research of Alfred Kinsey in the mid-twentieth century. Earlier models that understood homosexuality to be related to hereditary degeneration and constitutional defects were eventually supplanted by models of the body as a surface upon which perverse practices left their marks and by psychoanalytic theories focused on psychosexual development and social relations. Mid-twentieth-century developments in the history of the homosexual body reveal a significant shift away from assuming homosexuality to be the result of innate and immutable defects toward seeing it as a result of certain circumstantial, social, and psychological factors. Kinsey's statistical studies went one step further, by discrediting the whole idea that homosexuality was an attribute of a psychologically discrete or physically distinct group of people. Historically, the notion that homosexuals were in some way distinct was central to efforts of policing deviance. The greatest contribution of Kinsey's work—and, indeed, its most threatening quality—is that it made the border between homosexuality and heterosexuality permeable and highly contingent. By placing all sexual behavior on a continuum, his research brought to light the pervasive nature of homosexuality and, paradoxically, rendered it invisible, at least within the terms of the earlier constitutional framework. Thus it is possible to read the popular and professional backlash against Kinsey as a desire to make homosexuality visible again as an aberration to be contained rather than accepted as a common sexual practice.

Looking back over this history, we can say that, for an anxious moment, the homosexual body vanished not only because of methodological contradictions in such projects as the Sex Variant study, but also because rigorously conducted empirical science revealed that perhaps homosexuality was in every body. Indeed, following the Sex Variant study and Kinsey's research, constitutional studies were for the most part discontinued; but the quest for finding homosexuality in the body was far from exhausted, taking new forms which focused on the vicissitudes of the hormonal system and on patterns of sexual response.[64] Since the 1950s, the desire to resurrect clear boundaries between the nominal categories of heterosexual and homosexual has invited, if not required, the body to be a site for grounding some mythical essential difference.

Indeed, the question of whether homosexuality results from a rare biological trait or is a widespread cultural practice has animated scientific studies on the subject since 1869. Neither the Sex Variant researchers nor Kinsey resolved this question, and yet its answer has very large political implications for how those who engage in homosexuality are treated in the larger society. The issue has emerged once again in the twilight of the twentieth century, as banner headlines announce scientific evidence establishing the "biology of homosexuality." Alas, contemporary arguments on the subject repeat a familiar dichotomous structure, following two main directions: If homosexuality is an acquired or socially produced condition, presumably it could be curable or preventable, whereas if it is innate or due to biological factors, it should be tolerated as a natural variation. Historically, the body has been approached as an important source of evidence for debating this issue, just as it continues to be in recent controversies over whether there is a biological basis for sexual orientation. Then and now, arguments about whether or not sexual orientation is governed by biological factors generate a variety of agendas, not only about the definition of homosexuality, but also about whether those who practice it should be criminally prosecuted, medically treated, or legally protected as a minority.

The progressive position in the first half of the twentieth century was very much like a gay rights position articulated today: If biology determines homosexuality, those who engage in it should neither be punished nor be forced to change, because nature dictates their desires, and therefore homosexuality is not their fault. It is interesting to note that a new wave of "gay science," practiced and supported to a great degree by liberal gay-rights advocates, echoes this position by seeking to prove that sexual orientation can be correlated to the body.[65]

The key players in this new wave of "gay-positive" biology are mainly political liberals, and many are gay and lesbian scientists who defend their research by stating that scientific data can be used to argue that those who have homosexual desires should be tolerated because they are "hardwired" to do so.[66] An even more optimistic claim made by some supporters of this recent research is that proving the biological basis of homosexuality could protect lesbians and gay men against discrimination. In this construction, biology is presumed to be the strongest grounds for making equal protection claims under the law; lesbians and gay men are considered analogous to women and to racial minorities who are purportedly protected by law from discrimination on the basis of their biological race or sex.

It is curious that making a case through biology is being proposed again as a strategy for winning equal protection arguments, given that biologically based arguments historically have been used to support the elimination of whole groups of people under the rubric of social hygiene.[67] Many who are skeptical about making arguments of this sort but who support gay and lesbian rights prefer instead to make their arguments by drawing analogies from constitutional case law that protects the rights of individuals to form associations and to practice the religion of their choice.[68] In their view, one's sacred beliefs are not a matter of biological difference and yet they ought to be protected by the Bill of Rights. This latter approach stresses cultural affiliation, not biological destiny, and in so doing offers a much broader set of possibilities for appreciating and respecting differences among and within people.

Of course, the motives of those who advocate this new wave of scientific research reflect particular cultural and political circumstances of the present, and we must take into account what distinguishes their interests from apparently similar perspectives in the 1890s, the 1930s, and the 1950s. "Gay biology" today in many ways defies the popular critique of biological determinism from the 1960s and 1970s, and occurs in the context of several major historical developments: (1) a resurgence of right-wing political power in the United States during the 1980s, coupled with the reemergence of biological explanations for such culturally complex issues as crime, intelligence, unemployment, and oppression (to name just a few); (2) the elaboration of a far-reaching political movement based on the assertion of gay and lesbian identity and rights that has been fighting for rights to bodily self-determination; and (3) the emergence of a deadly epidemic that was originally named Gay Related Immune Deficiency and thus linked homosexuality with the body and disease in new and very urgent ways. These profoundly transformational developments in U.S. culture contribute to the recent trend in which homosexual bodies have taken on a mysterious, magical promise of identity, authenticity, and liberation for self-identifying gays and lesbians, who at the same time face escalating conflict, violence, and mortality. In this moment, bodies have become a means for both denouncing and affirming sexual difference in highly politicized terms.

Indeed, the AIDS epidemic shapes the cultural appropriation of rather tentative scientific findings about the biology of sexual orientation. For while gay rights advocates see scientific proof of biological difference as a protective shield, phobic detractors have a different kind of psychical and political investment in this research that can be tied to AIDS anxiety and homophobia. Since 1982, when the public began to hear about AIDS, one could see signs of a tenacious subterranean hope on the part of some people who identify as heterosexual (including those who engage in homosexual sex but would never call themselves gay) that there is some kind of essential difference between gay people and straight people.[69] From the official halls of Congress to local neighborhood schools and clinics, we see constant and willful attempts to fuel the popular belief that there is some irreducible difference between homosexuals and heterosexuals. This belief allows heterosexuals to continue to blame queers for AIDS and, simultaneously, be convinced that they themselves couldn't possibly embody the virus. In spite of epidemiological evidence to the contrary, this hope persists and is stoked by the likes of Jesse Helms, William Dannemeyer, Sam Nunn, and Donald Wildmon, who insist that there is a difference between "those people" over there doing grotesque things and "us" over here who would never do those bad things. Contemporary well-meaning attempts to locate homosexuality in the body share a tendency to believe that the difference between "us" and "them" is biologically ingrained and politically useful.

So in spite of its inability, refusal, or failure to produce distinct signs of homosexuality, the body continues to occupy a central role in scientific and popular understandings about lesbianism and homosexuality. Even after a history of baroque and futile attempts, scientists apparently refuse to give up the search for signs of homosexuality in the body. Again, the body is seen as simultaneously truthful and determinative, binding the person with whom it is associated to a set of proficiencies, weaknesses, and desires. As a product of "nature," the body rules the subject who submits to its needs and who, according to a rather peculiar reading of civil rights doctrine, is free of blame and released from the pressure to reform. Thus, while it may no longer be probed for evidence of degeneracy, the body continues to be treated as an important source of information for speculating about the sexual practices of certain individuals and for categorizing those individuals accordingly. Evidently, as long as we find a cultural urge to single out the homosexual as a specific type of person—whether that be under the banner of homophobic contempt or its rhetorical opposite, gay rights—the idea of the homosexual body persists. It would appear that a century-old tendency toward binary thinking that separates a friendly "us" from a dangerous "them" makes great use of the body as a site wherein difference is imagined to materialize. For all its high-tech qualities, the new wave of gay biology and genetics repeats the basic mistakes Kinsey first pointed out in 1941: It takes for granted that individuals are either homosexual or heterosexual, and that these are mutually exclusive, real categories, rather than nominal social constructs or political identities. But perhaps it is time for us to take Kinsey seriously and consider that "we" are everywhere and, at least as far as biology goes, "us" = "them."

Notes

1. On the basis of his observations of a girl who liked to dress like a boy and who acquired sexual satisfaction with other girls, Westphal concluded that her abnormality was congenital and thus should not be prosecuted by the police. Karl Friedrich Otto Westphal, "Die kontrare Sexualempfindung: Symptom eines neuropatholgischen (psychopathischen) Zustandes," *Archiv für Psychiatrie und Nervenkrankheiten* 2 (1869): 73–108. For other early and important sources linking homosexuality with nervous disease, see Jean-Martin Charcot and Valentin Magnan, "Inversions du sens génital et autres perversions génitales," *Archives de Neuroloigie* 7 and 12 (January–February 1882; November 1882): 55–60, 292–322; James G. Kiernan, "Psychical Treatment of Congenital Sexual Inversion," *Review of Insanity and Nervous Disease* 4 (4) (June 1894); James G. Kiernan, "Sexual Perversion," *Detroit Lancet* 7, no. 11 (May 1884): 483–484; G. Frank Lydston, "Sexual Perversion, Satyriasis, and Nymphomania," *Philadelphia Medical and Surgical Reporter* 61, no. 11 (September 14, 1889): 285; Albert Moll, *Perversions of the Sex Instinct* (trans. Maurice Popkin 1891; Newark, N.J.: Julian Press, 1931); and Cesare Lombroso and Guglielmo Ferrero, *The Female Offender* (London: T. Fisher Unwin, 1895).

2. Richard von Krafft-Ebing, *Psychopathia Sexualis, mit besonderer Berucksichtigung der contraren Sexualempfindung: Eine klinisch-forensische Studie* (Stuttgart: Enke 1886). All further citations to Krafft-Ebing in the text refer to *Psychopathia Sexualis: A Medico-Forensic Study* (New York: Samuel Login, 1908), which is an English translation of the tenth German edition.

3. Havelock Ellis, *Sexual Inversion.* 6 vols. Vol. 2, *Studies in the Psychology of Sex.* (Philadelphia: F. A. Davis Company, 1901). The six-volume set was revised twice, in a 1915 edition and a 1926 edition; all subsequent references in this chapter are to the latter.

4. E. Ray Lankester, *Degeneration—A Chapter in Darwinism* (London: Macmillan, 1880); and Eugene S. Talbot, *Degeneracy: Its Causes, Signs and Results* (London: Scott, 1898).

5. Sigmund Freud, " 'Civilized' Sexual Morality and Modern Nervous Illness," in *Standard Edition of the Complete Psychological Works of Sigmund Freud.* 24 vols. James Strachey, ed. (London: Hogarth, 1953), vol. 9:177–204.

6. Krafft-Ebing, *Psychopathia Sexualis,* 274.

7. Havelock Ellis, *The Criminal* (New York: Scribner and Welford, 1892).

8. Ellis, *Sexual Inversion,* 268–69.

9. Ellis contradicted himself on this point when he described the incidental lesbianism among otherwise "normal" women, precipitated by the unusual conditions of factory work and the entertainment industry. He observed that certain workplaces fostered perversion, as lace

makers and seamstresses, cooped up together in hot quarters, commonly engaged in lesbian practices. Working in close, hot quarters, factory girls talked about sex to a point of intense arousal, which could only be relieved through masturbation. This "vague form of homosexuality" occurred when, in the summer, some girls wore no underwear and worked with their naked legs crossed while other girls inspected them. The heat of the midday led to sensual stimulation which led to mutual masturbation among the aroused women. Similarly, actresses, chorus girls, and ballet dancers grew excited about performing as they waited in the wings and in crowded dressing rooms. Ellis suggested that the mere presence of women in close proximity to each other could trigger a state of nervous exhilaration that would culminate in a sexual frenzy of lesbianism. But these women could be restored to normalcy once removed from such arousing circumstances. (*Sexual Inversion,* 213–15.)

10. Hubert C. Kennedy, "The 'Third Sex' Theory of Karl Heinreich Ulrichs," in Salvatore J. Licata and Robert P. Petersen, eds., *Historical Perspectives on Homosexuality* (New York: Haworth Press and Stein & Day, 1981).

11. Karl Heinreich Ulrichs, *Forschungen 129ber das Ratsel der mannmannlichen Liebe* (1898; New York: Arno Press, 1975).

12. Edward Carpenter, *The Intermediate Sex* (London: Sonnenschein, 1908); Edward Carpenter, "The Intermediate Sex," in *Love's Coming of Age* (London: Allen and Unwin, 1923).

13. Magnus Hirschfeld, "The Homosexual as Intersex," in *Homosexuality: A Subjective and Objective Investigation,* ed. Charles Berg and A. M. Krich (London: Allen and Unwin, 1958).

14. Krafft-Ebing, *Psychopathia Sexualis;* and Albert Wilson, *Unfinished Man* (London: Greening, 1910).

15. George Chauncey, Jr., "From Sexual Inversion to Homosexuality: Medicine and the Changing Conceptualization of Female 'Deviance'," in *Passion and Power: Sexuality in History,* ed. Kathy Peiss and Christina Simmons (Philadelphia: Temple University Press, 1989).

16. Sigmund Freud, "The Psychogenesis of a Case of Homosexuality," *Standard Edition,* vol. 18:155–72.

17. See, for example, Richard Green, *The "Sissy Boy Syndrome" and the Development of Homosexuality* (New Haven: Yale University Press, 1987); John Money, *Gay, Straight, and In-Between: The Sexology of Erotic Orientation* (New York: Oxford University Press, 1988); Richard C. Friedman et al., "Hormones and Sexual Orientation in Men," *American Journal of Psychiatry* 134 (1977): 571–72; Richard C. Friedman and Leonore O. Stern, "Juvenile Aggressivity and Sissiness in Homosexual and Heterosexual Males," *Journal of the American Academy of Psychoanalysis* 8 (1980): 427–40; and Robert C. Stoller, "Boyhood Gender Aberrations: Treatment Issues," *Journal of the American Psychoanalytic Association* 26 (1978): 541–58.

18. Ellis was not as adamant as Krafft-Ebing about the pathological nature of the condition. To him, inversion was a result of an abnormal congenital organic variation. Like colorblindness, which deprived a person of the ability to distinguish between certain colors, the invert lacked the ability to see and feel normal emotional associations with the opposite sex. Ellis believed that, "strictly speaking, the invert is degenerate; he has fallen away from the genus." But he avoided the term degenerate because its meaning had become vulgarized and vague, and thus unfit for scientific use. Although all inverts had some form of congenital predisposition to homosexuality, he claimed that "inversion is rare in the profoundly degenerate." (*Sexual Inversion,* 320.)

19. Magnus Hirschfeld, *Die Homosexualität des Mannes und des Weibes* (Berlin: Louis Marcus, 1914); and Hirschfeld, *Sex in Human Relationships,* trans. John Rodker (New York: AMS Press, 1975); and Hirschfeld, *Sexual Anomalies and Perversions* rev. ed. (New York: Emerson Books, 1948).

20. A. Brill, "The Conception of Homosexuality," *Journal of the American Medical Association* 61 no. 5 (1913): 335–40; Sandor Ferenczi, "More about Homosexuality," (1909) in Ferenczi, *Final Contributions to the Problems and Methods of Psychoanalysis,* ed. Michael Balint, trans. Eric

Mosbacher (New York: Basic Books, 1955); and Kenneth Lewes, *The Psychoanalytic Theory of Male Homosexuality* (New York: New American Library, 1989), 55.

21. Freud, "The Psychogenesis of a Case of Homosexuality"; and Freud, " 'Civilized' Sexual Morality."

22. Sigmund Freud, *Three Essays on the Theory of Sexuality*, in *Standard Edition*, vol. 7, 123–246; Freud, "Analysis of a Phobia in a Five-Year-Old Boy ("Little Hans"), in *Standard Edition*, vol. 10, 1–147; Freud, "Leonardo da Vinci and a Memory of His Childhood," in *Standard Edition*, vol. 11, 59–138; Freud, "Certain Neurotic Mechanisms in Jealousy, Paranoia, and Homosexuality," in *Standard Edition*, vol. 18, 221–34.

23. Freud, *Three Essays on the Theory of Sexuality*; Lewes, *The Psychoanalytic Theory of Male Homosexuality*, 35.

24. Freud, *Three Essays on the Theory of Sexuality*.

25. Lewes, *The Psychoanalytic Theory of Male Homosexuality*, 24–68; Herman Nunberg and Ernst Federn, eds., *Minutes of the Vienna Psychoanalytic Society*, vol. 1, *1906–1908*, trans. M. Nunberg (New York: International Universities Press, Inc., 1962).

26. Krafft-Ebing mapped the homosexual body in a number of ways: while he observed that the genital anatomy of inverts was neither hermaphroditic nor unique, he believed that the sexual response of inverts registered symptoms of nervousness. He wrote that neuroses "awakened and maintained by masturbation" could result in extreme cases of *neurasthenia sexualis*, which caused afflicted men to ejaculate spontaneously and experience "abnormal feelings of lustful pleasure." But Krafft-Ebing primarily believed that sexual inversion was a disturbance of the normal development of the cerebral center, which, in turn, controlled the genitals and reproductive organs. Interestingly, he argued that this damage was not visible in organic tissue, but was perceptible only through watching the invert's behavior. (Krafft-Ebing, *Psychopathia Sexualis*, 327.)

27. Ellis, *Sexual Inversion*, 254.

28. In the early 1930s, psychiatrists George Henry and Hugh Galbraith set out to investigate whether homosexuals were somatically distinct. Their intention was to weigh strict psychogenic explanations of homosexuality against "constitutional" data derived from examining subjects' bodies; they concluded that homosexuals had distinct physical, as well as psychological, features that separated them from the rest of the general population. (George W. Henry and Hugh M. Galbraith, "Constitutional Factors in Homosexuality, *American Journal of Psychiatry* 13 [1934]: 1249–70.) See also George W. Henry, "Psychogenic and Constitutional Factors in Homosexuality; Their Relation to Personality Disorders," *Psychiatric Quarterly* 8 (1934): 243–64.

29. In 1937, psychiatrist Joseph Wortis refuted the relationship of constitutional factors to male homosexuality and called into question the common assumption that homosexuals were a kind of intersex. Citing several studies, including Henry and Galbraith's 1934 study, Wortis argued that no existing research offered sufficient evidence to support the idea that homosexuals had a distinct body build. Favoring psychogenic explanations, Wortis suggested that homosexuality was due mostly to early childhood identifications with parents of the opposite sex. Joseph Wortis, "A Note on the Body Build of the Male Homosexual," *American Journal of Orthopsychiatry* 8 (1937): 1121–25.

30. Bearing the influence of Magnus Hirschfeld's extensive research in Germany, the study incorporated methods for locating signs of homosexuality on or in certain bodies, alongside an interview method for identifying psychical processes at play in this form of variance. (Hirschfeld, *Die Homosexualität*.)

31. The published research appears in two volumes as George W. Henry, *Sex Variants: A Study of Homosexual Patterns* (New York: Paul B. Hoeber, 1941). For other discussions of the Sex Variant study, see Jennifer Terry, "Lesbians under the Medical Gaze: Scientists Search for Remarkable Differences," *Journal of Sex Research* 27, no. 3 (1990): 317–40; Terry, "Theorizing Deviant Historiography," *differences* 3, no. 2 (1991): 55–74; and Henry L. Minton, "Femininity in Men

and Masculinity in Women: American Psychiatry and Psychology Portray Homosexuality in the 1930s," *Journal of Homosexuality* 13, no. 1 (Fall 1986): 1–21.

32. CSSV members Robert Latou Dickinson, Adolph Meyer, Earnest Hooton, and Lewis Terman were members of the American Eugenics Society, which formed in 1923 with the aim of organizing eugenics campaigns on a national basis. Daniel J. Kevles, *In the Name of Eugenics: Genetics and the Uses of Human Heredity* (Berkeley: University of California Press, 1985), 59–60, 252.

33. Diane Paul, "Eugenics and the Left," *Journal of the History of Ideas* (October 1984): 567–90.

34. The alleged increase of homosexuality among robust, affluent white people alarmed many eugenicists, who feared that good stock was declining and bad stock taking over. Much of the popular anxiety, particularly about lesbianism, in the 1930s was tied to fears that certain women would prefer women lovers and thus refuse to do their part to reproduce the race. Of course, a great deal of cultural ambivalence surrounded this issue: on the one hand, some white middle-class homosexuals were perceived to be people of otherwise good racial stock, who could replenish the best of the population were it not for their perverse tendencies. On the other hand, since it was generally assumed that homosexuality was hereditary, their refusal to reproduce might be a very good thing because it would lessen the risk of homosexuality in future generations. For more on the relationship between the scientific study of homosexuality and eugenics, see Jennifer Terry, Siting Homosexuality: A History of Surveillance and the Production of Deviant Subjects (1935–1950) (Ph.D. diss., University of California, Santa Cruz, 1992).

35. Robert Latou Dickinson and Lura Beam, *A Thousand Marriages: A Medical Study of Sex Adjustment* (Baltimore: Williams and Wilkins, 1931); Dickinson and Beam, *The Single Woman: A Medical Study in Sex Education* (Baltimore: William and Wilkins, 1934); Dickinson, *Human Sex Anatomy* (Baltimore: Williams and Wilkins, 1933); Dickinson, "Marital Maladjustment: The Business of Preventive Gynecology," *Long Island Medical Journal* 2 (1908): 1–4; and Dickinson and Louise Steven Bryant, *Control of Contraception: An Illustrated Medical Manual* (Baltimore: Williams and Wilkins, 1932).

36. Dickinson consulted for the California-based Human Betterment Foundation on the subject of sterilizing mental patients in state hospitals and founded the Committee on Maternal Health in 1923, which, among other things, articulated an explicit goal of bringing birth control to poor women, who were seen to be most in need of contraception for their own good and the good of the race. Many of his ideas about marriage counseling techniques were developed in dialogue with other eugenicists in Germany and Britain, including Havelock Ellis. And while he appeared to be a moderate when it came to more egregious policies of involuntary sterilizations, as late as 1948, well after most American scientists publicly denounced Nazi eugenics campaigns, Dickinson applauded German policies of premarital screening to prevent hereditary disorders and feeblemindedness. Letter to Karl Bruegger, 20 March 1948, Robert Latou Dickinson Papers, Countway Library, Harvard University Medical School; and James Reed, *From Private Vice to Public Virtue: The Birth Control Movement and American Society since 1830* (New York: Basic Books, 1978), 195.

37. These were based on ideas developed in Lewis M. Terman and Catharine Cox Miles, "Sex Differences in the Association of Ideas," *American Journal of Psychology* 41 (1929): 165–206; *Attitude-Interest Analysis Test* (New York: McGraw-Hill, 1936); and *Sex and Personality: Studies in Masculinity and Femininity* (New York: McGraw-Hill, 1936).

38. After completing the Sex Variant study, Dickinson proposed a male counterpart study to the gynecology of homosexuality, which was never completed, in part because Dickinson's younger colleague and friend, Alfred Kinsey, gently pointed out its myriad methodological flaws. "Masturbation, Physical Signs in Males," unpublished notes dated December, 1946, Robert Latou Dickinson Papers, Countway Library, Harvard Medical School; and letter from Kinsey to Dickinson, dated 10 January 1947, Miscellaneous Correspondence, Robert Latou Dickinson Papers, Countway Library.

39. Lombroso and Ferrero, *The Female Delinquent;* Pauline Tarnowsky, *Étude Anthropométrique sur les prostitutuées et les voleuses* (Paris: Lecrosnier and Babe, 1889); David Horn, "This Norm

Which Is Not One," in this volume; and Laura Engelstein, *The Keys to Happiness* (Ithaca: Cornell University Press, 1992), 128–64.

40. Robert Latou Dickinson, "Premarital Examination as Routine Preventive Gynecology," *American Journal of Obstetrics and Gynecology* 16 (November 1928): 631; and Dickinson, "Marital Maladjustment: The Business of Preventive Gynecology," *Long Island Medical Journal* 2 (1908): 1–4.

41. Dickinson produced several textbooks of vivid images of female and male genitalia. See *Human Sex Anatomy: A Topographical Hand Atlas,* 2d ed. (Baltimore: Williams and Wilkins, 1949).

42. "Dickinson Collection Comes to Cleveland Health Museum," *Museum News of the Cleveland Health Museum* (June–July 1945): 2.

43. Ellis observed in 1901 that, among his own small number of subjects, many had a tendency toward arrested development and physical infantilism. He discerned small external genitalia and small uteri and ovaries among inverted women, while "women with a large clitoris seem[ed] rarely to be a masculine type" (*Sexual Inversion,* 256). Ellis compared his findings with those of colonial doctors, anthropologists, and European travel writers who claimed to identify the signs of lesbianism around the world. One doctor with the British Indian Medical Service observed swollen vulvas on the bodies of female inmates in the central gaol of Bengal, reporting that these women, known to have engaged in tribadism, had "indelible stigmata of early masturbation and later sapphism," including very large clitorises, which were "readily erectile," and elongated and erectile nipples. Ellis concurred with colonial officials in assuming that these signs were the *result,* rather than the precipitating cause, of lesbianism (*Sexual Inversion,* 210).

44. Dickinson and Beam, *A Thousand Marriages,* 54.

45. Robert Latou Dickinson, "Hypertrophies of the Labia Minora and Their Significance," *American Gynecology* 1 (1902): 225–54. In 1931, he wrote that "the story of years of active self-excitation both as contemporary practice and as historical development" could be deduced from "certain labial contours, enlargements and varicosities, the range of excursions of the clitoris, reactions of the pelvic floor, and thickening of the tissues of the mammary glands." Dickinson, "Medical Analysis of a Thousand Marriages," *Journal of the American Medical Association* 97, no. 8 (August 22, 1931): 531.

46. Miscellaneous Sketch Books (1935), Robert Latou Dickinson Papers, Countway Library, Harvard Medical School.

47. This list of characteristics resembled those reported earlier by Havelock Ellis, who found that inverted women tended to have firm muscles and deeper voices, swaggered when they walked, and were usually good whistlers, in contrast to inverted men, who mostly were incapable of whistling at all. Apparently, by the 1930s there was a growing consensus of what would count as masculine characteristics in women. (Ellis, *Sexual Inversion,* 256.)

48. Henry, *Sex Variants,* 1023.

49. Henry, *Sex Variants,* 1096. This basic premise was supported by endocrinological research that, although originally developed as a way to distinguish the sexes, had recently revealed that "female hormones" were present in men and "male hormones" were common in women. (Diana Long Hall, "Biology, Sex Hormones and Sexism in the 1920s," *Philosophical Forum* 5 [Fall–Winter 1973–74]: 82.)

50. Alfred Kinsey, "Homosexuality: Criteria for a Hormonal Explanation of the Homosexual," *The Journal of Clinical Endocrinology* 1, no. 5 (May 1941): 424.

51. Ibid., 425.

52. Ibid., 424.

53. Alfred Kinsey, Wardell B. Pomeroy, and Clyde E. Martin *Sexual Behavior in the Human Male* (Philadelphia: Saunders, 1948); and Alfred C. Kinsey et al., *Sexual Behavior in the Human Female* (Philadelphia: Saunders, 1953).

54. Kinsey, "Homosexuality," 425.

55. Ibid., 425.
56. Ibid., 425.
57. Ibid., 426.
58. Ibid., 427.
59. Edmund Bergler and William S. Kroger, *Kinsey's Myth of Female Sexuality: The Medical Facts* (New York: Grune and Stratton, 1954).
60. For articles on official, state-sanctioned homophobia during the years just following World War II, see John D'Emilio, "The Homosexual Menace: The Politics of Sexuality in Cold War America," in *Passion and Power: Sexuality in History,* ed. Peiss and Simmons, 226; Allan Berube and John D'Emilio, "The Military and the Lesbians during the McCarthy Years," *Signs* 9, no. 4 (1948): 759–75; Jonathan Ned Katz, *Gay American History: Lesbians and Gay Men in the U.S.A.* (New York: Crowell, 1976); and Jennifer Terry, "Purging the Perverts: A History of Hysteria and Homophobia in the Loyalty-Security Programs of Postwar America (1946–1960)," unpublished manuscript.
61. See, for example, Edmund Bergler, *Counterfeit-Sex: Homosexuality, Impotence, Frigidity,* 2d ed. enl. (New York: Grune and Stratton, 1958); Bergler, *Homosexuality: Disease or Way of Life?* (New York: Hill and Wang, 1956); Bergler, *One Thousand Homosexuals: Conspiracy of Silence, or Curing and Deglamorizing Homosexuals?* (Paterson, N.J.: Pageant Books, 1959); Irving Bieber, "Clinical Aspects of Male Homosexuality," in *Sexual Inversion: The Multiple Roots of Homosexuality,* ed. Judd Marmor (New York: Basic Books, 1965); Charles Socarides, *The Overt Homosexual* (New York: Grune and Stratton, 1968); Louis S. London and Frank S. Caprio, *Sexual Deviations* (Washington, D.C.: The Linacre Press, Inc., 1950); and Frank Caprio, *Female Homosexuality* (London: Peter Owen, 1955).
62. See A. Kardiner, "How the Problem Has Been Studied: Freud and Kinsey," in *Sex and Morality* (New York: Bobbs-Merrill, 1954); Albert Deutsch, ed., *Sex Habits of American Men: A Symposium on the Kinsey Report* (New York: Grosset and Dunlap, 1948); L. Kubie, "Psychiatric Implications of the Kinsey Report," *Psychosomatic Medicine* 10 (1948): 95–106; Karl Menninger, "One View of the Kinsey Report," *GP* 8 (1953): 67–72; B. Wortis, "The Kinsey Report and Related Fields: Psychiatry," *Saturday Review of Literature* 31 (1948): 19, 32–34; Bergler, "The Myth of a New National Disease: Homosexuality and the Kinsey Report," *Psychiatric Quarterly* 22 (1948): 66–88; and Bergler, "Homosexuality and the Kinsey Report," in *The Homosexuals: As Seen by Themselves and Thirty Authorities,* ed. Aron Krich (New York: Citadel Press, 1954).
63. For later examples of this, see Irving Bieber et al., *Homosexuality: A Psychoanalytic Study* (New York: Basic Books, 1962); Charles W. Socarides, "Theoretical and Clinical Aspects of Overt Male Homosexuality," *Journal of the American Psychiatric Association* 8 (1960): 552–66; Socarides, "Homosexuality," *International Journal of Psychiatry* 10 (1972): 118–25; and Socarides, *Homosexuality* (New York: Aronson, 1978).
64. See, for example, John Money, "Sin, Sickness, or Status?: Homosexual Gender Identity and Psychoendocrinology," *American Psychologist* 42, no. 4 (April 1987): 384–99; Lee Ellis and M. Ashley Ames, "Neurohormonal Functioning and Sexual Orientation: A Theory of Homosexuality-Heterosexuality," *Psychological Bulletin* 101, no. 2 (1987): 233–58; Heino F. L. Meyer-Bahlburg, "Sex Hormones and Female Homosexuality: A Critical Examination," *Archives of Sexual Behavior* 8, no. 2 (1979): 101–119; Meyer-Bahlburg, "Psychoendocrine Research on Sexual Orientation: Current Status and Future Options," in *Progress in Brain Research,* ed. G. J. DeVries et al. (Amsterdam: Elsevier, 1984); P. D. Griffiths et al., "Homosexual Women: An Endocrine and Psychological Study," *Journal of Endocrinology* 63 (1974): 549–56; Jennifer Downey et al., "Sex Hormones in Lesbian and Heterosexual Women," *Hormones and Behavior* 21 (1987): 347–57; G. Dorner et al., "A Neuroendocrine Predisposition for Homosexuality in Men," *Archives of Sexual Behavior* 4 (1975): 1–8; Dorner, et al., "Stressful Events in Prenatal Life of Bi- and Homosexual Men," *Experiments in Clinical Endocrinology* 81 (1983): 83–87; William Masters and Virginia Johnson, *Human Sexual Response* (Boston: Little, Brown, 1966); and Masters and Johnson, *Homosexuality in Perspective* (Boston: Little, Brown, 1979).

65. For recent examples of scientific research purporting to find the origins of sexual orientation in biology, see Simon LeVay, "Evidence for Anatomical Differences in the Brains of Homosexual Men," *Science* 253 (1991): 1034–37; Simon LeVay, *The Sexual Brain* (Cambridge: MIT Press, 1993); D. F. Swaab, L. J. Gorren, and M. A. Hofman, "Gender and Sexual Orientation in Relation to Hypothalamic Structures," *Hormonal Research* 38 (supp. 2) (1992): 51–61; Swaab et al. "The Human Hypothalamus in Relation to Gender and Sexual Orientation," *Progress in Brain Research* 39 (1992): 205–19; Laura Allen and Roger Gorski, "Sexual Orientation and the Size of the Anterior Commissure in the Human Brain," *Proceedings of the National Academy of Sciences* 89 (August 1992): 7199–202; J. Michael Bailey and Richard C. Pillard, "A Genetic Study of Male Sexual Orientation," *Archives of General Psychiatry* 48 (1991): 1089–96; J. Michael Bailey et al., "Heritable Factors Influence Sexual Orientation in Women," *Archives of General Psychiatry* 50 (1993): 217–23; J. M. Bailey and D. S. Benishay, "Familial Aggregation of Female Sexual Orientation," *American Journal of Psychiatry* 150, no. 2 (February 1993): 272–77; Angela M. L. Pattatucci and Dean H. Hamer, "The Genetics of Sexual Orientation: From Fruit Flies to Humans," paper presented at Wenner-Gren Foundation Symposium no. 116, *Theorizing Sexuality: Evolution, Culture and Development* (March 19–27, 1993); and Dean H. Hamer, et al., "Evidence for Homosexuality Gene," *Science* 261 (July 16, 1993), 291–92.

66. Joe Dolce, "And How Big Is Yours?" (interview with Simon LeVay) *The Advocate,* June 1, 1993: 38–44; Chandler Burr, "Homosexuality and Biology," *Atlantic Monthly* 271, no. 3 (March 1993): 47–65; Natalie Angier, "Study of Sex Orientation Doesn't Neatly Fit Mold," *New York Times,* July 18, 1993; Michael Bailey and Richard Pillard, "Are Some People Born Gay?" *New York Times,* December 17, 1991; "Just What Do Gay Twins Reveal?" (interview with Richard Pillard), *The Guide* 12, no. 2 (February 1992): 24–28; Angela Pattatucci, "On the Quest for the Elusive Gay Gene: Looking within and beyond Determinism," manuscript, 1993; and "Dr. Simon LeVay Discusses Gay Brains," *San Francisco Sentinel,* February 18, 1993, 11.

67. See Robert N. Proctor, "The Destruction of 'Lives Not Worth Living,' " chapter 6 in this volume.

68. Nan D. Hunter, comments from Lavender Law Conference, Chicago (October 1992); Ruth Ann Robson, "Discourses of Discrimination and Lesbians as (Out)Laws," *Radical America* 24, no. 4 (April 1993): 39–46; and Meredith F. Small, "The Gay Debate: Is Homosexuality a Matter of Choice or Chance?" *American Health,* March 1993, 70–76. For general critiques of recent biological research on sexual orientation, see Kathryn E. Diaz, "Weird Science: An Interview with Jennifer Terry," *Gay Community News* 19, no. 14–15 (October 20–27, 1991); William Byne and Bruce Parsons, "Human Sexual Orientation: The Biologic Theories Reappraised," *Archives of General Psychiatry* 50 (March 1993): 228–39; Anne Fausto-Sterling, *Myths of Gender: Biological Theories about Women and Men.* 2d ed. (New York: Basic Books, 1992); and Ruth Hubbard and Elijah Wald, *Exploding the Gene Myth* (Boston: Beacon Press, 1993).

69. Michael Fumento, *The Myth of Heterosexual AIDS* (New York: Basic Books, 1990); Robert E. Gould, "Reassuring News about AIDS: A Doctor Tells Why You May Not Be at Risk," *Cosmopolitan,* January 1988, 146; John Langone, "AIDS: The Latest Scientific Facts," *Discover,* December 1985, 28–53; John Langone, *AIDS: The Facts* (Boston: Little, Brown, 1988); and Helen Singer Kaplan, *The Real Truth about Women and AIDS: How To Eliminate the Risks without Giving Up Love and Sex* (New York: Simon and Schuster, 1987).

6

The Destruction of "Lives Not Worth Living"[1]

Robert N. Proctor

HISTORIANS EXPLORING THE origins of the Nazi destruction of "lives not worth living" have only in recent years begun to stress the ties between the destruction of the mentally ill and handicapped, on the one hand, and the Jews on the other. And yet the two programs were closely linked in both theory and practice. Ethnic and physical deviants were linked through a logic that traced the inferiority of both groups to deficiencies in their physical bodies. Both groups of peoples (unwanted ethnic minorities and the physically or mentally handicapped) were stigmatized as diseased, fouling the purportedly healthy German populace; both were cast as parasites, living off the productive lives of others. To see how this worked, we must first explore the euthanasia operation and its associated productivist or performance ethic, by which Nazi physicians sought to rid the nation of its "lives not worth living."

The Euthanasia Operation

The details of the euthanasia operation are by now well known, as a result of more than a decade of intensive scholarship in Germany and elsewhere. In early October 1939, in the year designated by the government as the year of "the Duty to be Healthy," Adolf Hitler issued a secret memo certifying that "Reichsleiter Bouhler and Dr. med. Brandt are hereby commissioned to allow certain physicians to grant a mercy death (*Gnadentod*) to patients judged incurably sick by critical medical examinations."[2] By August 24, 1941, when the first phase of the operation was brought to an end, more than 70,000 patients from 130 German hospitals had been killed,[3] in an operation which provided the stage rehearsal for the subsequent destruction of Jews, homosexuals, Communists, Gypsies, Slavs, and prisoners of war.*

Despite occasional posturings of humanitarianism, the essence of the Nazi argument for the destruction of the insane was economic. Euthanasia was defended as a means of cutting costs, or ridding society of "useless eaters." Racial hygienists such as Gustav Boeters pointed out that, during the First World War, when food and medical supplies were rationed after the British blockade of German ports, nearly half of all patients in German psychiatric hospitals perished from starvation and disease.[4] People such as these were simply too low on the list to obtain those supplies. It was in this context that Hoche and Binding called for the killing of the mentally ill and other "defectives."** On August 10, 1939, when German physicians met to plan the euthanasia operation, Philipp Bouhler, head of the Nazi Party Chancellery, made it clear that the purpose of the operation was not only to continue the "struggle against genetic disease" but to free up hospital beds and personnel for the coming war.[5] The philosophy behind this was simple: patients were to be either cured or killed.[6]

The original intent of those who planned the euthanasia operation was that

*The idea of the destruction of "lives not worth living" had been discussed in professional legal and medical literature long before the Nazi rise to power; the most influential text prior to the Nazi period was Alfred Hoche and Karl Binding's *Die Freigabe zur Vernichtung lebensunwerten Lebens* (Berlin, 1920). For background, see Michael Burleigh, *Death and Deliverance: 'Euthanasia' in Germany c. 1900–1945* (Cambridge, Eng., 1994).

**Germany was not the only society to discuss euthanasia. American discussion peaked (predictably) in the period 1936–1941. As in Britain, most American advocates of euthanasia were motivated by a desire to establish the right of individuals to have a "death with dignity." Many also argued, however, that euthanasia would help save on medical costs. See, for example, W. A. Gould, "Euthanasia," *Journal of the Institute of Homeopathy* 27 (June 1933): 82. In 1935, the French-American Nobel Prize winner Alexis Carrel suggested that the criminal and the insane should be "humanely and economically disposed of in small euthanasia institutions supplied with proper gasses." See his *Man the Unknown* (London, 1936), p. 296. W. G. Lennox in a 1938 speech to Harvard's Phi Beta Kappa proposed euthanasia for "the congenitally mindless and for the incurable sick who wish to die"; Lennox was also astute enough to perceive that "The principle of limiting certain races through limitation of off-spring might be applied internationally as well as intranationally. Germany, in time, might have solved her Jewish problem this way." See his "Should They Live? Certain Economic Aspects of Medicine," *American Scholar* 7 (1938): 454–66. American support for the concept of forcible euthanasia evaporated after rumors of German extermination operations began to filter into the American media. The issue was not entirely dead, however: in 1942, as Hitler's psychiatrists were sending the last of their patients into the gas chambers, Dr. Foster Kennedy, Professor of Neurology at Cornell Medical College, published an article in the official journal of the American Psychiatric Association calling for the killing of retarded children age five and older—"those hopeless ones who should never have been born—Nature's mistakes." See his "The Problem of Social Control of the Congenitally Defective: Education, Sterilization, Euthanasia," *American Journal of Psychiatry* 99 (1942): 13–16; also his "Euthanasia: To Be or Not to Be," *Colliers* 103 (May 20, 1939): 15.

the scale of the operation should be dictated by the formula 1000:10:5:1—that is, for every one thousand Germans, ten needed some form of psychiatric care; five of these required continuous care, and among these, one should be destroyed.[7] Given a German population of 65–70 million, this meant that 65,000–70,000 individuals were to be killed. And in fact, the program kept closely to this schedule, despite occasional and (later) much celebrated protests from the Catholic Church. By the end of August 1941, when the gassing phase of the operation was stopped, 70,273 individuals had been killed. We know these figures exactly, because the committee responsible for overseeing the operation kept meticulous records.[8] We also know that Nazi officials celebrated the economic savings produced by the operation. One rather extraordinary account of the mid-war years calculates that, altogether, euthanasia had saved for the German economy an average of 245,955.50 Reichsmarks per day (88,543,980 RM per year). This same account figured that, assuming an average institutional life expectancy of ten years, the Reich had been saved expenses in excess of 880 million Reichsmarks. Other records show that, by the end of 1941, 93,521 beds had been freed up by the operation.[9]

It is important to recognize the banality of the program: in 1941, the psychiatric institution at Hadamar celebrated the cremation of its ten thousandth patient in a special ceremony, where everyone in attendance—secretaries, nurses, and psychiatrists—received a bottle of beer for the occasion.[10] Even after the end of the gas-chamber phase of the operation, the killings continued, albeit in different forms. Whereas earlier killings had been primarily by means of gas chambers, killings after the summer of 1941 were performed through a combination of injections, poisonings, and starvation. Euthanasia took on less the character of a single Reich-wide "operation" and more the character of normal hospital routine.[11] Equally disturbing is the fact that doctors were apparently never *ordered* to murder psychiatric patients and handicapped children. They were *empowered* to do so, and fulfilled their task without protest, often on their own initiative. In the abortive euthanasia trial at Limburg in 1964, Hans Hefelmann testified that "no doctor was ever ordered to participate in the euthanasia program; they came of their own volition." Himmler himself noted that the operations undertaken in psychiatric hospitals were administered solely by medical personnel.[12]

The Medicalization of Anti-Semitism

How, though, does one explain the extension of the euthanasia program to the destruction of the Jewish people through the final solution? One of the key ideological elements in this was what might be called the *medicalization of anti-Semi-*

tism—the view developed by Nazi physicians that the Jews were a diseased race, and that the "Jewish question" might be solved by medical means.[13]

According to Walter Gross, head of the Office of Racial Policy and one of the period's foremost racial activists, it was first with the Nuremberg Laws of 1935 (especially the Blood Protection Law) that the explicit link was made between the "genetically healthy" (*Erbgesunden*) and the "German-blooded" (*Deutschblütigen*). The 1933 Civil Service Law excluding non-Aryans from government employment was, in Nazi jargon, primarily a "socio-economic" and not a "medical or biological" measure; according to Gross, it was not until the Nuremberg Laws banned marriage and sexual relations between Jews and non-Jews that legislation for the protection of Germans against the Jews was put on a biological basis. All subsequent legislation in the sphere of race and population policy, Gross claimed, was based upon this distinction between "healthy" and "diseased" races.[14]

The Nazi concept of healthy and diseased races was at one level expressed in medical metaphors of the Jew as "parasite" or "cancer" in the body of the German Volk. One physician phrased this in the following terms: "There is a resemblance between Jews and tubercle bacilli: nearly everyone harbors tubercle bacilli, and nearly every people of the earth harbors the Jews; furthermore, an infection can only be cured with difficulty."[15] Nazi physicians also argued, however, that Jews actually suffer from a higher incidence of certain metabolic and mental diseases. In his speech before the 1935 Nazi Party Congress at Nuremberg (the same meeting at which he proposed the euthanasia of the mentally ill), Gerhard Wagner, Führer of the Nazi Physician's League, argued that for every 10,000 inhabitants in the Reich there were 36.9 mentally infirm among the Germans, 48.7 among the Jews. Wagner cited the "interesting figures" of the "Jewish doctor Ullmann" documenting that, between 1871 and 1900, the relative proportions of Germans and Jews in psychiatric institutions rose from 22:29 in 1871 to 63:163 in 1900. Interesting as well was the fact that Jews showed a higher rate of sexual deficiency, expressed, for example, in the blurring of secondary sexual characteristics. This, Wagner affirmed (citing the "Jewish doctor Piltz" [*sic*]), is what explains not only the higher incidence of homosexuality among the Jews, but also the prominence of female Jews in "masculine pursuits," such as revolutionary political activism. Wagner concluded from this that the Jews were a diseased race; Judaism was "disease incarnate."[16]

Wagner was not of course the first to make such claims. As Sander Gilman has recently shown, conceptions of the Jewish body as inherently diseased reach back well into the nineteenth century.[17] Jewish physicians were also often contributors to this literature, sometimes critically, sometimes not. Maurice Sorsby wrote a whole book in 1931 just to refute notions that cancer was a disease suffered by Jews in different ways than from other races of the world; Rabbi Max Reichler wrote a somewhat less critical book expounding the virtues of a "Jewish eugenics."[18] Nazi medical theorists exploited and refined this literature, especially those parts documenting differential racial susceptibilities to disease. A. Pilcz, for example, in the 1902 *Wiener klinische Rundschau* had reported that Jews suffer disproportionately from acute psychosis and insanity, and are especially susceptible to psychoses of a "hereditary-degenerative nature."[19] A Dr. Rajanski noted that the 1871 German census showed that Jews were nearly twice as likely as Christians to be mentally ill, and in Vienna, a Dr. M. Engländer presented statistics to demonstrate a higher incidence of idiocy, myopia, glaucoma, diabetes, and tuberculosis among Jews than among non-Jews, and attributed these higher rates to poor nutrition and to inbreeding.[20] Much of this research was published in Jewish journals: the 1902 *Jüdisches Volksblatt,* for example, reported a higher incidence of nervous disorders, gallstones, bladder and kidney stones, neuralgia, chronic rheumatism, and brain malfunction among Jews, but also noted a lower susceptibility of Jews to lung infections, typhus, various fevers, syphilis, and alcoholism.[21] Gerhard Wagner cited these and other articles as evidence for his claim that Jews suffer many diseases that non-Jews do not. Wagner concluded from this that the interbreeding of Jews and non-Jews posed grave risks to German public health: if Germans continued to allow the mixing of "Jewish and non-Jewish blood," this would result in the spread of the "diseased genes" of the "already bastardized" Jewish race into the "relatively pure" European stocks.[22]

The study of the racial specificity of disease was in fact to become one of the leading priorities of biomedical science under the Nazis. Otmar Freiherr von Verschuer, founder of the journal *Der Erbarzt* and director of the Frankfurt Institute for Racial Hygiene, was one of the leading figures in this effort. In his 1937 book on "Genetic Pathology" (*Erbpathologie*), Verschuer identified more than fifty different ailments suspected of being genetic in origin; he also classified diseases according to how common they are among particular racial groupings. Measles, Verschuer argued, is rare among Mongols and Negroes; myopia and difficulties associated with giving birth are more common among civilized than among "natural" peoples of the world. This latter phenomenon, he suggested, was a consequence of the fact that "the more advanced peoples of the world" have suffered a loss of selective pressure as a result of the progress of medical science. Verschuer maintained that tuberculosis is rare among Jews, though Jews suffer more from diabetes, flat feet, hemophilia, xeroderma pigmentosum, deafness, and nervous disorders. Both Jews and "coloreds" (*Farbige*) suffer higher rates of muscular tumors. Interestingly, tuberculosis is the only disease Verschuer lists as less prevalent among Jews than non-Jews. To explain this he adduces an adaptive story: Jews have for centuries lived in the cities. The long-standing urban life of the Jews has led them to develop resistance against diseases commonly found in cities (such as tuberculosis); this same urban existence has led Jews to become susceptible to ailments such as flat feet.[23]

Nazi physicians sometimes speculated that Jewish racial degeneracy might be explained in terms of the supposedly hybrid origins of the Jewish race. In 1935, for example, Dr. Edgar Schulz of the Office of Racial Policy published an article demonstrating higher rates of insanity, feeble-mindedness, hysteria, and suicide among Jews than among non-Jews. Schulz claimed that these and other disorders arose from the fact that Jews were not, strictly speaking, a single race, but rather an amalgam of Negro and Oriental blood. As a result of this impure racial constitution, Jews suffered "tensions and contradictions" that became manifest as disease. This was a phenomenon supposedly observed not just among the Jews, but among any population that had suffered racial mixing.[24] The anthropologist Theodor Abel in 1937 thus described the *Rheinlandbastarde* (offspring of black French soldiers and German women from World War I) as suffering from a host of racial maladies—including tuberculosis, rickets, bad teeth and gums, weak musculature, gout, flat feet, bronchial problems, and nervous disorders, such as nail-biting, eye-twitching, speech defects, and crying in the night.[25] Eugen Fischer's early study of Rehobother half-castes—sometimes held up as the first successful application of genetics to human populations—was supposed to have shown that children of interracial marriages score lower on standardized tests than their classmates.[26]

Racial scientists were remarkably creative in their attempts to explain the odious effects of racial miscegenation. Wilhelm Hildebrandt, for example, in his 1935 *Racial Mixing and Disease*, argued that the maladies produced by racial miscegenation were a product of the fact that different races have different life spans and that bodily organs therefore mature and degenerate at different rates. If a long-lived person were to mate with a short-lived individual, then the various organs might mature and die at different rates, disturbing the "equilibrium" found in relatively pure races.[27]

Interestingly, differential racial susceptibility to disease was one topic Jews were permitted to write about, even at the height of the Nazi regime. In 1937, Drs. Franz Goldmann and Georg Wolff presented statistics demonstrating lower rates of infant mortality and tuberculosis among Jews than non-Jews. These authors, writing for Germany's officially sanctioned Jewish body (the Reichsvertretung der Juden in Deutschland), also pointed out that Jews suffered higher rates of mortality from diabetes, diseases of the circulatory system (especially arteriosclerosis), and suicide. From 1924 to 1926, suicide rates among Jewish men were 20 percent higher than for non-Jewish men (the difference between

Jewish and non-Jewish women was even higher—30 percent); by 1932–34, the Jewish suicide rate was 50 percent higher among Jews than non-Jews.[28]

There is evidence that some Nazis did recognize that differences in racial susceptibility to disease might be due to social rather than "racial" causes. In 1940, for example, Martin Stämmler and Edeltraut Bieneck analyzed demographic shifts among Jewish and non-Jewish inhabitants of Breslau in the period 1928–1937. Stämmler and Bieneck noted that Jewish birthrates had declined considerably over this period, and that consequently there was a higher proportion of elderly among Jews than non-Jews. This helped account for the higher rates of mortality Jews suffered from disorders such as cancer, diabetes, and circulatory failure; it also helped explain the lower death rates for tuberculosis and infectious diseases, ailments that commonly strike the young.[29] Stämmler and Bieneck also noted the rise in Jewish mortality rates over this period: from 14/1,000 in 1928 to more than 21/1,000 in 1937. The *British Medical Journal,* reviewing the work of these authors, pointed out that nowhere in their analysis did Stämmler and Bieneck discuss the role of state violence in producing these statistics.[30]

The interpretation of the "Jewish problem" as a "medical" problem was to prove useful in Nazi policy and propaganda. In the early months after the invasion of Poland in 1939, Nazi police officials were able to turn to medicine to justify the "concentration" and extermination of the Jews. Just how this was done illustrates not only something about the role of medical ideology in the persecution of Germany's minorities, but also how the concept of the Jew as "disease incarnate" began to take on the character of a self-fulfilling prophecy.

Genocide in the Guise of "Quarantine"

On September 1, 1939, Hitler's armies invaded Poland on the pretext of retaliation for an attack on a German border station by Polish troops—an attack which we now know to have been staged by SS guards disguised as Polish officers. Shortly after the occupation of Poland, SS officers were ordered to confine all of Poland's Jews into certain ghettos, including, first and foremost, the traditional Jewish ghetto of Warsaw.

In territories occupied by the German army, Nazi medical authorities used the pretext of danger of disease to justify a series of repressive measures against the Jewish population. In Warsaw, when the Nazis established a separate section for Germans on the city's streetcars, the Nazi-controlled *Krakauer Zeitung* justified this decision as "a hygienic necessity."[31] When the Nazis banned Jews from unauthorized railway travel throughout occupied Poland, Nazi newspapers announced this under the headlines "Germ-carriers Banned from the Railways."[32] One of the most brutal forms of persecution using hygiene as a pretext was the confinement of Jews to the ghettos. In February 1940, for example, more than 160,000 Jews from the areas surrounding the industrial town of Lodz (renamed Litzmannstadt after occupation by the Germans) were rounded up and forced into one small part of the town. The original intention was to remove all Poles and Jews from the town, leaving it entirely German; when this proved impractical, Nazi authorities decided to confine the Jews to the northern part of the town and to regulate all trade or interchange of any kind between the Jewish and non-Jewish sectors. On April 30, the Jewish quarter was sealed off and surrounded by a wall, similar to that being erected around the Jewish ghetto in Warsaw. German newspapers reported that the ghetto in Lodz was the "most perfect" of all the settlements established by the Germans in occupied Poland; one author called it the "purest temporary solution to the Jewish question anywhere in Europe."[33] Similar ghettos were subsequently established on a smaller scale in Cracow, Lublin, Radom, and other parts of Poland.

In each of these cases, hygiene was preferred as one of the leading grounds for concentration. The establishment of the Jewish ghetto at Lodz, for example, was justified on the grounds that this was necessary to protect against the dangers of epidemic disease.[34] And soon after confinement, of course, the people in these ghettos did begin to suffer from higher rates of infectious disease. These outbreaks of disease allowed Nazi medical philosophers to justify the (continued) concentration of the Jews in terms of a medical quarantine. The concept of quarantine was effectively used by Nazi authorities to confine, transport, and deport individuals throughout the war.

It was in the Warsaw ghetto that the Nazis were able to realize to the fullest their prophecies of "Jewish disease." Shortly after the invasion of Poland, German radio stations carried a report of an associate of Goebbels who had recently returned from a visit to Warsaw and Lodz. The author of this report described the Jews of the ghettos as "ulcers which must be cut away from the body of the European nations"; he claimed that if the Jews of the ghettos were not completely isolated the "whole of Europe would be poisoned."[35]

Before the war, the population of Warsaw was approximately 1.2 million, including 440,000 Jews, two-thirds of whom lived in the ghetto in the northwest part of town. When Nazi occupation forces began forcibly concentrating Poland's rural Jews into the ghetto, one effect was to create a breeding ground for disease. The crowded living conditions were exacerbated by shortages of food and clean water. In 1940 and 1941, as the number of Jews arriving in the ghetto grew from 500 per day to over 1,000 per day, diseases began to break out, soon reaching epidemic proportions. The world medical press was not unaware of these conditions. The July 6, 1940, issue of the *British Medical Journal* reported that "typhoid fever is still raging in Warsaw, where there are from 200 to 300 cases every day. Fully 90 percent of the victims are Jews. The German authorities have increased the number of disinfecting stations from 212 to 400, but have made no attempt to eradicate the source of the disease by clearing out the worst part of the ghetto, where tens of thousands of Jews are confined under pestilential conditions."[36]

The situation was to become much worse in subsequent months, as can be seen from official mortality figures for the early years of occupation. Before the war, mortality in the ghetto due to all causes had been about 400 per month. By January 1941 nearly 900 people were dying every month, and death rates were increasing week by week. By March, the number of deaths had grown to 1,608 per month, and, in June of 1941 alone, 4,100 people died from infection and disease, compounded by starvation, murder, and lack of adequate medical supplies. According to Wilhelm Hagen, a German doctor working in the ghetto at this time, the commissar of the ghetto (a Dr. Auersbach) sabotaged efforts on the part of well-meaning German doctors to alleviate the situation (Auersbach reportedly blocked the shipment of medical supplies to the hospitals and food to the city). By the end of 1941, official rations had been reduced to bread worth about 2,000 calories per person per day, and most people were receiving even less than this. Hunger and epidemic disease reinforced each other as official mortality rates from tuberculosis alone rose from 14/100,000 in 1938 to more than 400/100,000 in the first quarter of 1941. The case was even more dramatic with typhus. In the single month of October 1941, health authorities responsible for the Warsaw ghetto recorded 300 deaths from typhus—nearly as many as from all causes combined before German occupation.[37] The very existence of such statistics betrays the regime's nightmarish mix of murderous policy and bureaucratic fact-grubbing.

Epidemics that raged inside the Warsaw ghetto in 1941 and 1942 provided Nazi occupation forces with a medical rationale for the isolation and extermination of the Jewish population. On October 29, 1940, the *Hamburger Fremdenblatt* noted that 98 percent of all cases of typhoid and spotted fever in Warsaw were to be found in the ghetto. In the spring of 1940, non-Jewish doctors were barred from treating Jewish patients; on March 12, the *Krakauer Zeitung* explained this ban as follows:

> This decree is based on the fact that infectious diseases, particularly spotted fever and typhoid, are widespread especially among the Jewish population. When Jews suffering from those diseases are treated by non-Jewish doctors—doctors who are at the same time treating the sick of other races—there is a danger of their transmitting diseases from the Jews to the non-Jewish population.

In the first months of the German occupation, traffic between Jews and non-Jews in and outside the Warsaw ghetto was not restricted. Germans and non-Jewish Poles were allowed to enter the ghetto. After 1940, however, contact with Jews was declared a "threat to public health." Jews trying to escape from the ghetto were shot on the grounds that they were violating the "quarantine" imposed by the Nazis. Years later, long after the war, when the Nazi chief of police for Warsaw (Arpad Wigand) was brought before a West German Court and accused of ordering the shooting of Jews trying to leave the ghetto, defense attorneys argued that this had been a "necessary precaution" to preserve the quarantine.[38]

Criminal Biology

Criminal biology was to forge a further link in the medical solution to the *Judenfrage*. According to professor of psychiatry and psychology Robert Ritter, the urgent task of criminal biology was "to discover whether or not certain signs can be found among men which would allow the early detection of criminal behavior; signs which, in other words, would allow the recognition of criminal tendencies *before* the actual onset of the criminal career."[39] Such efforts were not, of course, an invention of the Nazis. Criminal biologists had tried, since Cesare Lombroso's *L'uomo delinquente* in the late nineteenth century, to construct a medical-forensic system linking moral, criminal, and racial degeneracy. Crime, in this view, was a literal disease; disease in turn represented a sign of moral degeneracy. Each reflected a form of the other: mental deviation could be found in moral and social criminals; criminality might be discovered through certain signs or physical manifestations on the body.

In the twentieth century, spurred by advances in genetics and hopes for eugenics, criminal biology became an important research priority for both government and academia. Concerns on the part of criminal biologists were in many ways close to those of the racial hygienists. Criminal biologists argued that crime was both genetically determined and racially specific; criminal biologists attempted to prove that the incidence of mental disease was higher among criminals than among the noncriminal population. Criminal biologists also worried that criminals were reproducing at a faster rate than noncriminal elements of the population. In the Nazi period, government statistical offices tried to determine the proportion of murderers who were "genetically defective": in 1938, government statisticians provided data to show that, whereas in 1928–30, only 14.5 percent of all murderers were genetic defectives (*erblich belastet*), by 1931–33, this proportion had supposedly grown to 20.1 percent.[40] Criminal biologists, such as Johannes Lange, argued in the Weimar period that twin studies indicated that crime was the product of hereditary disposition, rather than social environment.[41] Racial hygienists at the fourth Congress of Criminal Biology at Hamburg in June 1934 predicted that a thorough implementation of the Sterilization Law would lower general criminality by 6 percent, and "moral" (that is, sexual) crimes by 30 percent.[42]

The institutionalization of German criminal biology accelerated with the rise of the Nazis. By 1935, leading legal and medical journals propounded the theory that crime and "antisocial behavior" in general were inborn racial characteristics. J. F. Lehmann's *Monthly Journal of Criminal Biology and Penal Reform* set itself the goal of analyzing the genetics of crime; other journals followed suit. Under the Nazis, criminal biology became an important part of university teaching, and by the end of the 1930s most German universities offered instruction in this area—often in conjunction with courses on racial hygiene.[43] The German government supported research in this area in a number of ways. In October of 1936, Justice Minister Franz Gürtner ordered the establishment of fifty Examination Stations throughout Germany to explore the genetics and racial specificity of crime. The main targets of these stations were individuals under the age of twenty-five serving long-term prison sentences; everyone serving a three-month sentence or longer was required to be examined. In addition, larger Criminal Biology Research Stations were established at Munich, Freiburg-Breslau, Cologne, Münster, Berlin, Königsberg, Leipzig, Halle, and Hamburg to evaluate the effects of various biological measures on the incidence of crime (especially the castration allowed by the 1935 "Law against Dangerous Career Criminals").[44] In 1939, the *Deutsches Ärzteblatt* reported SS Reichsführer Heinrich Himmler's order that, henceforth, the examination of the genetic and family background of criminal suspects would become a routine part of criminal investigations.[45]

Criminal biologists also addressed the "Jewish question." The conceptual link here, as one might imagine, was the idea that Jews were racially disposed to certain forms of crime, in the same way they were racially disposed to certain kinds of disease. One should recall that disease for the Nazis was often broadly construed to cover not just physical disorders, but also behavioral and cultural maladies. Johannes Schottky, for example, in his 1937 book *Race and Disease* argued that Jews were racially predisposed to suffer not just from flat feet or gout, but also from mental disorders such as feeblemindedness, neurasthenia, hysteria, various sexual disorders, and pathological drives for recognition and power.[46] It was not such a large step from here to argue that Jews were innately criminal. Fritz Lenz put forth this thesis in his influential textbook on human genetics, first published in 1921 and reissued in numerous editions over the next twenty years.[47] Nazi medical authorities followed this lead. Gerhard Wagner, in the same speech in which he proposed the "euthanasia" of the mentally ill, declared that the incidence of criminality was higher among Jews than non-Jews, as was the incidence of bankruptcy (fourteen to thirty times higher for Jews), distribution of pornography ("exceptionally Jewish"), prostitution (pimping), drug smuggling, purse snatching, and general theft.[48]

Science thus conspired in the solution to the Jewish question: Jews were racially disposed to commit crime, as they were racially disposed to suffer from a rash of other diseases. By the late '30s, German medical science had constructed an elaborate worldview equating mental infirmity, moral depravity, criminality, and racial impurity. This complex of identifications was then used to justify the destruction of the Jews on medical, moral, criminological, and anthropological grounds. To be Jewish was to be both sick and criminal; Nazi medical science and policy united to help "solve" this problem.

The Final Solution

The German census of May 17, 1939, revealed that there were 330,892 Jews in Germany; there were in addition 72,738 "half-breeds" (*Mischlinge erster Grad*) and 42,811 quarter-Jews (*Mischlinge zweiter Grad*). In six years of Nazi rule, the number of Jews in Germany had fallen by 390,000. For those remaining, a variety of "solutions" were proposed, and medical doctors were among those leading the way.[49] Germany's medical journals made it clear that Jews had no place in the New German Order. Dr. W. Bormann, for example, writing in the *Ärzteblatt für den Reichsgau Wartheland,* declared that the "retrieved" German territories of the occupied east were to be settled "exclusively with Germans."[50] On November 23, 1939, when laws were passed requiring Jews in occupied Poland to wear the yellow star, Germany's leading medical journal justified this as a necessary measure "to create an externally visible separation between the Jewish and Aryan population." The journal proposed that in order to establish a geographical separation between the races, there were two possible solutions: the creation of a separate Jewish state, and confinement to a ghetto. This latter was preferable, the journal argued, because "it could be implemented more rapidly and with greater effect." The *Deutsches Ärzteblatt* reported with satisfaction that, as a result of Nazi policies, areas of mixed Polish-Jewish population had already begun to disappear, and one no longer saw the names of Jewish businesses. Furthermore,

> for the first time in centuries, the Jew has been forced to change his lifestyle; for the first time, he is required to work. For this purpose, Jews have been organized into forced labor brigades. At the head of each of these

> brigades there is a Jew who supervises his racial comrades, and is responsible to German authorities for insuring that the work assigned to his brigade is carried out in an orderly manner. This procedure has proven to work exceptionally well in the *Generalgouvernement* [occupied Poland].[51]

Germany's medical journal claimed that this arrangement was intended to last for two years, after which a working committee of Jews was to organize accommodations and work for Polish Jews—work primarily "of a handicraft nature."[52]

Privately, more radical measures were suggested. The agronomist Hans Hefelmann suggested exporting all Jews to Madagascar;[53] others in the Nazi regime suggested the establishment of a huge Jewish reservation near Lublin for permanent settlement.* Philipp Bouhler, head of the Party chancellery, proposed sterilizing all Jews by X-rays. Dr. Victor Brack recommended sterilization of the two to three million Jews capable of work, who might therefore be spared from extermination. In postwar testimony at Nuremberg, Brack recounted the alternatives pondered by Nazi authorities:

> In 1941, it was an "open secret" in higher Party circles that those in power intended to exterminate the entire Jewish population in Germany and occupied territories. I and my co-workers, especially Drs. Hefelmann and Blankenburg, were of the opinion that this was unworthy of Party leaders and humanity more generally. We therefore decided to find another solution to the Jewish problem, less radical than the complete extermination of the entire race. We developed the idea of deporting Jews to a distant land, and I can recall that Dr. Hefelmann suggested for this purpose the island of Madagascar. We drew up a plan along these lines and presented it to Bouhler. This was apparently not acceptable, however, and so we came up with the idea that sterilization might provide the solution to the Jewish question. Given that sterilization is a rather complicated business, we hit upon the idea of sterilization by X-rays. In 1941 I suggested to Bouhler the sterilization of Jews by X-rays; this idea was also rejected, however. Bouhler said that sterilization by X-rays was not an option, because Hitler was against it. I worked on this program further and finally came up with a new plan. . . .[54]

*This latter option was said to have the additional advantage that it would allow German scholars to pursue "Jewish studies." The official medical journal for the newly conquered *Reichsgau* of Wartheland thus defended proposals to establish "Jewish studies" at the new Reich University of Posen on the grounds that "as a German metropole in the middle of an area where we find Jews in their more or less 'natural condition', those in the new university are in the unique position of being able to record and ascertain [*festhalten*] the Jewish question for all times." See "Gründung der Reichsuniversität Posen," *Ärzteblatt fü den Reichsgau Wartheland* 2 (1941): 59.

Brack's "new plan" was simply another version of the plan to induce sterilization by X-rays. Jews were to be forced to have their reproductive organs (ovaries or testes) irradiated to such an extent (500–600 roentgens for men; 300–350 roentgens for women) that their reproductive capabilities would be irreparably destroyed. This, Brack suggested, might be best achieved by having individuals stand in front of a counter, where they would be asked to fill out a form; an official standing behind the counter would operate the apparatus, administering to the genitals a dose of radiation for the time it took to fill out the form (two to three minutes). This, Brack noted, would effectively amount to the castration of the individual. With twenty such setups, Brack figured that one could sterilize perhaps 3,000–4,000 people per day.[55]

Sterilization was ultimately rejected as a solution to the Jewish question. The decision to destroy Europe's Jews by gassing them in concentration camps emerged from the fact that the technical apparatus already existed for the destruction of the mentally ill. In the early phases, both the children's and the adult "euthanasia" operations were administered at first only to non-Jews: Jews were explicitly declared not to "deserve" euthanasia.[56] But the actions were soon extended to Jews, and in mass fashion. On August 30, 1940, the Bavarian minister of the interior ordered that all Jewish patients in psychiatric hospitals were to be transported to the hospital at Eglfing-Haar, near Munich; a memo of December 12 that year justified this move on the grounds that mixing of German and Jewish patients posed "an intolerable burden on both nursing personnel and the relatives of patients of German blood." In fact, Jewish psychiatric patients had begun to be rounded up and sent to gas chambers at Berlin-Buch since earlier that summer (June 1940).[57] In early September 1940, 160 Jewish patients held at Eglfing-Haar were filmed as part of the propaganda film *Scum of Humanity* (*Abschaum der Menschheit*). Later that month, on September 20, these patients were sent to Brandenburg, where they were gassed on September 22, 1940.

In early 1941, the Reich Ministry of the Interior ordered all Jews in German hospitals killed—not because they met the criteria required for euthanasia, but because they were Jews.[58] The Jews were not the first group to be singled out for extraordinary euthanasia. Criminals in Germany's hospitals had already been disposed of by this time; and, in the course of the year 1941, a number of other groups would fall within the shadow of the program. On March 8, 1941, Dr. Werner Blankenburg wrote to local *Gauleiter,* asking that all "asocials" and "antisocials" in Germany's workhouses be registered with euthanasia officials. In April, Germany's concentration camps began a new program designed to destroy camp inmates no longer capable of or willing to work. This project, within which Jews were also to be included, was code-named "14 f 13."

Operation 14 f 13 typifies the transition under way at this time from the systematic destruction of the handicapped and the psychologically ill to the systematic destruction of the ethnically and culturally marginal. In 1941, Buchenwald Commandant Koch announced to the SS officers in his camp that he had received secret orders from Himmler that all feeble-minded and crippled inmates were to be killed. Koch also announced at this time that all Jewish prisoners were to be included in operation 14 f 13.[59] In December of 1941, SS Lieutenant Colonel Liebehenschel notified the concentration camps at Sachsenbuch, Gross-Rosen, Neuengamme, Mauthausen, Auschwitz, Flossenbürg, Niedernhagen, Sachsenhausen, and Dachau that a medical commission would soon arrive to select prisoners for "special treatment." Camp officials were instructed to prepare the necessary paperwork; forms were to be filled out indicating the diagnosis of the inmates, including information on race, and whether the individuals were suffering from incurable physical ailments. SS Lieutenant Colonel Friedrich Mennecke, head of the Eichberg State Medical Hospital, arrived in Gross-Rosen in mid-January 1942, to begin selecting prisoners for destruction. By this time, selection had moved some distance from what we today would consider criteria for health. Psychopaths, criminals, asocials, antisocials, and individuals "foreign to the community" (*Gemeinschaftsfremde*) were now included; people were being taken from tuberculosis hospitals, workhouses, and homes for the elderly; racial and social "inferiors" were also being selected for elimination—all along guidelines established by the euthanasia experts Nitsche, Heyde, and Brack.[60]

Important to realize today is that, for the Nazi doctors, there was no sharp line dividing the destruction of the racially inferior and the mentally or physically defective. Many of the physicians responsible for administering the euthanasia operation in German hospitals were also responsible for formulating criteria and administering the first phases of the destruction of the Jews and other "lives not worth living" in Germany's concentration camps. Physicians cross-examined after the war at the Nuremberg Trials pointed out that they often did not distinguish whether certain exterminations had been for racial, political, or medical reasons. The testimony of Dr. Mennecke, questioned by the defense attorney for Karl Brandt, made this clear:

Attorney: When was the decision made to exterminate individuals based upon racial and political considerations? Had this already been decided by the time you first visited a concentration camp?

Mennecke: No.

Attorney: When was it then?

Mennecke: As far as I can remember it first began in Buchenwald or Dachau.

Attorney: How was it done prior to this? What was your task in the concentration camps?

Mennecke: The examination of certain prisoners with respect to the question of psychosis or psychopathology.

Attorney: So it was first a question of mental illness?

Mennecke: A medical question.

Attorney: And later it became a political and racial question?

Mennecke: Yes. That is, alongside the political and racial question I also had to make purely medical judgments.

Attorney: So, you had two kinds of cases: the mentally ill, which had to be evaluated according to medical criteria, and those which had to be evaluated according to political and racial criteria?

Mennecke: One simply cannot distinguish the two, Herr Attorney. The two cases were simply not divided and clearly separated from one another.[61]

Consistent with this conception of the Jews as a pathological people, many of Germany's leading medical journals debated what to do about "the Jewish question." *Deutsches Ärzteblatt,* Germany's leading medical journal (then as now), carried a regular column in the early war years on "Solving the Jewish Question"; the column also reviewed ongoing efforts along these lines in Romania, China, Japan, Italy, and other countries.[62] Popular medical journals discussed the history and fate of the Jews, in some cases replete with pictures portraying the exodus of the Jews from Egypt, the problems of Zionism, and so forth.[63] The *Deutsches Ärzteblatt* worried that Britain was building an army of 100,000 Jews in Palestine, an army that "could and would be used against the Germans." The journal reprinted Hitler's words (in boldface) that if another war should come to Europe, this would mean not the "Bolshevization" of the earth, but rather the destruction of the Jewish race in Europe.[64]

Racial hygienists also carried forth this banner. Otmar von Verschuer, successor to Eugen Fischer as head of the Kaiser Wilhelm Institute for Anthropology, Human Genetics, and Eugenics, described in his textbook on racial hygiene the need for a "complete solution of the Jewish question" (*Gesamtlösung der Judenfrage*).[65] During the March 27–28, 1941, opening ceremonies for the Institute for Research on the Jewish Question (Institut zur Erforschung der Judenfrage) in Frankfurt, Eugen Fischer and Hans F. K. Günther were guests of honor at a meeting at which a host of possible "solutions" to the Jewish question were discussed.[66] Here, Walter Gross, head of the Office of Racial Policy, reviewed the history and shortcomings of previous attempts to solve "the Jewish question" (emancipation, persecution, partial annihilation, etc.), concluding that a "final solution" (*Gesamtlösung, endgültige Lösung*) to the Jewish question could come only with the "removal of Jews from Europe."[67] In late 1941 or early 1942, Eugen Fischer carried the racial message to occupied Paris, where he delivered a speech in which he declared that Bolshevist Jews were of "such a mentality, that one can only speak of inferiority and of beings of another species."[68] And in 1944, the same year that he accepted Dr. Josef Mengele as scientific assistant, Otmar von Verschuer proudly claimed that the dangers posed by Jews and Gypsies to the German people had been "eliminated through the racial-political measures of recent years." Verschuer also reminded his readers, though, that the purification of Germany from "foreign racial elements" would require an even more ambitious effort, extending across the entirety of Europe.[69]

In November 1942, the *Informationsdienst des Hauptamtes für Volksgesundheit der NSDAP,* a newsletter published by the Reich Health Publishing House, noted that in the "Confidential Information of the Party Chancellery" there had appeared a paper entitled "Preparatory Measures for the Final Solution of the European Jewish Question."[70] Copies of this newsletter were circulated among the entire medical faculties of certain universities (e.g., Giessen); one can only wonder what crossed the minds of these professors upon seeing the title of this paper.[71]

The continuities linking the various phases of the Nazi's program to destroy "lives not worth living" were both practical and ideological. In the fall of 1941, with the completion of the first major phase of the euthanasia operation, the gas chambers at psychiatric institutions throughout southern and eastern Germany were dismantled and shipped east, where they were reinstalled at Belzec, Majdanek, San Sabba (Trieste), Treblinka, and Sobibor. The same doctors and technicians and nurses often followed the equipment, bringing with them skills in how to murder covertly, extract teeth, recycle valuables, and cremate bodies—all in assembly-line fashion. Germany's psychiatric hospitals forged the most important practical link between the murder of the handicapped classed as "lives not worth living" in Germany's hospitals and of Germany's Jewish and Romani (Gypsy) populations.

Sexual and Racial Pathologies

Jews, Gypsies, and the physically or mentally handicapped were not the only groups stigmatized as sick and degenerate by Germany's racial scientists. Communists, homosexuals, the feeble-minded, the tubercular, and a wide class of "antisocials" (vagabonds, beggars, alcoholics, prostitutes, drug addicts, the homeless, and other groups) were also marked for destruction. In each case, medical professionals were involved in both the theory and practice of this destruction.

Consider the case of homosexuals. By the 1930s, Nazi medical authorities could draw upon a sizable literature documenting the supposed degeneracy of (male) homosexuals. In the very first (1904) volume of the *Archiv für Rassen- und Gesellschaftsbiologie,* Ernst Rüdin argued that homosexuality was a genetically determined "diseased form of degeneracy."[72] Dobrovsky's "Dental Investigations of Homosexual Men" in the 1924 *Zeitschrift für Konstitutionslehre* postulated that gay men show abnormal gum and tooth development; Weil's article of the same year in the *Archiv für Frauenkunde* purported to discover distinctive bodily deformities in homosexual men.[73] Research along these lines was encouraged during the Nazi period, and not just in Germany. In the summer of 1934, scholars convening under the rubric of the International Federation of Eugenics Organizations debated whether the "disease" of homosexuality was genetically inherited or socially acquired. Dr. J. Sanders of the Hague concluded from his studies of the sexual preferences of twins that there were "neither medical nor scientific grounds for treating homosexuals differently from heterosexuals, either legally or socially." German scientists at the congress disagreed with Sanders's conclusions, however. Professors Lothar Tirala of Munich and Otmar von Verschuer of Frankfurt argued that homosexuality was a "moral pathology" and advised that Germans use "all possible means to suppress such sick perversions in the body of our people."[74]

By the mid-1930s, physicians in Nazi Germany were united in arguing that homosexuals posed a threat to public health. Physicians writing in the nation's leading public health journal regularly described homosexuality as a "pathology" and homosexuals as "psychopaths."[75] The magnitude of the threat was not something to be taken lightly: in 1938, the Office of Racial Policy reported that Germany was faced with an "epidemic of some 2 million homosexuals, representing 10 percent of the entire adult male population."[76]

The most common theory of homosexuality advanced at this time was that it was an inborn, biologically determined disorder. In 1939, for example, a physician by the name of Deussen published an article titled "Sexual Pathologie" supporting this view, in the journal *Progress in Genetic Pathology.* Deussen cited Theobald Lang's work purportedly showing that the sisters of male homosexuals tend to exhibit particularly "masculine" characteristics; this led him to accept Richard Goldschmidt's theory of male homosexuals as "genetic females."[77] Interestingly, belief in a genetic basis of homosexuality was not confined to the political right. In 1937 the socialist physicians' *Bulletin* in exile published an article claiming that homosexuality was "inborn, and hence not subject to the free will of the individuals who come into the world with this inversion."[78] The journal advocated abolishing Section 175 of the Prussian criminal code criminalizing homosexuality.

In light of such arguments, some Nazi physicians disputed the claim that homosexuality was a genetic disorder. In June of 1938, for example, a physician writing for the Office of Racial Policy characterized homosexuals as "weak, unreliable, and deceitful"; they were typically "servile and yet power hungry . . . incapable, in the long run, of functioning in a positive manner in society as a whole." This author argued that homosexuality was *not*, however, a genetic disease: the view that homosexuality was inherited simply played into the hands of those who wished to believe that homosexuality was not a matter of choice. This was the argument "upon which the entire ideology of homosexuality is based"—namely, that these people "cannot do otherwise." This author estimated that, although perhaps only two percent of all homosexuals were actually "genetically sick," this two percent exerted an enormous influence in society: "40,000 abnormals—whom one might well expel from the community—are in a position to poison two million citizens." Homosexuals, in this author's view, were like the Jews: they build a "state within a state; they are state criminals: not 'poor, sick' people to be treated, but enemies to be eliminated!"[79]

Homosexuality was illegal in the Weimar Republic; the Nazis, however, imposed new and far more repressive measures against this group. On the night of June 30, 1934, most of the leaders of the SA (many of whom were homosexuals—most notably, Ernst Roehm) were assassinated in what subsequently became known as "the Roehm putsch" or "night of long knives." The Roehm putsch marked only the first phase of the Nazi persecution of homosexuals. Beginning soon after the *Machtergreifung* (Nazi seizure of power), Nazi officials had begun to construct an inventory of all representatives of the so-called third sex; these lists were subsequently used to "cleanse" the population of these people. In the mid-1930s, thousands of individuals identified as homosexuals were arrested and sent to concentration camps, where they were detained so as not to "infect" the broader population. Thousands of camp inmates wearing the pink triangle were ultimately sent to the gas chambers, as part of the attempt to extirpate this "pathology" from Germany.[80]

Other groups were singled out for destruction, and, here again, the cooperation of medical personnel on both ideological and practical levels was crucial. Beginning early in the Nazi regime, the Reich Health Office began constructing elaborate genealogical tables of all Gypsies in Germany.[81] In 1938, public health authorities were asked to register with the police all Gypsies and "Gypsy half-breeds," based upon information gathered from genealogical tables or genetic registries.[82] This information was then used by police authorities to round up Gypsies for deportation. An article in the *Deutsches Ärzteblatt* described some of the tasks faced by the physician in sorting out the Gypsy question:

> Experience gathered thus far in the struggle against the Gypsy plague reveals that the half-breeds are responsible for the largest fraction of criminal offenses among the Gypsies. On the other hand, it has been shown that attempts to make the Gypsies settle down have failed, especially among the purest strains of this race; this is because of their strong migratory instinct. It has thus become necessary to separate pure and half-breed Gypsies, for the purpose of coming to a final solution of the Gypsy problem. Toward this end, the SS Reichsführer and Chief of German Police [Heinrich Himmler] has issued elaborate instructions. In order to achieve this goal, it will be necessary to determine the racial affiliation of all Gypsies living in the Reich, also that of all people living like Gypsies.[83]

This article specified that all Gypsies were to be registered with the Reich Criminal Police Bureau, in a special division created for this purpose (the Reichszentrale zur Bekämpfung des Zigeunerwesens).

Gypsies, like Jews and gays, were often described by Nazi medical authorities as a health risk to the German people. Otmar von Verschuer claimed that 90 percent of Germany's 30,000 Gypsies were "half-breeds" and that most Gypsies were "asocial and genetically inferior."[84] In 1944, medical authorities in Bulgaria ascertained that Gypsies, by virtue of their migrant lifestyle, were responsible for spreading infectious diseases; Bulgarian authorities ordered all Gypsies to give up wandering and settle in a single location.[85] Medical involvement in the destruction of the Gypsies was also more direct. In the winter of 1941–42, Dr. Robert Ritter, one of Germany's foremost criminal biologists, participated in a conference at which the drowning of 30,000 Gypsies by bombardment of their ships in the Mediterranean was proposed.[86] Ritter ultimately distinguished himself as one of the chief architects of the genocide of this group. Beginning in the mid-1930s, he received funds from the German Research Council (DFG) to research the Gypsy question at his Racial Hygiene and Population Biology Research Division within the Berlin Health Office. On January 20, 1940, Ritter reported to the DFG that the Gypsy question could be solved only "if the majority of asocial and useless Gypsies can be rounded up and put to work in special camps, where they can be prevented from any further reproduction." He helped prepare evaluations (*Gutachten*) used for identifying Gypsies to be destroyed; on January 31, 1944, the German professor reported that he had recently completed evaluation of another 23,822 Gypsy "cases."[87]

The campaign against tuberculosis also took on a similar—if more controversial—character under the Nazis. Even before their rise to power in January 1933, the Nazis claimed that the eradication of tuberculosis was one of their highest goals. Opinions were divided, however, on what the appropriate attitude toward the disease should be. Some complained about the exorbitant costs associated with treating the disease: Dr. F. Koester, for example, writing in the country's leading tuberculosis journal, stated in 1938 that care for Germany's 400,000 tubercular cost the government four or five billion Reichsmarks annually. Others, however, expressed doubts that it was indeed a good idea to eliminate tuberculosis entirely, given that this would mean the loss of an important means of "natural selection."[88] Tuberculosis had been recognized as an infectious disease since Koch's discovery of the tubercle bacillus at the end of the nineteenth century, yet many also recognized that infection with the bacterium was not a sufficient condition for someone coming down with the disease. People knew that one had to be in a weakened physical state—but opinions differed on the nature and origins of this state. Many argued that the "white plague" was largely the product of poor nutrition, living space, or working conditions; others argued that one's physical or racial constitution was the crucial variable. Followers of Ernst Kretschmer's *Konstitutionslehre,* for example, argued that those with the leptosome body type were especially susceptible and that such individuals should be counseled not to go into nursing or medicine (where they were more likely to be infected). Fritz Lenz suggested that members of the Nordic race (or light-skinned individuals more generally) were particularly resistant to the disease; Hans Luxenburger postulated a correlation between TB and schizophrenia.[89]

Even prior to 1933, the widespread conception of tuberculosis as a genetically inherited disease[90] prompted many to call for the isolation of those with the disease from the "genetic stream" of the population. Alfred Grotjahn as early as 1915 advocated celibacy for the afflicted, and by the 1930s TB was one of the most common grounds given for abortion. Health officials also stressed the importance of isolating the tubercular from contact with the larger population. The government initiated a campaign for early detection of the disease, and instituted obligatory chest X-rays for the entire SS, members of the army, the SA, and workers in the armament industries.[91] Proposals were put forward that the tubercular should be sterilized, but this particular suggestion was never put into practice (nor was a proposed ban on marriage between the tubercular and nontubercular). Instead, on December 1, 1938, an "Order for the Struggle against Contagious Diseases" required that all cases of tuberculosis be registered with state health authorities; the order also allowed the forcible confinement of individuals with the disease.

During the Second World War, the struggle against tuberculosis took on a more urgent nature, especially in occupied territories. On May 1, 1942, the *Gauleiter* for the Wartheland region in occupied Poland wrote a letter to Himmler suggesting that Poland's 35,000 incurable tuberculars be exterminated, and that preparations for this operation be made as soon as the region's remaining 100,000 Jews were destroyed. Reinhard Heydrich, head of Germany's state Security Police (SD), approved the operation in a letter of June 9, 1942, on the condition that health authorities could determine which cases of the disease were incurable. Kurt Blome, head of Germany's medical postgraduate education program and now deputy chief of the Nazi party's Office for Public Health, was asked to explore the alternatives open to German authorities to deal with the matter. Blome distinguished three possible options: "special treatment" (*Sonderbehandlung* = Nazi language for extermination); isolation of the severely infected; or creation of a special reserve for all tuberculars. Blome estimated that the first of these options—the one favored by Heydrich and Himmler—would take six months to complete. He also cautioned, however, that the operation would have to remain secret; if the operation were to become public, Germany's enemies would be able to mobilize the "doctors of the world," and opposition would be even greater than to the euthanasia operation. Blome therefore recommended the creation of a reservation, comparable to a leper colony.[92] It remains unclear today how far plans for such a colony were ever put into practice.

Part of our revulsion for medical involvement in Nazi racial crime stems from the fact that it violated a relationship that is supposed to involve a great deal of intimacy and trust. Patients generally place themselves at the mercy of their physicians, and the relationship is easy to exploit. In the Nazi period, the trust implicit in the doctor-patient relationship was exploited in order to achieve goals that would have been difficult to achieve by other means. At Buchenwald, for example, 8,000 Russian prisoners of war were executed in the course of supposed "medical exams"; unsuspecting prisoners were taken to a medical examination room, where they were told to stand in front of a device apparently designed to measure their height. Prisoners were then shot in the neck from a cavity secretly built into the apparatus. (This device can still be seen as part of the exhibit the former East German government established in the former concentration camp at Buchenwald.) The traditional doctor-patient relationship was exploited in other ways as well. Physicians and medical technicians operated the gas chambers; physicians were responsible for the selection of people to be gassed in concentration camps. Physicians also supervised the medical experiments in concentration camps—perhaps the most notorious aspect of medical complicity in Nazi racial crime.[93] It is possible of course, in hindsight, to separate analytically the sterilization program (eugenics), the destruction of the mentally ill (euthanasia), and the destruction of Germany's racial minorities (the final solution). The fact is, however, that each of these programs was seen as part of a larger program of racial purification. Medical journals used the term "life not worth living" to refer to those sterilized under the 1933 Sterilization Law, to those killed in psychiatric hospitals, and to those killed in concentration camps.

If we want to understand the logic of the Nazi racial program, then it is not possible to draw a sharp line between what happened before and after 1939. Nor is it possible to maintain that Germany's biomedical community restricted its participation to only the earliest or more "theoretical" phases of this process. Physicians played an active role in both the theory and the practice of each phase of the Nazi program of racial hygiene and racial destruction. Central in each case was an image of the unwanted person as alien, unclean, or unproductive. Medical metaphors of "parasites" or "bacilli" allowed enemies of the state to be murdered in the guise of quarantine; medico-legal institutions provided the personnel and ideological legitimacy for operations that will hopefully forever stand as a low watermark for professional ethical responsibility.

In this light, we can appreciate the conclusion reached by Max Weinreich in his *Hitler's Professors:*

> It will not do to speak in this connection of the "Nazi gangsters." This murder of a whole people was not perpetrated solely by a comparatively small gang of the Elite Guard or by the Gestapo, whom we have come to consider as criminals . . . the whole ruling class of Germany was committed to the execution of this crime. But the actual murderers and those who sent them out and applauded them had accomplices. German scholarship provided the ideas and techniques which led to and justified this unparalleled slaughter.
>
> [Those involved] were to a large extent people of long and high standing, university professors and academy members, some of them world famous, authors with familiar names. . . .[94]

Notes

1. This chapter is a slightly revised and shortened version of a chapter from my book, *Racial Hygiene: Medicine under the Nazis* (Cambridge, 1988), reprinted by permission of Harvard University Press, Copyright © 1988 by the President and Fellows of Harvard College.

2. *Der Prozess gegen die Hauptkriegsverbrecher vor dem internationalen Militärgerichtshof,* vol. 26 (Nuremberg, 1947), p. 169. The order was issued in early October, but backdated to Sep-

tember 1, 1939, to coincide with the beginning of the invasion of Poland. The text of the memo was dictated by Dr. Max de Crinis, professor of neurology and psychiatry at the University of Berlin.

3. Figures for the first phase of the adult euthanasia operation are recorded quite precisely in German archives: 70,273 were killed in six institutions from October 1939 to September 1941. See Götz Aly, "Medizin gegen Unbrauchbare," in Götz Aly et al., *Aussonderung und Tod, die klinische Hinrichtung der Unbrauchbaren* (Berlin, 1985), p. 23. Not counted in this figure, however, are the related "operations" in the territories occupied by German armies; these are sometimes difficult to distinguish from extermination operations carried out in concentration camps under less "scientific" auspices. In Germany itself, according to the statistics of the Gesellschaft für Deutsche Neurologie und Psychologie, there were 158,164 mentally ill in 253 German hospitals (public and private) at the beginning of 1936. By the end of the war, German psychiatric hospitals were virtually empty. For background on the exterminations, see Ernst Klee's *"Euthanasie" im NS-Staat* (Frankfurt, 1983); also Karl-Heinz Hafner and Rolf Winau, "Die Freigabe der Vernichtung lebensunwerten Lebens," *Medizinhistorisches Journal* 9 (1974): 227–54; Friedrich Kaul, *Nazimordaktion T-4* (Berlin, 1973); Frederic Wertham, *A Sign for Cain* (New York, 1966), especially the chapter on the "Geranium in the Window." A bibliography containing more than 2,000 entries on the question of euthanasia has been published by Gerhard Koch: see his *Euthanasie, Sterbehilfe, Eine dokumentierte Bibliographie* (Erlangen, 1984), published as part of the *Bibliographica genetica medica*, vol. 18.

4. Benno Müller-Hill, *Tödliche Wissenschaft* (Reinbek, 1984), p. 12.

5. Kaul, *Nazimordaktion T-4*, p. 58; Alexander Mitscherlich and Fred Mielke, *Medizin ohne Menschlichkeit* (Frankfurt, 1978), pp. 191–92.

6. Aly, "Medizin gegen Unbrauchbare," p. 23.

7. Kaul, *Nazimordaktion T-4*, p. 64.

8. Aly, "Medizin gegen Unbrauchbare," p. 23. Records for these figures can be found in Roll 18, T-1021, National Archives, Washington.

9. Klee, *"Euthanasie,"* pp. 340–41.

10. Wertham, *A Sign for Cain*, p. 157.

11. Aly, "Medizin gegen Unbrauchbaren," pp. 19–31.

12. Wertham, *A Sign for Cain*, p. 167.

13. For an early Nazi presentation of this argument, see Theobald Lang, "Die Belastung des Judentums mit Geistig-Auffälligen," *Nationalsozialistische Monatshefte* 3 (1932): 23–30.

14. Walter Gross, "Die Familie," *Informationsdienst*, September 20, 1938. Gross here announced that the work of the Office of Racial Policy would be finished "only with the disappearance of the last Jew from our Reich."

15. Peltret, "Der Arzt als Führer und Erzieher, "*Deutsches Ärzteblatt* 65 (1935): 565–66.

16. Gerhard Wagner, "Unser Reichsärzteführer Spricht," *Ziel und Weg* 5 (1935): 432–33.

17. Sander Gilman, *The Jew's Body* (New York, 1992).

18. Maurice Sorsby, *Cancer and Race: A Study of the Incidence of Cancer among Jews* (London, 1931); Max Reichler, *Jewish Eugenics* (New York, 1916). In 1918, the *American Journal of Physical Anthropology* reported Maurice Fischberg's assertions that "the Jews are physically puny—a large proportion are feeble, undersized; their muscular system is of deficient development with narrow, flat chests, and of inferior capacity. They make the appearance of a weakly people, often actually decrepit" (pp. 106–107). Fischberg suggested that both the talents and the infirmities of Jews were to be explained by the impact of their habits upon their breeding, as in their "giving preference in marriage to the scholar," but also in their "dysgenic" custom of distributing relief to the poor.

19. A. Pilcz, "Die periodischen Geistesstörungen," *Wiener Klinische Rundschau* 16 (1902): 490.

20. Martin Engländer, *Die auffallend häufigen Krankheitserscheinungen der jüdischen Rasse* (Vienna, 1902).

21. *Jüdisches Volksblatt* 50 (1902). In 1919, a Dr. Rafael Becker published evidence that, in the period 1900–1904, Swiss Jews were more than twice as likely to be mentally ill as Swiss Protestants or Catholics. See his "Die Geisteserkrankungen bei den Juden in der Schweiz," *Zeitschrift für Demographie und Statistik der Juden*, April, 1919, pp. 52–57.

22. Wagner, "Unser Reichsärzteführer Spricht," p. 432. Wagner's discussion of the Jew in terms of an "impure race" recalls the theory occasionally put forward by Nazi racial theorists that Jews represented a cross between Oriental and Negro races. Hans F. K. Günther, for example, presented this theory in his *Rassenkunde des Jüdischen Volkes* (Munich, 1929).

23. Otmar Freiherr von Verschuer, *Erbpathologie* (Munich, 1937), pp. 86–182.

24. Edgar Schulz, "Judentum und Degeneration," *Ziel und Weg* 5 (1935): 349–55. Many within Germany's racial hygiene community rejected the notion that inbreeding invariably caused a decline in racial fitness. Fritz Lenz, for example, supported marriage between cousins and even siblings, arguing that such inbreeding was dangerous only where deleterious recessive alleles were present.

25. Rainer Pommerin, *Die Sterilisierung der Rheinlandbastarde* (Düsseldorf, 1979), p. 47.

26. See my "From *Anthropologie* to *Rassenkunde:* Concepts of Race in German Physical Anthropology," in *Bones, Bodies, Behavior: Essays on Biological Anthropology,* ed. George W. Stocking, Jr. (Madison, 1988), pp. 138–79.

27. Wilhelm Hildebrandt, *Rassemnischung und Krankheit* (Leipzig, 1935). For a review of comparable Anglo-American views, see William Provine, "Geneticists and the Biology of Race Crossing," *Science* 182 (1973): 790–96.

28. Franz Goldmann and Georg Wolff, *Tod und Todesursachen unter den Berliner Juden* (Berlin, 1937); reported in the *British Medical Journal,* October 2, 1937, p. 663.

29. Martin Stämmler and Edeltraut Bieneck, "Statistische Untersuchungen über die Todesursachen der deutschen und jüdischen Bevölkerung von Breslau," *Münchener medizinische Wochenschrift* 87 (1940): 447–50.

30. "German Medicine, Race, and Religion," *British Medical Journal,* August 17, 1940, p. 230.

31. *Krakauer Zeitung,* January 14/15, 1940.

32. This was the headline of the *Krakauer Zeitung* of February 8, 1940.

33. *Kölnische Zeitung,* April 5, 1941.

34. Wilhelm Hagen, *Auftrag und Wirklichkeit, Sozialarzt im 20. Jahrhundert* (Munich-Gräfelfing, 1978), p. 166.

35. *Manchester Guardian,* November 3, 1939.

36. *British Medical Journal,* July 6, 1940, p. 36.

37. Hagen, *Auftrag und Wirklichkeit,* pp. 171, 179–81.

38. On Hamburg attorney Jürgen Rieger's 1978 defense of Arpad Wigand, see p. 382n. 82 of my *Racial Hygiene.* On the threat to public health, see *Der Stürmer,* no. 13, March, 1941, p. 3.

39. Robert Ritter, "Kriminalität und Primitivität," *Monatsschrift für Kriminalbiologie* 31 (1940): 197–210.

40. *Informationsdienst,* October 10, 1938.

41. Johannes Lange, *Verbrechen als Schicksal, Studien an kriminellen Zwillingen* (Leipzig, 1929).

42. "Der 4. Kriminal-Biologische Kongress in Hamburg," *Archiv für Bevölkerungswissenschaft und Bevölkerungspolitik* 4 (1936): 51.

43. For a comprehensive listing of courses offered in the fields of criminal biology at German universities, see the *Monatsschrift für Kriminalbiologie und Strafrechtsreform* 32 (1941): 53–73.

44. *Reichs-Gesundheitsblatt* 12 (1937): 118. See also "Kriminalbiologische Sammelstelle in Berlin," *Ärzteblatt für Berlin* 42 (1937): 623. The Criminal Biology Stations, established nationwide in 1936, were modeled on similar institutions already existing since 1923 in Bavaria.

45. *Deutsches Ärzteblatt* 68 (1938): 858.

46. Johannes Schottky (ed.), *Rasse und Krankheit* (Munich, 1937).

47. I discuss the science and ideology of the Baur-Fischer-Lenz textbook at length in my *Racial Hygiene,* chapter 2.

48. Wagner, "Unser Reichsärzteführer Spricht," p. 433.

49. See, for example, Rudolf Ramm, "Die Aussiedlung der Juden als Europäisches Problem," *Die Gesundheitsführung* 11 (1941): 175–78; also Walter Gross, "Rassenpolitische Voraussetzungen einer europäischen Gesamtlösung der Judenfrage," *Rassenpolitische Auslands-Korrespondenz,* March, 1941, pp. 1–6.

50. W. Bormann, "Grundsätze der deutschen Ostraumpolitik," *Ärzteblatt für den Reichsgau Wartheland* 2 (1941): 168.

51. "Die Juden im Generalgouvernement," *Deutsches Ärzteblatt* 70 (1940): 430–31.

52. Ibid., p. 432.

53. Mitscherlich and Mielke, *Medizin ohne Menschlichkeit,* p. 241.

54. Ibid., pp. 240–41.

55. Ibid., p. 242.

56. Kaul, *Nazimordaktion T-4,* pp. 90–95.

57. Ibid., pp. 97–99.

58. Aly, "Medizin gegen Unbrauchbare," p. 28.

59. *14 f 13* was the code word used for the secret operation ordered by Himmler to kill all feeble-minded and crippled inmates of Germany's concentration camps. From Buchenwald at this time, three to four hundred Jews were sent to Bernburg for extermination. After the war, Viktor Brack denied any connection between the euthanasia program (of which he was in charge) and this special "Aktion 14 f 13." See Mitscherlich and Mielke, *Medizin ohne Menschlichkeit,* p. 213.

60. Aly, "Medizin gegen Unbrauchbare," pp. 30–31. The best history of links between the killings of the Jews and the Gypsies, on the one hand, and of handicapped children and adults, on the other, is Henry Friedlander, *The Origins of Nazi Genocide; From Euthansia to the Final Solution* (Chapel Hill, 1995).

61. Mitscherlich and Mielke, *Medizin ohne Menschlichkeit,* p. 216.

62. See, for example, "Zur Lösung der Judenfrage," *Deutsches Ärzteblatt* 72 (1942): 16–17; also "Löst Rumanien die Judenfrage?" in the same journal, vol. 70 (1940): 359–60.

63. See, for example, Bernhard Hörmann, "Zur Judenfrage," *Die Volksgesundheitswacht,* Erntin, 1933, pp. 3–10.

64. "Das Ende der jüdischen Kriegsziele," *Deutsches Ärzteblatt* 70 (1940): 312–14.

65. Otmar von Verschuer, *Leitfaden der Rassenhygiene* (Leipzig, 1941), pp. 125–30; compare also his "Rassenbiologie des Juden," *Forschungen zur Judenfrage* 3 (1938): 137–51.

66. Müller-Hill, *Tödliche Wissenschaft,* pp. 18–19.

67. Walter Gross, "Rassenpolitische Voraussetzungen," pp. 1–6.

68. Müller-Hill, *Tödliche Wissenschaft,* p. 20.

69. Otmar von Verschuer, "Bevölkerungs- und Rassefragen in Europa," *Europäischer Wissenschaftsdienst* 1 (1944): 3.

70. "Vorbereitende Massnahmen zur Endlösung der europäischen Judenfragen," listed as paper no. 881 in the "Vertrauliche Informationen" of the Parteikanzlei, mentioned in the *Informationsdienst des Hauptamtes für Volksgesundheit der NSDAP,* November, 1942, p. 69.

71. NSD 28/8, Bundesarchiv Koblenz.

72. See the *Archiv für Rassen- und Gesellschaftsbiologie* 1 (1904): 99–100.

73. Dobrovsky, "Gebissuntersuchungen an homosexuellen Männer," *Zeitschrift für Konstitutionslehre* 10 (1924); Arthur Weil, "Sprechen anatomische Grundlagen für das Angeborensein der Homosexualität?" *Archiv für Frauenkunde* 10 (1924): 23–51.

74. "Bericht über die 11. Versammlung der Internationalen Föderation Eugenischer Organisationen," *Archiv der Julius Klaus-Stiftung* 10 (1935): 40. In 1937, Otmar von Verschuer reported further research suggesting that homosexuality was a genetically inherited trait; see his *Erbpathologie,* p. 90.

75. See, for example, J. Lange, "Die Feststellung und Wertung geistiger Störungen im Ehegesundheitsgesetz," *Öffentlicher Gesundheitsdienst* 4 (1938): 533; also Miesbach's article, "Aus gerichtsärztliche gutachterlichen Praxis" in the same journal, vol. 4 (1938): 672.

76. "Staatsfeinde sind auszumerzen!" *Informationsdienst,* June 20, 1938.

77. Deussen, "Sexualpathologie," *Fortschritte der Erbpathologie* 3 (1939): 90–99; compare also Theobald Lang, "Beitrag zur Frage nach der genetischen Bedingtheit der Homosexualität," *Zeitschrift für Neurologie* 155 (1936): 5; 157 (1937): 557; and 162 (1938): 627. For Goldschmidt's theory, see his *Die sexuelle Zwischenstufen* (Berlin, 1931).

78. "Zum Problem der Homosexualität," *Internationales Ärztliches Bulletin,* December 1937, p. 114.

79. "Staatsfeinde sind auszumerzen!" *Informationsdienst,* June 20, 1938. This article also contains a discussion of certain "natural affinities" between Jews and homosexuals.

80. For the broad outlines of this history, see Richard Plant, *The Pink Triangle: The Nazi War against Homosexuals* (New York, 1986).

81. For a Nazi activist's bibliography on Germany's "Gypsy question," see Robert Ritter, "Zeitschriftenartikel über Zigeunerfragen," *Fortschritte der Erbpathologie* 3 (1939): 2–20. For a postwar history of the persecution of Gypsies in the Third Reich, see Joachim S. Hohmann, *Zigeuner und Zigeunerwissenschaft* (Marburg, 1980).

82. *Öffentlicher Gesundheitsdienst* 4 (1938): 995–96.

83. "Rassische Erfassung der Zigeunern," *Deutsches Ärzteblatt* 68 (1938): 901.

84. Verschuer, *Leitfaden der Rassenhygiene* p. 130.

85. *Die Gesundheitsführung* 14 (1944): 28.

86. Müller-Hill, *Tödliche Wissenschaft,* p. 20.

87. Ibid., p. 23.

88. See, for example, Kurt Klare, *Tuberkulosefragen* (Leipzig, 1939), p. 51; also F. Koester, "Zeitgemässe Tuberkulosefragen und ihre Auswirkungen für die praktische Tuberkulosebekämpfung," *Deutsche Tuberkulose-Blatt* 12 (1938): 25–29, 49–53.

89. See Kristin Keltin, Das Tuberkuloseproblem im Nationalsozialismus. Med. diss., University of Kiel, 1974.

90. The most important defense of the heritability of a predisposition toward tuberculosis is Otmar von Verschuer and Karl Diehl, *Zwillingstuberkulose. Zwillingsforschung und erbliche Tuberkulosedisposition* (Jena, 1933).

91. Susanne Hahn, "Ethische Grundlagen der faschistischen Medizin, dargestellt am Beispiel der Tuberkulosebekämpfung," in *Medizin im Nationalsozialismus,* ed. Achim Thom and Horst Spaar (East Berlin, 1983), pp. 139–40.

92. Mitscherlich and Mielke, *Medizin ohne Menschlichkeit,* pp. 231–36.

93. The best single account of the experiments is Mitscherlich and Mielke's 1949 *Medizin ohne Menschlichkeit,* translated into English as *The Death Doctors* (London, 1962).

94. Max Weinreich, *Hitler's Professors* (New York, 1946), pp. 6–7.

7

Domesticity in the Federal Indian Schools

The Power of Authority over Mind and Body

K. Tsianina Lomawaima

AMERICAN INDIAN CHILDREN have been coerced or recruited into federally run off-reservation boarding schools from 1879 until the present. The publication of native memoirs and film and radio productions periodically incite public interest or excite public outrage, an outrage that occasionally translates into government investigations, reports (Kennedy 1969; Meriam et al. 1928), and reforms. Scholars have detailed the history of federal educational policy and schools for American Indians, but only rarely have they portrayed boarding schools as arenas for a reciprocating exercise of power (Foucault 1979, 1980; Sheridan 1980) between school staff and students—in other words, as an interaction Indian students helped to create.[1]

In this chapter I analyze federal educational policy, federal educational practice, and Indian students' actions in order to comprehend that policy and practice (and their seeming contradictions) and to reveal the strategies Indian children devised to undermine federal objectives. I focus on federal policies of domestic training for girls; the practices of regimenting, training, and clothing female students' bodies; and the strategies girls invented to avoid wearing government-issue (GI) clothing. The focus on one gender's boarding-school experiences is motivated by the strict sex segregation enforced in the schools—which generated different school experiences for boys and girls—and by the feminist critique of the Victorian "cult of domesticity." Domesticity training for Indian girls was a clear surface manifestation of the gender- and race-defined fault lines segmenting American society at the turn of the century (and later as well).

In the early 1900s, federal boarding schools forbade native language use and religious practice, and they separated families. Policy makers calculated these practices to achieve far-reaching social goals, to civilize and Christianize young Indian people and so draw them away from tribal identification and communal living. The government's ambitious goals of individual transformation mobilized, in the boarding schools, a rigid and detailed military discipline that scheduled every waking moment, organized classrooms and work details, and even mandated a "correct" physical posture, "correct" ways of moving and exercising, and "correct" details of dress.

In order to mold young people's minds, nineteenth-century educators bent first to mold their bodies, according to gender- and race-specific notions of capacities and inclinations. The federal emphasis on physical training reflected racist conceptions of the intrinsic link between uncivilized minds and undeveloped bodies. The boarding school exemplified Foucault's assertion that in Western, industrial societies' systems of "corrective" detention, "it is always the body that is at issue—the body and its forces, their utility and their docility, their distribution and their submission" (Foucault 1979:25). As Foucault points out, power does not operate on bodies primarily in repressive, negative ways (see, for example, Foucault 1980:59; Sheridan 1980:184). Power is strong because it is creative—it "produces" effects, knowledge, habits, discourses. A new image of the female Indian body was created according to the dictates of Victorian decency and domesticity. This study of the ways the "body itself is invested by power relations" (Foucault 1979:24) depends upon a critical analysis of the boarding school as an institutional training ground where the colonized were to learn subservience. The practices of military regimentation, uniform dress, and domesticity training flowed from the federal vision of boarding school as a complete transformative experience, training Indians for their place as a detribalized social and economic underclass.

Boarding schools were not, however, perfect laboratories in which to fulfill federal intentions. Students adapted to regimentation and resisted authority in creative ways.[2] Peer groups both bound and subdivided a diverse student population along lines of gender, tribe, blood, age, and family background. Narrative comments from boarding-school alumni recall and reconstruct student life at the Chilocco Indian Agricultural School in north-central Oklahoma in the years from 1920 to 1940.[3] The Chilocco Indian Agricultural School began with a congressional order of 17 May 1882 that authorized construction of a school for Indian children on 8,000 acres near the Kansas state line, and it operated from 1884 to 1980 (National Archives 1931). School administrators, teachers, and staff engaged in an often adversarial dialogue with students, who were creative (although dominated) participants in the culture of boarding-school life.[4] The wearing of a uniform was a battleground for the contest of power between students and authorities. Alumni tell many stories about their school days, and one—the bloomer story—documents both female student resistance to uniform dress and the complexity of the peer group cooperation and competition essential to student resistance.

Here is the bloomer story as told by Maureen, a Choctaw who entered Chilocco in 1931 at age fourteen:

> We wore gray sateen bloomers and black cotton stockings and GI shoes the first year I was up there. Well, when we went to the dances, we could wear what we called "home clothes" [the personal clothes that the girls brought from home], but if we put home clothes on over those big old sateen bloomers, it looked terrible. So all the girls hated that and some of them would get brave. . . . They pulled their home pants on and then put their bloomers over them so when we had inspection—they'd inspect for that before we left—we'd raise our dresses and show we had our bloomers on. Then when we'd get out of the building we'd pull those bloomers off. So some of 'em got a real clever idea and they made legs, just cut the legs off the bloomers, so when Miss M [the head matron] would come along to inspect, they'd just raise their dress so far [to show the piece of leg]. But they got caught with that, too; we got caught with everything eventually.

Today both women and men tell subtly disparate versions of the bloomer story to family members, at alumni reunions, to interested listeners. The story responds directly to the themes of domesticity training and proper dress that saturated federal education for young Indian women, and it provides a student counterpoint to the discussion of federal policy and practice reconstructed from documentary evidence.

In the 1870s, Congress authorized army officer Richard H. Pratt to convert military barracks at Carlisle, Pennsylvania, into the Carlisle Indian School, the first federal off-reservation boarding school for Indian youth. Carlisle had been preceded by, and was for many years contemporaneous with, the Indian School at Hampton Institute, a normal and industrial training school for black Americans in Hampton, Virginia. Pratt ran Carlisle in a military fashion, with issued uniforms, close order drill, and students organized by company and by rank.[5] The educational experiment at Carlisle met with federal approval, and within five years schools had been established in Genoa, Nebraska; Lawrence, Kansas; and Chilocco, Oklahoma. Using a curriculum that emphasized piety, obedience, and manual labor, these schools aimed to transform the Indian child. The essential transformation would be internal, a matter of Christian belief, nontribal identification, mental discipline, and moral elevation. For female students, that meant training for domesticity; for male students, it meant instruction in semiskilled trades and agriculture. The regimentation of the external body was the essential sign of a new life, of a successful transformation. The famous "before and after" pictures of Carlisle students are as much a part of American iconography as the images of Custer's Last Stand.[6] "Savages" shed buckskin, feathers, robes, and moccasins; long black hair was shorn or bobbed or twisted into identical, "manageable" styles; pinafores, stiff starched collars, stockings, and black oxfords signified the "new woman."

The school policies and practices of domestic education, together with their intense surveillance, control, regimentation, and restriction of Indian girls, warrant close examination. Why did the girls have to labor so long and hard? Why did they have to wear a uniform? Why was every moment of their waking and sleeping hours monitored so carefully? Why were they strictly forbidden to step outside their own yards? Analysis of the roots of domestic education for all American women makes clear the underlying federal agenda, which was to train Indian girls in subservience and submission to authority. (Surely this agenda also underlay the manual labor requirements for the boys, but that instruction was at least minimally linked with employment in life after school.)

The degree of surveillance of Indian students dwarfed any possible requirements of straightforward education. The Indian Service demanded such minute documentation of the Indians' personal lives and backgrounds that it is remarkable that the school staff had any time to educate. In 1917, Commissioner of Indian Affairs Cato Sells requested the following information on each student proposed for graduation from a federal school that year: name, age, sex, tribe, degree of Indian blood, physical condition, course and length of study, and handling of personal funds. He also asked for a biographical sketch; a statement on the student's industry, reliability, apparent business qualifications, and general character and habits; a recommendation as to his or her competency; and a statement from the student's reservation superintendent as to all of his or her property holdings (National Archives 1917a).

Beyond the surveillance built into a federal bureaucracy devoted to the total control of Indian people, the acute, piercing focus on Indian girls' attire, comportment, posture, and hairstyles betrays a deep-seated, racially defined perception of Indian people's physical bodies as "uncivilized." Late-nineteenth- and early-twentieth-century racist ideology linked physical and mental competencies to genetic inheritance. It is not surprising, then, that an education appropriate to Indians' ostensibly lower intellectual capacities would so vigorously stress the concomitant need to develop physical skills and habits. In the years after Pratt enrolled the first Kiowa and Cheyenne prisoners of war at the Hampton Institute, Booker T. Washington was appointed the Indian boys' "house-father." Steeped in Hampton's racial "scale of civilization," Washington immediately commented on these young savages' physical ineptitude:

> The untutored Indian is anything but a graceful walker. Take off his moccasins and put shoes on him, and he does not know how to use his feet. When the boys and girls are first brought here it is curious to see in what a bungling way they go up and down stairs, throwing their feet in all sorts of directions as if they had no control over them. (cited in Adams 1977:174)

Federal policy makers intended to create a new kind of Indian through the moral, spiritual, and physical training of the boarding schools. For Indian girls, that meant a process of civilization derived from the Victorian model of middle-class, white domesticity, a template requiring alteration to fit the lower level of physical organization and mental capacity inherited by Indians.

Domesticity: Training for Subservience

After the Civil War, federal and public schools turned to popular manuals, such as Catharine Beecher's *Treatise on Domestic Economy* (1846), as they created departments of "domestic science."[7] Beecher spoke clearly to women: every civilized society required a system of laws to sustain social relations, and it was a fact of nature that those social relations entailed duties of subordination. In her view, freedom in a democratic state meant that every individual was free to choose his or her own superior, to whom obedience was owed (1846:25–26). A 1938 promotional brochure printed at Chilocco repeated this message in the student Code of Conduct: "I believe that intelligent obedience is necessary to leadership. Therefore, I will have respect for authority" (U.S. Indian Service 1938:8). Respect would be naturally engendered by productive manual labor, an assumption that also underlay practices in French prisons[8] and schools for American blacks (Anderson 1978; Hultgren and Molin 1989).

John Stuart Mill recognized the relationship between domesticity and subservience in his criticism of white women's "superficial" education, "calculated to render women fit for submission, vicarious experiences and a service ethic of largely ineffective philanthropy" (cited in Millett 1970:70). We can see similarities in the links between domesticity and subservience training for all women, but certainly the experiences of white and nonwhite women were quite different. Feminist authors point out that the cult of domesticity enshrined white women's fragility and invalidism and demanded that someone be strong enough to work. Addressing the National Congress of Mothers in 1893, Sylvanus

Stall contorted the cultural evolutionary stance of racist ideology to suit the challenges of finding good domestic help:

> At war, at work, or at play, the white man is superior to the savage, and his culture has continually improved his condition. But with woman the rule is reversed. Her squaw sister will endure effort, exposure, and hardship which would kill the white woman. (Ehrenreich and English 1978:114)

Habituation to simple labor clearly superseded any truly vocational goals—that is, training for employment—for Indian girls. In the 1920s, even school officials recognized the gap between vocationalism and mindless labor in their distinction between "vocational" and "prevocational." Most vocational girls worked in the sewing room and the laundry; a few were detailed to the kitchen, dining room, and dormitories. The drudgery in the kitchen and dining room was assigned to the prevocational girls, who could someday look forward to moving up in the ranks, promoted from scrubbing pots to darning socks by the bushelful (National Archives 1925a). Students spent half of each day on "vocational" or work details, sewing hundreds of shirts, darning thousands of socks, polishing miles of corridor. The work detail seems a too literal translation of Bourdieu and Passeron's "pedagogic work (PW), a process of inculcation which must last long enough to produce a durable training, i.e., a habitus" (1977:31). In reality, little distinguished prevocational from vocational, itself a misnomer, since the girls' productive labor was more important to sustaining the institution than as training for employment.

We know that a few schools placed Indian women in domestic service, but documentation in this area is incomplete. In his history of the Phoenix Indian School, Trennert (1988) indicates that the school provided domestic help to the white households of Phoenix over the years. Chilocco placed some girls in summer domestic service in the early 1900s but never mobilized this placement widely or permanently. For young Indian women graduating from boarding schools, the proper place in domestic service was often available only in the very institutions that had trained them. In 1917 the Matrons' Report on the skills and aptitudes of the senior girls graduating from Chilocco ranked their character and disposition as well as their competencies in sewing, washing, ironing, and housekeeping, all with an eye toward their suitability as assistant matrons (National Archives 1917b). Only a fraction of the female graduates could expect employment in the Indian Service. An economic rationale of placing Indian women in domestic employment does not account for the centrality of domesticity training in their education.

An ideological rationale more fully accounts for domesticity training. In this regard, there are striking parallels with the education for black girls in South Africa in this century, an education that Gaitskill (1988) characterizes as "vocational, domestic, and subservient," enforced in boarding schools that were "very much rival domestic establishment[s], giving intimate daily contact with alternative 'maternal' figures and Western cultural norms" (1988:153, 159). As in America, this was training in dispossession under the guise of domesticity, developing a habitus shaped by messages about subservience and one's proper place.

The struggle to reform the Indian home targeted the education of young women. They would serve as the matrons of allotment households,[9] promoting a Christian, civilized lifestyle and supporting their husbands in the difficult climb up the cultural evolutionary staircase (from hunter or pastoralist to farmer). The Victorian vision of Woman as Mother affirmed women's capacity to bear this burden, to influence society and shape the future by nurturing their children (Richards 1900). An epigraph by Helen Hunt prefaced a description of "Home Economics Class Instruction" at Chilocco: "A woman who creates and sustains a home, and under whose hand children grow up to be strong and pure men and women, is a Creator, second only to God" (U.S. Indian Service 1938:31). To create this godly creature, Indian schools had to convince or force Indian girls to renounce the teachings of their own mothers. Estelle Reel, superintendent of Indian schools from 1898 to 1910, encouraged the matrons to assure Indian girls who were being trained in housekeeping that "because our grandmothers did things in a certain way is no reason why we should do the same" (Cheney-Cowles Museum 1901:440).[10]

School training in acquiescence to federal authority was more important than the details of needlework, laundry, or food preparation. Chilocco's 1938 "Requirements for Graduation" ranked the acquisition of proper codes of behavior above academic or vocational skills. Students had to demonstrate vocational achievement, complete all projects, and remove all Fs and Incompletes from their academic records. However, doing so was not sufficient for graduation: "Inasmuch as a purpose of the school is to train Indian youth for citizenship, THE STUDENT WHOSE RECORD IS NOT SATISFACTORY AS TO PERSONAL CONDUCT WILL NOT BE RECOMMENDED FOR GRADUATION" (U.S. Indian Service 1938:47). This was a wonderfully flexible system for the exercise of power. Students who fulfilled all the requirements except "personal conduct" might not achieve the distinction of "graduating" from the school. Correspondingly, hard work and a good attitude were rewarded. Students with commendable conduct and "satisfactory skills in a chosen profession" earned a "special vocational certificate" without completing the "related [academic] requirements" (U.S. Indian Service 1938:47).

The Battleground of the Body

To construct the ideal Indian woman, educators had to teach Indian girls new identities, new skills and practices, new norms of appearance, and new physical mannerisms. Dormitory personnel, matrons and disciplinarians,[11] academic teachers and trades instructors, all enforced the rigid code of appearance for Chilocco students. Ellen, a Creek who entered the school in 1927 at age 13, recalls the impact of that scrutiny:

> I remember [two of the teachers], they were on the side of our matron when it came to strictness. In Mrs. S's class, sometimes the girls would keep their red sweaters on because it was cold. She wouldn't say anything, she would just stand there and look at them. They'd like to try and ignore her, avert their eyes, look at the ground. Finally she'd say, "Well, when these grandmas take their sweaters off, we can start class." That's what she did, she'd call us grandmas. I can still remember seeing [one of the boys] on his way to class in the morning, trying to button his shirt collar, button his cuffs, tie his shoes. Because she was his first class. All the boys would really button up, then after class they would unbutton their collars, roll up their cuffs, even roll up their pants if they thought they could get away with it.

Records and reminiscences attest that authority's gaze focused more intensely on the girls' appearance than the boys'. Boys were allowed more latitude in dress on campus and on their work details.[12] Official correspondence and inspection reviews mention girls' clothing more frequently and in more detail. The girls had to pass inspections to determine whether they had donned all the required undergarments. Clothing was a clearly marked terrain of power in the boarding school, especially in the girls' dormitories. And the surveillance of female students penetrated well beyond their classroom appearance. One alumna recalls the "blue bag" full of rags issued to girls during their menstrual periods. A matron dispensed the rags and in the process kept track of each girl's cycle. Hospital records from the 1920s corroborate this record keeping in the list of students hospitalized or treated: dysmenorrhea was one of the categories of cases catalogued each month (National Archives 1921).

Bureaucratic correspondence in the 1920s illustrates the discourse on girls' clothing and appearance. In 1923, Chilocco's Superintendent Merritt wrote to Washington to protest inspecting Field Supervisor Spalsbury's remarks that Chilocco girls had an "uncouth appearance" and that their hair styles should be improved. The superintendent noted that nearly half the girls had arrived that fall with short hair. He regretted to see the older girls' hair "bobbed in the latest fashion [as] Indians' hair is very straight and some of our girls look rather unattractive." He questioned the policy of total uniformity of appearance, and assured the head office that "most of the girls at Chilocco . . . give considerable attention to their personal appearance, and the matrons give the matter of supervising them along this line considerable attention" (National Archives 1923). The superintendent's weak plea for "some individuality" was not often repeated in his correspondence with Washington.

In 1925 Supervisor Spalsbury suggested the Indian Service discontinue the olive drab uniform for students (since they were not likely to require the camouflage, in his words, unless the Service had plans for military action) and replace it with a neat gray trimmed in black braid. The superintendent was noncommittal on this point, no doubt aware the Service was not so free with funds or uniforms, but he responded at length to the supervisor's comments on the girls' shoes. Spalsbury felt the high-cut, heavy shoes issued to all students, who called them "bullhides," ought to be relegated to the "realms of the departed": "They simply are not worn by well dressed girls or women any more. . . . [W]hy should we impose them on our Indian girls (and not our own daughters)?" (National Archives 1925b). The superintendent's reply to this enlightened concept of applying the standards of white society to Indian students was "I do not believe in adopting fads too quickly" (National Archives 1925c). He was willing to agree that a change in uniform dress was warranted: "Any legitimate means of overcoming the mental apathy of the average Indian pupil, which is undoubtedly the greatest obstacle to his intellectual advancement, is worthy of serious consideration" (National Archives 1925d).

The superintendent explicitly tied the "mental apathy" of Indian students to their physical bodies: spruce up the uniform and you will stimulate the mind. Booker T. Washington had earlier remarked on the accommodation of racially inferior intellects to retarded physical development. Estelle Reel spelled it out in excruciating detail in an interview at the turn of the century:

> Allowing for exceptional cases, the Indian child is of lower physical organization than the white child of corresponding age. His forearms are smaller and his fingers and hands less flexible; the very structure of his bones and muscles will not permit so wide a variety of manual movements as are customary among Caucasian children, and his very instincts and modes of thought are adjusted to this imperfect manual development. . . . In short, the Indian instincts and nerves and muscles and bones are adjusted one to another, and all to the habits of the race for uncounted generations, and his offspring cannot be taught like the children of the white man until they are taught to do like them. (Cheney-Cowles Museum 1900)

Reel developed a standard curriculum for the federal schools, one that privileged manual training, teaching to do. She stressed the moral development inherent in manual development, but never as strongly for boys as for girls, the future uplifters of Indian home life. Reel's encouragement of native cultural crafts, especially basketmaking and Navajo rug weaving, was one interesting offshoot of her educational philosophy:

> All civilized nations have obtained their culture through the work of the hand assisting the development of the brain. Basketry, weaving, netting, and sewing were the steps in culture taken by primitive people. A knowledge of sewing means a support for many. Skill in the art of using the needle is important to every woman and girl as an aid to domestic neatness and economy and as a help to profitable occupation. (Cheney-Cowles Museum 1901:450–51)

Civilized women, particularly boarding-school matrons and teachers, had to take the lead in inculcating the correct handling of a needle and thread: "Never permit sewing without a thimble. Do not let children make knots in thread. See to it that all sit in an erect position, never resting any part of the arm on the desk" (Cheney-Cowles Museum 1901:452). Female students were drilled not only in the correct motion of the arm in taking stitches but also in marching, breathing, calisthenics, and games. Reel rarely mentioned all the mental and moral benefits accruing from manual labor when she referred to boys' vocational training. In the same report where she spelled out correct sewing form and posture she simply mentioned that shoemaking was good training for boys so that they might be able to do it "for themselves . . . and, if any desire, to follow the trade" (Cheney-Cowles Museum 1901:453).

Whether or not school employees consciously recognized the link between enforced uniformity, regimentation of the body, and subservience training, some students did recognize it. Florence, a Choctaw who entered the ninth grade in 1933, makes the connection explicit in her remarks:

> One thing I think that figures in prominently into this lack of warmth I'm talking about is the loss of individuality that comes from those damn GI issue striped denim drawers, gray sweaters. If we were going to have sweaters, why did they all have to be gray? In that cold climate, you know? . . . That was just some of the kind of things that—well, it was just, I guess some kind of a feeling, at least on the people who controlled the purse . . . to encourage submission. I don't know what else you'd call it.[13]

Alumni Voices

Like feminist critiques of the Victorian cult of domesticity, theories of power relations contextualize the practices of surveillance, domination, and subservience training common to federal off-reservation boarding schools early in this century. These practices stemmed from racist notions of mental and physical capabilities and from historical and social forces that militated against Indian assimilation. What this framework does not directly illuminate is how students responded and resisted. Critical theory contextualizes the institution within the society that produced it, but it has not as yet contextualized the Indian student in her struggle for survival within the institution.

We are at the juncture where Foucault calls for more than the "affirmation, pure and simple, of a 'struggle' ": we must establish "concretely . . . who is engaged in struggle, what the struggle is about, and how, where, by what means and according to what rationality it evolves" (1980:164). The view of boarding-school life presented so far—constructed from documentary evidence, supplemented by student narrative—is external and incomplete. Additional narrative evidence from Chilocco alumni reveals that federal practice did not succeed in training Indian students to be subservient and that those students successfully resisted both policy and practice. The external perspective of critical theory and the internal perspective of boarding-school alumni converge at the image of the regimented female body. Both see this image as a crucial embodiment of the net of power relations binding students and staff. Many alumni today encapsulate the essence of student creativity in what I call the "bloomer story," a tale of ladies' underwear. The bloomer stories that follow enrich our understanding of student activity and control in the boarding school. Juxtaposed with the analytic framework provided earlier, the bloomer story illuminates the student side of the delicate balance of forces within the school. As we shall see, it weaves together a number of themes: government-issue uniformity versus the individuality of home clothes; adult supervision and inspection versus student sleight of hand; strict segregation by gender versus the limited socializing possible at the dances. We can begin with the issue of home clothes.

By the early 1930s, the government was issuing to each female student a gingham school dress, a blue chambray work dress, and a "dress uniform" (for church and visits to town) composed of a white middy blouse, navy skirt, navy cape, and beret. Regulations had eased to allow black oxford shoes for the girls. The boys could wear sneakers if they purchased them, but the girls were not allowed such leeway. One alumna recalls that her father mailed her black oxfords adorned with a small patch of gray; the matron refused to let her wear them. Another alumna's shoes were judged unfit because they had two eyelets, not the regulation four. The girls' undergarments were also government issue: black cotton stockings, woolen long underwear for winter, and bloomers, black sateen for the older girls, gingham or hickory for the younger (National Archives 1926).

When students arrived at Chilocco in the 1920s and 1930s, the boys stored their own clothes in footlockers by their beds, and the girls' "home clothes" were locked in trunk rooms, open only a few hours each month. Girls wore their own dresses to the Saturday night dances, which alternated weekly with movie showings. The strictly chaperoned dances were one of the few times when girls and boys mingled.

According to Vivian, a Choctaw who came to Chilocco in 1929 at age twelve:

> [At the dance] you had to be held a certain way; the boys weren't allowed to put their hands high on you, or, you know, low on you because you would be punished for that when you got back to the building. Oh, it was something else [laughter].

The omnipresent eye of authority had to mitigate the seeming freedom of "home clothes," especially in what authority perceived to be one of the most dangerous moments of school life: the sexually charged mingling of adolescent boys and girls. Rules dictated that the bulky, shapeless GI bloomers be worn under one's home clothes, and matrons inspected the girls as they lined up in the dorms before leaving for the dances.

Barbara, a Cherokee who entered the school in 1928 at age twelve, recalls:

> And I remember this old matron, one time, said something about us arousing the boys' passions [laughter]; that's the reason we had to wear these bloomers. . . . And I didn't know, I hardly knew what that meant, you know. Really, it seems strange this day and age, but the farthest thing from our minds was sex, I guess. And yet the matrons seemed to be concerned.

Educators attempted complete surveillance of and control over female Indian bodies within the schools, but students successfully exercised their own power in their resistance. The battleground of the body was strewn with the corpses of those bloomers, as the following stories attest. The narrators stress the solidarity of same-gender peer groups whose members shared risks and protected one another. The following versions of the bloomer story detail the students' cooperation, ingenuity, and flamboyant display of individual identity. The first is narrated by a male alumnus, the second by a woman.

Coleman, a Delaware/Isleta who came to the school at age fourteen, recollects the story thus:

> My sister came up the next year [1938] after I was there. She tells a story. . . . Those girls had what they called GI bloomers, and [their matron] there in Home 3,[14] she was a real strict disciplinarian. . . . [She] would require those girls to wear their bloomers to the dance. She had inspection; as they went out they had to pull their dress up to show that they had their bloomers on. And so then they just march[ed] on out. Well, almost all those girls had their own panties, and it just irked them to wear those bloomers. So they came up with a deal where they would take their bloomers off and hide'em behind the hedge there, and then when they'd come back [to the dorm], why they'd get their bloomers and go on in. . . . A whole bunch of the girls were doing that, but we didn't know about it, the boys didn't know about it. So those boys that had to work at the bakery had to go to work at five o'clock in the morning. The night attendant would always wake'em up so they would go to work. One time, Saturday night, they had a dance. And these girls pulled their little trick. Then it started raining. They took the school bus down there and brought the girls back so they wouldn't get wet. They just ran the school bus up, girls would get in, and it would take'em right on up to the dorm. And just in a row, they'd go right on into their room. It was raining pretty hard. When those girls got up the next morning, the boys that had gone to the bakery had found some of those pants, those bloomers, and they stacked'em all out front, on the hedge, on the edge of the trees, and [they were] just out there so pretty with all those girls' names on 'em [laughter]. They had to write their names across the seat. We teased those girls for so long after that.

Marian, a Creek who matriculated in 1934 at age eighteen, tells a slightly different tale:

> I remember, though, we got smart, about the time we were seniors. We had those old black satin bloomers, so we cut the leg off of one, we had elastic up here and we had elastic down there [at the top and bottom of the cut-off bloomer leg.] [During inspection] we had to pull up our dress and show we had them bloomers on. OK, we got outside, we pulled that leg off, and we put it in the bushes [laughter]. The wind came up—and we had our names in those legs, and you see, the first bunch back to the dorm would clean the bushes and take the legs upstairs. And then after lights out, they'd slip down and deliver your leg to you. Oh God, my leg wasn't there. I waited all night for the leg, and the leg never did come, and the next morning going to breakfast, there it was, laying right on [the superintendent's] walk. I got that old leg and put it in my pocket, and I tell you it was bulging out like this.

Settings and details of the bloomer story shift according to narrator and venue of performance. Girls outwitted the matron on the way to the dance, before playing basketball, on the way to classes, or on the way to church. Winter versions of the story feature the shapeless woolen long-underwear legs as the distasteful item of required apparel. The variety of permutations and the relish with which they are told signal the special status of the story among alumni today. Nearly all versions entail the girls' ruse being uncovered, by the wind or by a matron's sharp eye, but alumni tell the story as part of their Chilocco experience, an experience that cumulatively covers nearly twenty years. Some Chilocco girls at some point in time undeniably pulled a fast one on their matron or basketball coach or commanding officer, but the bloomer story today is more myth than personal anecdote. It sums up critical realities, tensions, and meanings that permeated student life at Chilocco.

The story reveals the complex network of bonds and divisions that simultaneously bound and segmented the large student population. Girls united in groups on the basis of dorm-room associations, shared hometowns, native language ties, company or work-detail assignments, or kindred "personalities." Loyalty to the group reigned supreme in the student code of ethical behavior, and groups worked to cooperate—signaling the approach of a matron after lights out, for example—or to compete, as when companies vied to win the special dinner prepared for the best dress-parade drill. Boys' gangs ruled Chilocco's 8,000 acres of fields and groves, as tribal heritage brought boys together and the distinction between mixed blood and fullblood sorted them out. The bloomer story reveals the solidarity within groups, as girls cooperated to outwit the matrons. It also illustrates another important practice, "trixing," which both crossed and highlighted group boundaries. Coleman recalls how the boys "teased" the girls about their errant bloomers. "Trixing" was student slang for elaborate joking and prank playing, within and between groups or gangs and between students and staff.

The bloomer story has a symbolic resonance for alumni because it marks a milestone in their memories, one of student triumph over a uniform(ed) existence. Whether it features a full bloomer or only a leg, the story drapes cloth scraps across the campus for all to see, incontrovertible evidence of the revolt organized in nameless, numbered "Homes." It was not a hidden insurrection, carried out by boys in dark of night in a field or grove where the adults' gaze never penetrated. The plan was executed under the noses of the matrons by the strictly guarded girls. The bloomers were not anonymous emblems, either; they bore the full and public script of their owners' names. A blowing rain from the north washed away the anonymity of institutional life for a moment, leaving in its place a gray or black sateen proclamation of independence, resting on the threshold of the superintendent's door.

There is nothing inherently "Indian" about the bloomer story; the isolated narrative does not reveal its origin in any striking way. Its status as a symbolic marker of Indian identity for Chilocco alumni today arises from the conditions of its creation and the contemporary contexts of its narration. The Chilocco Indian Agricultural School was not an educational institution created by or created to serve (in a productive sense) Indian people. Like other federal and mission boarding schools, it was created to destroy Indian tribal communities and erase individual Indian identities. It was created by, controlled by, and directed by non-Indian people. It was, however, an institution inhabited by Indian students, who created its everyday life. Every student knew Chilocco was an Indian school. They also knew, or soon learned, that as Indian students they were subordinate to non-Indian authority. The petty, tyrannical details of a regimented life were linked at every turn to their identity as Indians. It follows logically that the details of their resistance to regimentation are also now linked inextricably with their identity today, as alumni of an Indian school. Personal reminiscences and shared stories, such as the bloomer story, are powerful symbols of identity today, not because they convey some "Indian" cultural content (in some externally defined "ethnographic" sense), but because they are the chronicles of Indian experiences told by Indian people.

Conclusion

> [W]omen's actions were shaped by considerations of power, even though mediated through the language of morality. [Epstein 1981:149–150]

In the quest to "individualize" the tribal consciousness, federal Indian schools pressed Indian students into a strictly homogeneous mold of dress, appearance, and (limited) educational opportunity. The seeming contradiction is no real paradox: federal boarding schools did not train Indian youth to assimilate into the American "melting pot" but trained them to adopt the work discipline of the Protestant ethic and to accept their proper place in society as a marginal class. Indians were not being welcomed into American society, they were being systematically divested of their lands and other bases of an independent life.

As tribal sovereignty was attacked on the political front, personal individuality was attacked in the dormitory and classroom. Boarding schools attempted to divest Indian children of independence in an institutional setting. "Proper" training for young Indian women, and the emphasis on "proper" clothes for boarding school girls, exemplified the federal practice of organizing the obedient individual, while federal policy aimed to disorganize the sovereign tribe. Federal "vocational" and domestic education for the Indian woman was an exercise in power, a reconstruction of her very body, appearance, manners, skills, and habits. This exercise of power was mediated as well as cloaked by the language of morality, which also cloaked the basic dispossession of people from their tribal identities and lands. Federal educators hoped to manufacture civilized and obedient souls in civilized and obedient bodies, uniformly garbed in olive drab or snappy gray.

The federal government has not completely alienated Indian people from the land, or the Indian woman from herself, but the forces and ideologies powering past attempts are still present in contemporary American life. The government's failure to achieve these goals is due in great part to Indian people's commitment to the idea of themselves. As individuals and as community members, Indian people cling stubbornly to making their own decisions, according to their own values. In the process, they have created spaces of resistance within the often oppressive domains of education, evangelism, employment, and federal paternalism.

Notes

Acknowledgments. The research described in this article was supported by National Institute of Mental Health National Research Service Award No. 1 F31 MH 09016-01, the Phillips Fund of the American Philosophical Society, the L. J. and Mary C. Skaggs Foundation, and the Institute for Ethnic Studies in the United States. The Chilocco Alumni Association generously assisted me in contacting and meeting alumni and former employees. I would like to thank Tom Biolsi, Stevan Harrell, Clara Sue Kidwell, Laura Newell, Lorna Rhodes, Bill Simmons, Julie Stein, and the *American Ethnologist's* anonymous reviewers for their insightful criticism of earlier drafts of the article.

1. Two excellent book-length memoirs of student experience are Francis LaFlesche's 1978[1900] classic account of Presbyterian mission life in the late 1800s and Johnston's 1988 memoir of Canadian Catholic residential education in the 1930s. Many Native American autobiographies refer to school life: see Dover (1978), Fredrickson (1989), Giago (1978), and Manitowabi (1970); for an annotated bibliography, see Brumble (1981). Film and radio productions include the 1989 Canadian movie *Where the Spirit Lives* and the 1991 Minnesota Public Radio show "Learning the White People Way" (Amazing Spirit Productions 1989; Minnesota Public Radio 1991). U.S. boarding-school education is discussed in generic terms in overviews by Szasz (1974) and Reyhner and Eder (1989); Canadian residential schools are discussed by Bar-

man, Hébert, and McCaskill (1986, 1987). Prucha (1979) and Berkhofer (1965) offer excellent documentary overviews of mission education. Trennert (1988), Hultgren and Molin (1989), Williams and Meredith (1980), and King (1967) describe individual schools. Many school histories are unpublished theses (see Adams 1975). For detailed analyses of boarding-school life that include a student/alumni perspective, see Haig-Brown (1988), McBeth (1983), and Lomawaima (1987a, 1987b); for a compilation of narratives, see Rygg (1977).

2. See Ahern (1983) and Littlefield (1989) for the application of theories of resistance to American Indian boarding schools. Indian resistance to the creative use of power in the schools is clearly akin to resistance movements elsewhere. Cf. Dhan-Gadi aboriginal resistance to the Australian government's "pedagogic intervention" (Morris 1989:112); Tshidi resistance to historical and economic marginalization (Comaroff 1985); and Indian caste resistance to colonial British medical regulations concerning plague (Arnold 1988).

3. The narratives in this article were collected during field research (fall 1983–spring 1984) in eastern Oklahoma and central Kansas. They are excerpted from interviews conducted with sixty-five Chilocco alumni and former staff members. All but two of the alumni attended Chilocco between 1920 and 1940, and seven later had careers in the Education Division of the Indian Service. Names have been altered to protect the narrators' privacy. (More comprehensive transcripts may be found in Lomawaima 1987b.)

4. Comparisons of federal Indian schools with boarding schools for Euro-American elites have not been undertaken, so far as I know. I imagine that the similarities, such as military organization, are superficial and that student experiences are quite different. It seems likely that institutions devoted to cultural obliteration and transformation would be different from institutions devoted to cultural reproduction and the training of elites. Bourdieu and Passeron discuss the difference between schools aimed at radical conversion (with their propensity for regulation and military drill) and schools, such as military prep schools, that train the children of the elites:

> At the other extreme, the traditional institutions for young ladies of good family represent the paradigmatic form of all institutions which, thanks to the mechanisms of selection and self-selection, address themselves exclusively to agents endowed with a habitus as little different as possible from the one to be produced, and can therefore content themselves with ostentatiously organizing all the appearances of really effective training (e.g., the Ecole Nationale d'Administration). (1977:44)

5. Pratt's memoirs have been edited by Robert Utley and published by Yale University Press (Utley 1964). Foucault proposes the creation of the ideal body of the soldier as a model for "essential techniques" to produce docile bodies: the attention to detail, "the meticulousness of regulations, the fussiness of . . . inspections," the ranked distribution of individuals in space, and the control of activity according to a strict timetable (1979:139–40). Foucault's description of the earliest model for modern prisons, the Rasphuis (which opened in Amsterdam in 1596), might serve to describe the Indian boarding schools:

> A strict timetable, a system of prohibitions and obligations, continual supervision, exhortations, religious readings, a whole complex of methods "to draw towards good" and "to turn away from evil" held the prisoners in its grip from day to day. (1979:120–21)

6. Throughout Carlisle's existence (1879–1918) these pictorial records of Indian transformation were published widely in popular American journals, magazines, and newsletters (see, for example, Anonymous 1881).

7. Ideals of feminine domesticity also filtered through white women's reform associations, which profoundly affected federal Indian policy (Mathes 1990).

8. The philosophy of European societies' carceral systems aimed to produce "strong, skilled agricultural workers. . . . [In such] work, provided it is technically supervised, submis-

sive subjects are produced and a dependable body of knowledge is built up about them" (Foucault 1979:295).

9. Congress passed the General Allotment Act (also known as the Dawes Act) in 1887 to break up tribal, communal property ownership. Reservations were subdivided into individual "allotments" of 80 to 160 acres that were scheduled to convert to private property after a trust period had elapsed. See Adams (1988) for a discussion of the links between educational and allotment policies.

10. In November 1991, the Estelle Reel Collection was transferred from the Museum of Native American Cultures in Spokane, Washington, to the Cheney-Cowles Museum, also in Spokane. The filing system referenced here was that used at the Museum of Native American Cultures before the move.

11. In the 1930s the term used to refer to male attendants in the boys' dormitories was changed from "disciplinarian" to "advisor," but the personnel did not change.

12. The segregation of boys' from girls' experiences at the schools is too extensive to explore fully here. It was a practical consideration of the boys' trades (cattle raising, farming, carpentry, printing, and so on) that the boys ranged over the campus, doing much of the work necessary to support the institution. They were not under constant adult supervision, as were the girls, who worked in the confines of the dormitory, sewing rooms, laundry, and kitchen.

13. The discrepancy between this and an earlier account, which alludes to red sweaters, reflects the differing attendance dates of the narrators.

14. The dormitories were numbered, not named. Homes 3, 4, and 5 housed the girls, from youngest to oldest. Home 6 housed the senior boys; Homes 1 and 2, the younger ones. Constructed in 1885, Home 2 had originally housed all students and staff.

References Cited

Adams, David Wallace

1975 The Federal Indian Boarding School: A Study of Environment and Response, 1879–1918. Ed.D. dissertation, School of Education, Indiana University.

1977 Education in Hues: Red and Black at Hampton Institute, 1878–1893. South Atlantic Quarterly 76(2):159–176.

1988 Fundamental Considerations: The Deep Meaning of Native American Schooling, 1880–1900. Harvard Educational Review 58(1):1–28.

Ahern, Wilbert H.

1983 "The Returned Indians": Hampton Institute and Its Indian Alumni, 1879–1893. Journal of Ethnic Studies 10(4):101–124.

Amazing Spirit Productions

1989 Where the Spirit Lives. B. Pittman, director. K. R. Leckie, screenwriter. Canada: Amazing Spirit Productions.

Anderson, James D.

1978 The Hampton Model of Normal School Industrial Education, 1868–1900. *In* New Perspectives on Black Educational History. V. Franklin and J. Anderson, eds. Pp. 61–96. Lexington, MA: G. K. Hall.

Anonymous
1881 Indian Education at Hampton and Carlisle. Harper's New Monthly Magazine 42(371):659–675.

Arnold, David
1988 Touching the Body: Perspectives on the Indian Plague, 1896–1900. *In* Selected Subaltern Studies. R. Guha and G. C. Spivak, eds. Pp. 391–426. New York: Oxford University Press.

Barman, Jean, Yvonne Hébert, and Don McCaskill, eds.
1986 Indian Education in Canada. Vol. 1: The Legacy. Vancouver: University of British Columbia Press.
1987 Indian Education in Canada. Vol. 2: The Challenge. Vancouver: University of British Columbia Press.

Beecher, Catharine E.
1846 A Treatise on Domestic Economy, for the Use of Young Ladies at Home, and at School. New York: Harper and Brothers.

Berkhofer, Robert F., Jr.
1965 Salvation and the Savage: An Analysis of Protestant Missions and American Indian Response, 1787–1862. Lexington: University of Kentucky Press.

Bourdieu, Pierre, and Jean-Claude Passeron
1977 Reproduction in Education, Society and Culture. Beverly Hills, CA: Sage Publications.

Brumble, H. David
1981 An Annotated Bibliography of American Indians and Eskimo Autobiographies. Lincoln: University of Nebraska Press.

Cheney-Cowles Museum
1900 Teaching Little Reds. Journal (Kansas City, MO), 21 October. Estelle Reel Collection, folder 2, envelope 2, 8/6/00–9/29/00, item 17. Cheney-Cowles Museum, Spokane, WA.
1901 Report of the Superintendent of Indian Schools. 20 October. Estelle Reel Collection. Cheney-Cowles Museum, Spokane, WA.

Comaroff, Jean
1985 Body of Power; Spirit of Resistance. Chicago: University of Chicago Press.

Dover, Harriette Shelton
1978 Memories of a Tulalip Girlhood: Vibrations. Everett, WA: Everett Community College.

Ehrenreich, Barbara, and Deirdre English
1978 For Her Own Good: 150 Years of the Experts' Advice to Women. New York: Anchor Press/Doubleday.

Epstein, Barbara Leslie
1981 The Politics of Domesticity: Women, Evangelism, and Temperance in Nineteenth-Century America. Middletown, CT: Wesleyan University Press.

Foucault, Michel

1979 Discipline and Punish: The Birth of the Prison. A. Sheridan, trans. New York: Vintage Books.

1980 Power/Knowledge: Selected Interviews and Other Writings, 1972–1977. C. Gordon, ed. New York: Pantheon Books.

Fredrickson, Vera Mae, ed.

1989 School Days in Northern California: The Accounts of Six Pomo Women. News from Native California 4(1):40–45.

Gaitskill, Deborah

1988 Race, Gender, and Imperialism: A Century of Black Girls' Education in South Africa. *In* "Benefits Bestowed"? Education and British Imperialism. J. A. Mangan, ed. Pp. 150–173. Manchester, England: Manchester University Press.

Giago, Tim A., Jr.

1978 The Aboriginal Sin: Reflections on the Holy Rosary Mission School. San Francisco, CA: Indian Historian Press.

Haig-Brown, Celia

1988 Resistance and Renewal: Surviving the Indian Residential School. Vancouver: Tillacum Library.

Hultgren, Mary Lou, and P. Molin

1989 To Lead and to Serve: American Indian Education at Hampton Institute. Virginia Beach: Virginia Foundation for the Humanities and Public Policy.

Johnston, Basil H.

1988 Indian School Days. Norman: University of Oklahoma Press.

Kennedy, Edward M.

1969 Indian Education: A National Tragedy—A National Challenge. Report of the U.S. Senate Committee on Labor and Public Welfare, Special Subcommittee on Indian Education. Senate Report No. 91-501. Washington, DC: U.S. Government Printing Office.

King, A. Richard

1967 The School at Mopass: A Problem of Identity. Case Studies in Education and Culture. G. and L. Spindler, eds. New York: Holt, Rinehart, and Winston.

LaFlesche, Francis

1978[1900] The Middle Five. Lincoln: University of Nebraska Press.

Littlefield, Alice

1989 The B.I.A. Boarding School: Theories of Resistance and Social Reproduction. Humanity and Society 13(4):428–441.

Lomawaima, K. Tsianina

1987a Oral Histories from Chilocco Indian Agricultural School, 1920–1940. American Indian Quarterly 11(3):241–254.

1987b "They Called It Prairie Light": Oral Histories from Chilocco Indian Agricultural School, 1920–1940. Ph.D. dissertation, Anthropology Department, Stanford University.

Manitowabi, Edna
1970 An Ojibwa Girl in the City. This Magazine Is about Schools 4(4):8–24.

Mathes, Valerie Sherer
1990 Nineteenth-Century Women and Reform: The Women's National Indian Association. American Indian Quarterly 14(1):1–18.

McBeth, Sally
1983 Ethnic Identity and the Boarding School Experience. Washington, DC: University Press of America.

Meriam, Lewis, et al.
1928 The Problem of Indian Administration. Washington, DC: Institute for Government Research.

Millett, Kate
1970 The Debate over Women: Ruskin versus Mill. Victorian Studies 14(1):63–87.

Minnesota Public Radio
1991 Learning the White People Way. Aired 15 May. St. Paul: Minnesota Public Radio.

Morris, Barry
1989 Domesticating Resistance: The Dhan-Gadi Aborigines and the Australian State. Oxford: Berg Publishers.

National Archives
1917a Letter from Commissioner Cato Sells. 28 April. Record group 75, entry 1, box 2, unlabeled folder.

National Archives Regional Center, Fort Worth, TX.
1917b Matrons' Report on Graduating Senior Girls. Record group 75, entry 1, box 3, folder Dec. 1917-May 1918. National Archives Regional Center, Fort Worth, TX.
1921 Letter from Supt. to Commissioner. 22 October. Record group 75, entry 4, box 12, folder 1922. National Archives Regional Center, Fort Worth, TX.
1923 Reply to Supervisor Spalsbury's Report of Nov. 2. 27 November. Record group 75, entry 4, box 14. National Archives Regional Center, Fort Worth, TX.
1925a Letter from Supt. to Supervisor for Five Civilized Tribes, Muskogee. 17 November. Record group 75, entry 4, box 16, folder 1925. National Archives Regional Center, Fort Worth, TX.
1925b Report from Inspector Spalsbury to Commissioner. 24 October. Record group 75, entry 4, box 12, unlabeled folder. National Archives Regional Center, Fort Worth, TX.
1925c Letter from Supt. Merritt to Commissioner. 12 December. Record group 75, entry 4, box 16, folder 1925. National Archives Regional Center, Fort Worth, TX.
1925d Superintendent Merritt's Reply to Supervisor Groves' Report. 11 November. Record group 75, entry 4, box 16, folder 1925. National Archives Regional Center, Fort Worth, TX.

1926 Edna Groves' Report. Record group 75, entry 4, box 16, folder 1926. National Archives Regional Center, Fort Worth, TX.

1931 Annual Report of the Chilocco Indian School. Record group 75, entry 8, box 2. National Archives Regional Center, Fort Worth, TX.

Prucha, Francis P.

1979 The Churches and the Indian Schools. Lincoln: University of Nebraska Press.

Reyhner, Jon, and Jeanne Eder

1989 A History of Indian Education. Billings: Eastern Montana College.

Richards, Josephine E.

1900 The Training of the Indian Girl as the Uplifter of the Home. *In* Journal of Proceedings and Addresses, National Education Association, 39th Annual Meeting, 7–13 July 1900. Pp. 701–705. Chicago: University of Chicago Press.

Rygg, Larry

1977 The Continuation of Upper-Class Snohomish Coast Salish Attitudes and Deportment as Seen through the Life History of a Snohomish Coast Salish Woman. M. A. thesis, Western Washington University.

Sheridan, Alan

1980 Michel Foucault: The Will to Truth. London: Tavistock Publications.

Szasz, Margaret

1974 Education and the American Indian. Albuquerque: University of New Mexico Press.

Trennert, Robert

1988 The Phoenix Indian School: Forced Assimilation in Arizona. Norman: University of Oklahoma Press.

U.S. Indian Service

1938 Chilocco: School of Opportunity for Indian Youth. Chilocco, OK: Chilocco Indian School Press.

Utley, Robert, ed.

1964 Battlefield and Classroom: Four Decades with the American Indian: The Memoirs of Richard H. Pratt. New Haven, CT: Yale University Press.

Williams, John, and Howard L. Meredith

1980 Bacone Indian University. Oklahoma City, OK: Western Heritage Books.

8

Nymphomania

The Historical Construction of Female Sexuality

Carol Groneman

Introduction

THE TERM *nymphomania* resonates with a sense of the insatiable sexuality of women, devouring, depraved, diseased. It conjures up an aggressively sexual female who both terrifies and titillates men. Surrounded by myth, hyperbole, and fantasy, the twentieth-century notion of a nymphomaniac is embedded in the popular culture: referred to in films, novels, music videos, and sex-addiction manuals, as well as in locker rooms and boardrooms. In the nineteenth century, however, nymphomania was believed to be a specific organic disease, classifiable, with an assumed set of symptoms, causes, and treatments. Like alcoholism, kleptomania, and pyromania—diseases that were identified in the mid-nineteenth century—a diagnosis of nymphomania was based on exhibited behavior. "Excessive" female sexual desire is, however, a much more ambiguous concept than habitual drunkenness, shoplifting, or setting fires. Consider the following cases of nymphomania diagnosed in the second half of the nineteenth century.

"Mrs. B.," age twenty-four and married to a much older man, sought the help of Dr. Horatio Storer, a gynecologist and the future president of the American Medical Association, because of lascivious dreams. He reported that she "can hardly meet or converse with a gentleman but that the next night she fancies she has intercourse with him, . . . though thinks she would at once repel an improper advance on the part of any man" (Storer 1856, 384). In fact, she "enjoys intercourse greatly" (with her husband) and has had sex with him nightly for the seven years of their marriage. The husband "has of late complained that he found physical obstruction to intercourse on her part, though she thinks it rather an increasing failure by him in erection" (Storer 1856, 384). In this "Case of Nymphomania," Storer directed Mrs. B. to separate temporarily from her husband as well as to restrict her intake of meat and abstain from brandy and all stimulants to lessen her sexual desire, to replace the feather mattress and pillows with ones made of hair to limit the sensual quality of her sleep, and to take cold enemas and sponge baths and swab her vagina with borax solution to cool her passions. "If she continued in her present habits of indulgence," Storer argued, "it would probably become necessary to send her to an asylum" (Storer 1856, 385). At the time he presented the case before the Boston Society for Medical Observation in 1856, the woman's husband was still absent and her lascivious dreams had not occurred for several days. The doctor was "hopeful as regards the mental symptoms, which, however, will for some time require decided enforcement of very strict laws" (Storer 1856, 386).

In another case, the mother of a seventeen-year-old girl contacted Dr. John Tompkins Walton in 1856 because the girl, Catherine, was having "a fit." "This paroxysm" according to Walton, "was peculiar and specific . . . in the lascivious leer of her eye and lips, the contortions of her mouth and tongue, the insanity of lust which disfigured [her], . . . as well as in the positions she assumed and the movements which could not be restrained" (Walton 1857, 47). He judged her to be "in a condition of ungovernable sexual excitement" and was convinced that the primary cause of the disease was seated in her "animal organization," which he deduced from her small eyes, large, broad nose and chin, thick lips, and the disproportionate size of the posterior portion of her head (Walton 1857, 47). (An enlarged cerebellum was believed by some doctors to indicate increased "amativeness," or sexual desire.) Moralism and science—and class bias—combined in Walton's belief that Catherine was infected by "the exposure and contagion incident to several families living in one house, with a hydrant and watercloset shared by all the court, and [by] the immorality of the youths who lounged about the place" (Walton 1857, 48). Ultimately, the girl admitted that she was a "wanton" and that her sexual appetite was insatiable. Walton rendered her "emasculate for a time" (although he does not describe his method), prescribed a vegetable diet, various drugs, cold hip baths, and leeches to the perineum.[1]

The case description continued with the cryptic statement that Catherine was later intercepted *in coitu;* paroxysms and additional treatment followed. Although the doctor assured her that he would "render her sexually fit to assume the duties of a wife whenever such services were needed," Catherine denounced him more than once during this time for "having destroyed her virility" (Walton 1857, 50). Seven months after the beginning of treatment, Dr. Walton concluded that the girl "has no inclination to resume her old habits, and renew the disease" (Walton 1857, 50).

Yet another case of nymphomania, reported in 1895, was that of "Mrs. L.," thirty-five years old, married with three children. Reflecting the influence of Richard von Krafft-Ebing's ([1886] 1965) major work on the classification of psychopathology published in 1886, Dr. L. M. Phillips of Penn Yann, New York, diagnosed Mrs. L.'s problem as "acquired anaesthesia sexualis episodiac . . . [seemingly] . . . paranoia erotisa episodiac manifesting itself as nymphomania" (1895, 469). It seems that Mrs. L. attended a New York theater party where a fashionable *tableau vivant* took place in which women posed seminude as living statues. Filled with disgust but also fascinated, Mrs. L. returned repeatedly that night to the room where the living statues displayed themselves. Following this experience, she lost interest in sex with her husband for two years, then recovered it for eighteen months, during which time "it burned with such intensity that it very nearly wrecked the physical well-being" of the couple (Phillips 1895, 468). She repeated these episodes of asexuality and "sexual pyrotechnics" over the ensuing six years before seeking the doctor's help. Phillips found no physical abnormalities—describing Mrs. L. as a "perfect woman"—and instead relied on a psychopathological diagnosis in which he argued that the psychosexual sphere of the brain and spinal cord had been indelibly imprinted with a hypnotic suggestive impression. In effect, Mrs. L. was compelled by this image to return to the attractive, yet repellent, scene, thus inducing nymphomania. Phillips did not describe the treatment he provided Mrs. L. but did state that she would have developed brain disease if she had not sought his help (Phillips 1895, 470). While his diagnosis foreshadowed the twentieth-century shift from a strictly organic to a more psychological explanation of nymphomania, it also hearkened back to the Renaissance belief that madness could occur when an impression of the unattainable beloved became seared upon the brain.[2]

This amazing confusion about the nature of nymphomania can be found throughout more than one hundred case studies—ranging from short references about a particular patient to full-scale examinations of specific instances of the disease—published in American and European medical journals and texts from the late eighteenth to the early twentieth century.[3] Nymphomania is variously described as too much coitus (either wanting it or having it), too much desire, and too much masturbation.[4] Simultaneously, it was seen as a symptom, a cause, and a disease in its own right. Its etiologies, symptoms, and treatments often overlapped with those of erotomania, hysteria, hystero-epilepsy, and ovariomania, despite doctors' attempts to classify each as a distinct "disease." For example, while symptoms such as convulsions, the appearance of strangulation, paralysis, and blindness were more likely to be associated with hysteria, a patient exhibiting these symptoms as well as "excessive" sexual desire or activity might have been diagnosed as a nymphomaniac subject to hysterical attacks or a hysteric with nymphomaniacal manifestations.

Indeed, nymphomania assumed a myriad of forms in these nineteenth-century medical reports, including a woman's desire for gynecological examinations, her introducing pins and other foreign objects into her urethra, vagina, or uterus, and her orgasm at the mere sight of a man. Nymphomania was also diagnosed by Krafft-Ebing in the case of a mother's incestuous desire for her son, while Chicago neurologist James G. Kiernan diagnosed nymphomania in cases of three schoolgirls who masturbated together and two women who lived together as "man and wife" (Krafft-Ebing [1886] 1965, 502; Kiernan 1891, 202). Cases were reported of puerperal nymphomania (relating to or occurring during childbirth), malarial nymphomania, mild or true nymphomania, homosexual nymphomania, platonic nymphomania, and nymphomania brought on by pulmonary consumption and by opium.[5] One doctor claimed that women with blond hair between the ages of sixteen and twenty-five were the most likely candidates (Howe 1883, 108), while others saw it as a disease of widows, virgins, or pubescent adolescents.

To further complicate the study of this disease, certain kinds of behavior were labeled nymphomania that today would be associated with psychosis, such as incessant and uncontrolled masturbation, lewd and lascivious tearing of clothes, and public display of the genitals. In numerous medical reports, physicians described nymphomaniacs in hospitals and mental institutions who made indecent proposals to almost anyone and who masturbated and exposed themselves openly.[6]

At the same time, however, and sometimes even in the same case studies, nineteenth- and early-twentieth-century European and American doctors also diagnosed as nymphomaniacs women whose "symptoms" consisted of committing adultery, flirting, being divorced, or feeling more passionate than their husbands.[7] Physicians writing for a popular audience diagnosed nymphomania in those women who actively tried to attract men by wearing perfume, adorning themselves, or talking of marriage (see, e.g., Talmey [1904] 1912, 112).

In the late nineteenth century, therefore, even minor transgressions of the social strictures that defined "feminine" modesty could be classified as diseased. A convergence of several factors helps to explain this medicalization of female behavior. Starting in the late eighteenth century, woman's nature was increasingly defined as inextricably bound up with her reproductive organs. This supposedly objective, scientific "fact" created the new framework within which physicians and other authorities found justifications for the limitations of women's social and economic roles. It was thought only natural that women would and should find their fulfillment solely in taking on the roles of wives and mothers.[8]

But the changing realities of women's lives in the second half of the nineteenth century contradicted this formulation. Contrary to their presumed natural passivity, modesty, and domesticity, women were demanding greater access to education, engaging in public debate over issues of prostitution and women's rights, joining the workforce in growing numbers, marrying later—or not at all—and having fewer children. Although medical and other authorities hoped to define femaleness as fixed and static, it was in fact unstable and fluid.

This paradox was augmented by the contradictions implicit in the Victorian construction of female sexual desire. Women—that is to say white, middle-class women—were supposed to be naturally modest and sexually passive (although not passionless), awaiting the awakening of desire in response to the approaches of men.[9] And yet, sympathetic medical observers recognized the reality of female heterosexual desire, occasionally bemoaning the effects of the necessary strictures placed on young girls, unmarried women, and widows.

These tensions and contradictions are highlighted in the physicians' case studies of nymphomania. The concept of nymphomania constructs a female sexuality that is totally out of control, both literally and figuratively: out of the control of Mrs. B., Catherine, and Mrs. L.; out of the control of their husbands, mothers, and doctors; and out of the control of the "natural laws" that supposedly determined women's passive response to male desire. This disease—defined as the extreme end of the sexual spectrum—embodied Victorian fears of the dangers of even the smallest transgressions, particularly among middle-class women whose conventional roles as daughters, wives, and mothers were perceived as a necessary bastion against the uncertainties of a changing society.

Within this theoretical context, gynecology emerged as a medical specialty in the second half of the nineteenth century by focusing particular attention on the generative organs as the source of most women's diseases (Moscucci 1990, 1–6, 10–13).[10] In addition, an increasing medical interest in perversion and deviance and a growing fear by many in the medical profession that these abnormal behaviors and desires were hereditary and incurable led to attempts to organize, classify, and thus gain some control over a myriad of newly defined psychopathologies, including diseases such as nymphomania.

An examination of the medical history of nymphomania allows us to observe the tensions and contradictions inherent in nineteenth-century assumptions about female sexuality during a period of contention over the nature of femaleness. Furthermore, the behavior described in case studies of nymphomania—masturbation, lascivious dreams, lesbian relationships, sexual intercourse, putting objects in the vagina and urethra, clitoral orgasm—however mediated through the doctors' presentations, permits us to glimpse a range of erotic activity of Victorian women that has generally been hidden.

In trying to unravel this history, I will look briefly at earlier notions of nymphomania and examine what is new in the nineteenth century. In particular, I will focus on the shifts and confusions in the medical profession's conceptualization and treatment of this disease to reveal the tensions in the attitudes toward female sexual desire and the nature of female sexuality at the end of the nineteenth century. Finally, I will look at the major shifts in perception, diagnosis, and treatment that begin to occur in the late nineteenth and early twentieth centuries as nymphomania was transformed from a biological to a psychological disease.[11]

From *Furor Uterinus* to Nymphomania

Notions of insane love—accompanied by symptoms of uncontrolled sexuality and/or pining away for love—are as old as medical theory. In *On the Diseases of Young Women,* Hippocrates described the melancholy madness that could consume young girls and recommended marriage as the cure (Ferrand [1623] 1990, 264, 505). The second-century Greek physician Galen believed that uterine fury occurred particularly among young widows whose loss of sexual fulfillment could drive them to madness (Ferrand [1623] 1990, 174). Clinical observations of nymphomania, or *furor uterinus,* as it was more likely to be called until the seventeenth century, were discussed by medical theorists as early as the fifteenth century, and numerous medical school dissertations and scholarly texts examining the disease appeared in the sixteenth and seventeenth centuries. Cases of *furor uterinus* were reported in Italy, France, Spain, Portugal, Germany, and England during these centuries (Louyer-Villermay 1819; Diethelm 1971, 62, 139).[12]

One of the clearest early definitions of the disease can be found in the works of the Italian physician Girolamo Mercuriale, a major contributor to sixteenth-century gynecological studies. According to him, *furor uterinus* was an "immoderate burning in the genital area of the female, caused by the surging of hot vapor, bringing about an erection of the clitoris. Because of this burning sensation women were thought to be driven insane" (Ferrand [1623] 1990, 385–86). Other physicians, such as Pieter van Foreest, whose works were often quoted during the sixteenth century, pointed to the "corrupted imagination" and to the brain alone as the seat of the furor caused by insane love (Diethelm 1971, 63–64). Some writers, such as Felix Platter—in an influential early-seventeenth-century textbook in which he described the case of a matron "who was in every other way most honorable, but who invited by the basest words and gestures men and dogs to have intercourse with her"—attributed the cause of this mania to possession by the devil (Diethelm 1971, 51). Although mainly discredited, demonology continued to be a factor for a few medical theorists even into the eighteenth century (Diethelm 1971; see also Eccles 1982, 82).

Traditionally, *furor uterinus,* specifically associated with the generative organs, was thought to differ in some way from the melancholia of insane love believed to be connected solely to the brain. These distinctions were never clear in practice, however; in his magnum opus, *A Treatise on Lovesickness,* the seventeenth-century French physician Jacques Ferrand included both under the rubric of lovesickness, claiming that they differed only in degree. This debate about the nature of insane love—about the differences between pining away for love and sexual fury (or pathological sexuality)—continued into the nineteenth century.

In addition, Ferrand argued that lovesickness in all its manifestations was more likely to affect women because they were less rational, more "maniacal," and more libidinous in their love ([1623] 1990, 311). Up to the Renaissance, physicians had contended that lovesickness was a disease almost exclusively afflicting noblemen; throughout the period, most of the victims discussed in the few detailed case studies of sexual excitement were men (Diethelm 1966, 238, 243, 246). (Literature, painting, and poetry, however, occasionally presented women as victims of lovesickness.) Gradually, the focus of the diseases of morbid love began to shift toward women. According to a recent study titled *Lovesickness in the Middle Ages,* the more specific connection made by some Renaissance medical writers between lovesickness and the sexual organs may have directed particular attention to women as sufferers "since women's ailments received special notice insofar as they were related to sexual physiology" (Wack 1990, 175). Furthermore, Enlightenment discourses on the rationality of man, as distinct from women's irrational nature, may have contributed to this shift toward locating love madness in women and femininity. Whatever the reasons for this change, by the nineteenth century insane love in its various forms was much more likely to be associated with women.

Changes in Attitudes toward Female Sexuality

Belief in female irrationality continued to inform the medical discussions of nymphomania into the nineteenth century; Renaissance notions of the normality of female sexual desire did not. Sweeping changes in the assumptions about female sexuality occurred in the Western world in the late eighteenth and nineteenth centuries. Well into the eighteenth century, both popular notions and medical understanding retained vestiges of the belief that women were as passionate, lewd, and lascivious as men were. While some doctors had begun to question whether female orgasm was necessary for pregnancy, the popular assumption that female pleasure and fertility were connected remained intact (Laqueur 1987). And yet by the nineteenth century, an ideology was firmly established: women by nature were less sexually desirous than men; the wifely and maternal role dominated their identity (Bloch 1978a; Friedli 1988, 235; Copley 1989, 84–85).

Some historians argue that the rise of evangelical Christianity in the late eighteenth century helped to transform attitudes about female sexuality, encouraging an ideology of female "passionlessness." The revitalized churches demanded moral restraint of women as evidence of their noble character. Women themselves, these theories suggest, adopt this link between passionlessness and moral superiority as a means of enhancing their status, gaining some control over their lives, and, ultimately, expanding their opportunities (Cott 1978; Smith-Rosenberg 1985, 302, n. 23).

Economic factors also contributed to this transformation. The development of urban industrial capitalism leading to a separation of work from home resulted in a hardening of the divisions between men's and women's roles, particularly among the middle classes. This growing sexual division of labor was underscored by medical-scientific theories that posited the naturalness of this divide by arguing that woman's passive nature left her ill equipped for the rough-and-tumble, competitive public world of work and politics. Thus, women's too delicate nervous systems, monthly "illness," smaller brains, and specific reproductive organs all made it unhealthy—indeed unnatural—for women to work, write, vote, go to college, or participate in the public arena (Smith-Rosenberg and Rosenberg 1973; Bloch 1978b; Digby 1989).

At the same time, according to several recent historical studies, a new representation of the female as profoundly different from the male was promulgated. From the Ancient period to the eighteenth century, they argue, the female body was seen simply as an inferior male body, one whose genitals had not descended because of lack of heat. This one-sex model mirrored the cosmological understanding of the social order. As that world view was transformed by revolutions, both scientific and political, a new model of the body that posited difference rather than sameness was created. Profoundly suspicious of passion, Enlightenment and post-Revolutionary writers argued that women had less sexual desire than men and thus were uniquely suited to be a civilizing force; male passion would be controlled by the strength of woman's moral virtue (Schiebinger 1987; Laqueur 1990; Moscucci 1990).[13]

Causes and Treatments of Nymphomania

Treatments prescribed for nymphomania also underwent major changes. Renaissance doctors, working within the context of humoral medicine, treated *furor uterinus* with bleeding, purges, emetics, and a variety of herbal medicines to restore equilibrium to the body's elements. Bleeding would draw off the noxious and excess humors or remove the "obstruction" caused by too much blood, restore harmony to the body, and cure the disease. One of the most famous cases, repeated in many of the early texts and still cited in the nineteenth century, described a young woman suffering from nymphomania who was bled over thirty times until she died (Diethelm 1971, 66). Later nineteenth-century physicians continued to maintain that menstrual problems were a major cause of diseases such as nymphomania (as well as hysteria and many other female diseases); the cure some of them suggested was to remove the ovaries and stop menstruation.

In the late eighteenth century, while a few physicians continued to recommend the age-old cures of "therapeutic intercourse" with prostitutes for the men who fell victim to lovesickness, virtuous living was more likely to be prescribed, for both men and women, as the necessary anodyne to the diseases of insane love. In a treatise titled *Nymphomania, or a Dissertation concerning the Furor Uterinus,* translated into English in 1775, an obscure French doctor, M. D. T. Bienville, stated emphatically that too much pleasure and high living, rich sauces, and spiced meat made the "blood too abundant," and thus indulgent women were much more likely to succumb to the disease of insane love (1775, 51). The emphasis on the consequences of luxurious living suggests that Bienville was particularly concerned about warning middle-class women not to yield to the excesses of the upper classes.

In keeping with the growing role of medical doctors as arbiters of morality, Bienville argued that physicians, not philosophers, must point out the moral dangers of these acts and show how they will lead to the grave (Bienville 1775, 42). These exhortations to moral behavior found in Bienville's writing, while not unknown in the Renaissance, would become even more strident in the decades to come. His work foreshadowed the Victorian conviction that the first false step into this "labyrinth of horrors" led the nymphomaniac-as-fallen-woman inexorably toward her death. Only thirty years earlier, the French physician Jean Astruc, in *A Treatise on the Diseases Incident to Women,* discussed quite dispassionately the pros and cons of recommending sexual indulgence as a cure for *furor uterinus:* "yes" to venereal action when it could be legitimate (i.e., marital intercourse) and "no" to masturbation because he had not seen cases in which it had done any good ([1740] 1743, 168). Bienville, however, appeared to be gripped by a growing and pervasive fear of the dangers of sexuality: "its [nymphomania's] progress becomes every day more rapid and alarming" (1775, xi).

Bienville believed that nymphomania struck a variety of women—young girls of marriage age who did not get the object of their passion, debauched girls who had lived a voluptuous life, married women when united to a husband of feeble or cold temperament, or young widows deprived of a vigorous man. Young girls of the middle classes, however, were his primary concern. Pubescent girls were particularly susceptible, he argued, both because their passions were easily stirred and because the first menses might bring on the disorder (Bienville 1775).

Nineteenth-century Conceptions of Nymphomania

Bienville's work provides a link between earlier notions of uterine fury and nineteenth-century conceptions of the disease in its greater focus on the nervous system, with less attention paid to the theory of an imbalance of the humors as the cause of insane love.[14] Later in the nineteenth century, a link between the genitals and the brain via the nervous system (through the spinal column) was posited as a more scientific explanation than the earlier notion of vapors rising from the uterus to the brain.

This still did not satisfactorily explain the "seat" or specific location of the disease—the brain or the genitals—and debates over this issue continued throughout the nineteenth century.[15] On the one hand, neurologists, anatomists, phrenologists, and others looked for an organic cause of nymphomania in the brain. Their attempts to establish somatic causes in cerebral lesions, changes in the brain's blood vessels, a thickening of the cranial bones, or overexcited nerve fibers generally came to naught. Autopsies of the brains of those who had been diagnosed with nymphomania led many to conclude that no significant morbid alteration of the brain in these cases was perceptible or that similar abnormalities were found in persons who had died of other diseases (see, e.g., Magendie 1836–37, 463–65, 505). In a particularly dramatic case, cited throughout the nineteenth century, the theory that a relationship existed between an enlarged or inflamed cerebellum and the sexual appetites was refuted by referring to the case of the nymphomaniacal girl whose autopsy revealed she had no cerebellum (Dunn 1849, 321; *Journal of Psychological Medicine and Mental Pathology* 1849b, 539; Shortland 1987).

On the other hand, especially in the second half of the nineteenth century, the relatively young and growing medical specialization of gynecology reversed the focus from the brain to the genitals. Diseased ovaries or disordered menstruation, gynecologists argued, could lead to injury of the nervous system and brain and thus to mental illness.[16] Furthermore, because the etiology of many female disorders, such as nymphomania, was so uncertain, gynecologists searched for a sign, or a symptom, that could clearly identify the disease. Redness, soreness, or itching of the genitals was often noted, but, in particular, enlargement of the clitoris or labia was believed to be the preeminent indicator of female lasciviousness.[17] A woman's body would yield evidence of behavior to the trained eye of the physician that the woman herself might deny. In this way, gynecology was able to lay claim to a unique role in the diagnosis and treatment of female disease. Although a study of six thousand French prostitutes, published in English in 1840, refuted the widely held belief that sexual excess would mark the genitals in some obvious way, gynecologists throughout the nineteenth century drew attention to the size of the clitoris and continued to diagnose hypertrophy of the clitoris among those women they labeled as nymphomaniacs (Meigs [1848] 1859, 151; Parent-Duchâtelet in Gilman 1985, 223).

The diets, drugs, bloodletting, cold baths, or moral treatments of the neurologists, alienists (psychiatrists), and other physicians had not provided a cure for nymphomania, hysteria, and the other diseases connected to women's sexuality in the nineteenth century. Gynecology, attempting to consolidate its professional status, offered a new and controversial treatment for certain of these diseases—gynecological surgery.[18] The underlying assumption that women were dominated by their reproductive organs led some physicians to blame virtually all women's diseases and complaints on disorders of these organs. The theory of reflex action posited that "irritation" of the sexual organs—which covered everything from the pain caused by ovarian cysts to an indeterminate "nervous excitability"—could affect the brain and lead to madness. Some argued that, because mental disturbances were connected to menstruation, stopping the bleeding by removing the ovaries would cure the illness. Except for this pseudohumoral explanation, the medical profession was not exactly sure how this "cure" worked. One contention was that the operation's shock value alone cured nymphomania.[19]

In addition to "normal ovariotomy" or oophorectomy (terms generally associated with removal of nondiseased ovaries), other gynecological surgery, such as excision of the clitoris and/or the labia, was also recommended in cases diagnosed as excessive sexual desire.[20] The removal of the clitoris was justified because—according to Isaac Baker Brown, one of the major, and ultimately discredited, British proponents of the operation—it supposedly "removed the abnormal peripheral excitement of the pudic nerve," which otherwise would probably lead to insanity and death (1866, 70).

The efficacy of gynecological surgery for nervous and mental disorders, including nymphomania, was both praised and condemned at major medical congresses and in the pages of leading medical journals throughout the last two decades of the nineteenth century. Not surprisingly, gynecologists were generally more enthusiastic about the operation than were neurologists, psychiatrists, and other physicians. In fact, one such specialist in female diseases argued that fully 75 percent of those women examined in mental hospitals showed pelvic disease or abnormality (Rohé 1892, 700). Within gynecological circles, however, the indiscriminate use of oophorectomy was condemned by many. One of the originators of the surgical removal of the ovaries, the eminent British surgeon T. Spencer Wells, argued against the use of the operation in cases of insanity and concluded that "in nymphomania and mental diseases it is, to say the least, unjustifiable" (Wells, Hegar, and Battey 1886, 470).

The Nature of Female Sexuality in the Nineteenth Century

The debates surrounding the treatment of nymphomania starkly reveal the tensions and ambiguities implicit in the representations of female sexuality and the nature of female desire. One part of the controversy centered on what effect the removal of ovaries would have on women as women. This question was fraught with contradiction: if the productive role in society for which women were biologically determined was motherhood, then removal of the ovaries essentially eliminated woman's reason for being. On the other hand, because the disease of nymphomania raised fears that the intensity of sexual desire would lead a woman to lose all self-control and modesty, some physicians argued that oophorectomy was justified (see, e.g., Stewart 1889). One doctor reported that a twenty-three-year-old unmarried woman, on whom he refused to operate, begged him to remove her ovaries because "she deplored the fact that anyone with sufficient opportunity could prevail over her scruples." Thus, some patients, and their fathers and husbands, assumed along with some doctors that this operation was a cure for the woman's disease of excessive sexual desire (Chunn 1887, 121; Cushing 1887, 441; Polack 1897, 302).

Another part of the argument over oophorectomy centered on the nature of sexual desire itself. Contrary to the twentieth-century popular notion that the Victorians perceived women as asexual, most nineteenth-century doctors assumed that women's sexual desire was natural, albeit more limited than men's.[21] What would happen to this sexual desire after the ovaries were removed? Some argued that oophorectomy did not eliminate desire, claiming that desire resided in the nervous system, not the ovaries. In a discussion led by the gynecologist B. Sherwood-Dunn and reported in the *Transactions of the American Association of Obstetricians and Gynecologists*, Rufus Hall of Cincinnati commented that he had tabulated more than four hundred cases (of removal of the ovaries), "and not one of this number has had a total loss of sexual feeling, and only three women, after a period of three years following the operation, have noticed any marked diminution in the sexual feeling" (Sherwood-Dunn 1897, 227). The gynecologist John M. Duff of Pittsburgh stated that he had performed a vaginal hysterectomy (although it was not clear whether the ovaries were also removed) on a woman who had had no sexual desire prior to the operation and who subsequently developed nymphomania. "After the operation," he continued, "she could hardly be satisfied in that direction. I have had two or three cases in which the sexual appetite developed, but not to the same extent as in this one" (Sherwood-Dunn 1897, 219). Other physicians argued strenuously that oophorectomy "unsexed" women, using terms like "spaying" for the operation. A scathing criticism of the surgical operation was offered by T. Spencer Wells: "But would anyone strip off the penis for a stricture or a gonorrhoea, or castrate a man because he had a hydrocele [an accumulation of fluid in the scrotum], or was a moral delinquent" (Wells, Hegar, and Battey 1886, 466).[22]

The medical establishment's lack of consensus makes clear that psychogynecological surgery was not based on a scientific, proven understanding of its effects. Rather, this method of treatment reflected the confluence of a particular construction of the female, the development of "safe" and anesthetized surgery, and the desire by gynecologists to consolidate their professional position by establishing themselves as the experts who could diagnosis, treat, and cure these elusive female disorders.

Satyriasis: Male Nymphomania?

Medical discussions of satyriasis, the presumed counterpart of nymphomania, provide additional insights into the construction of femaleness. Professional journals, medical textbooks, and encyclopedias often declared that satyriasis was the equivalent of nymphomania, but at the same time most doctors believed that satyriasis occurred far less frequently.[23] In addition, they were much more likely to assume that the vast majority of nymphomaniacs were severely diseased, while positing that many cases of satyriasis were very mild. The consequences predicted for the nymphomaniac were generally worse than those for the satyriasist: the outcome for nymphomaniacs was prostitution or the insane asylum, while satyriasists might go through life without getting into trouble if they learned to control themselves (see, e.g., Parke 1908, 346; Huhner [1916] 1920, 150–65). These "scientific" theories reflected the Victorian assumption that women, by nature, had less sexual desire than men; a "predominating sexual desire in woman arouses a suspicion of its pathological significance" (Krafft-Ebing [1886] 1965, 87).[24] Furthermore, as many doctors recognized, it was easier for men to fulfill their sexual desires in "illicit indulgences," which are "openly condemned, secretly practised, and tacitly condoned" (Maudsley 1867, 388). Thus, men have more sexual desire, but less disease of excess; women are less desirous, but more prone to morbid passion. We can see in these discussions that even within the biological framework posited by the medical profession, the social construction of the disease was tacitly recognized.

The case studies of satyriasists vary enormously both in mental institutions and in private treatment; they include a few examples of the use of castration as a cure.[25] But castration was never seen as a routine treatment for mental disorders in men, and men's nature—unlike women's—was never primarily defined by their genitalia. In addition, none of the cases articulated male behavior equivalent to the flirting, lascivious glances, wearing of perfume, or the other symptoms of "mild nymphomania" that garner the opprobrium of Victorian medical writers. The standards of behavior for women were, of course, much stricter than those for men. But there is more here than just a case of the Victorian double standard. Some doctors were surprisingly sympathetic to the strictures society placed on women's sexuality. For example, John Charles Bucknill, coauthor of an authoritative mid-nineteenth-century textbook on psychological medicine, held that nymphomania could result "from the struggle between mental purity and the physical impulses of sex," as well as from organic causes (Bucknill and Tuke [1858] 1968, 524). In a case discussed before the Boston Gynecological Society in 1869, one doctor went so far as to suggest, "If this woman could go masked as she is at the present moment [the woman patient was brought before the Society wearing a mask] to a house of prostitution, and spend every night for a fortnight at sexual labor, it might prove her salvation; such a course, however, the physician cannot advise" (Field in Storer 1869, 425).

Nymphomania in the Late Nineteenth Century

Such pragmatic understanding of the effects of the double standard, however, contrasted sharply with the age-old fears of the insatiable female that lurked just beneath the surface of some of the medical rhetoric. Krafft-Ebing epitomized the apocalyptical nature of this concern in *Psychopathia Sexualis* by warning, "woe unto the man who falls into the meshes of such an insatiable Messalina, whose sexual appetite is never appeased" ([1886] 1965, 403). In the latter part of the nineteenth century, the classification (and creation) by Krafft-Ebing and others of a wide variety of psychopathologies, such as sadism, masochism, lesbianism, nymphomania, and satyriasis—combined with the theory that these "perversions" were inheritable—contributed to a climate in which sexual dangers were thought to be rampant. Rather than focus on the sexually deviant act itself, Krafft-Ebing and other sexologists began to look to the very character of the person. A nymphomaniac, like a homosexual, was increasingly seen as a particular type of deviant, one whose pathology would be inherited by her daughters. Older notions of moderation-is-best gave way to anxiety that deviance and perversion would be passed on by defective genes, ineradicably marking the nervous system of the next generation (see, e.g., Maudsley 1873, 76; Krafft-Ebing [1886] 1965, 400–408).

Female sexual desire was believed to be particularly dangerous: women were more easily overwhelmed by the power of their sexual passion because they were closer to nature and thus more volatile and irrational than men. According to one doctor, "when they are touched and excited, a time arrived when, though not intending to sin, they lost all physical control over themselves" (Heywood Smith in Routh 1887, 505). Women's potential for explosive sexuality jeopardized the self-discipline and control of desire that the Victorian middle class asserted were the mainstays of civilization. Throughout these discussions, women were presented not only as metaphorically dangerous—to the family, to the moral order, to civilization itself—but literally dangerous as well. Some doctors argued that a nymphomaniac would not just seduce a man but would overpower him and actually force him to satisfy her sexual desires. Female sexuality was thus understood in terms of the male sexual act, a kind of reverse rape fantasy.

Physicians' own anxieties about the nature of sexuality—theirs and their patients'—were brought into sharp focus in these debates. One form this took was warnings to other doctors about the female patient as "seductress": a nymphomaniac who sought to entice the physician into a gynecological exam, demanding that a speculum or a catheter be inserted as a means of sexual gratification.[26] Physicians claimed that these women resorted to remarkable subterfuges to induce handling of the sexual organs, one of the most frequent of which was a pretense of urine retention. In addition, doctors described "hair pins, pencils, crochet needles, small keys, bits of bone, of tobacco pipes, of glass tubing, etc. etc." lodged in the bladder or urethra (Chapman 1883, 595). Many of the doctors assumed the women were using these objects, and the subsequent gynecological examination, as a means of sexual excitement.[27] None of the physicians suggested that sexual abuse or attempted abortion might be responsible; they saw these women as temptresses, not victims.

Decoding these cases is difficult. It is not implausible that a few women did approach doctors in this manner; others likely came seeking treatment for gynecological problems and were unfairly perceived to be acting lasciviously. Certainly the gynecologists' representation of the female patient as seducer reflected anxiety on the part of the physicians. Like male midwives in an earlier period, gynecologists were perceived by husbands and fathers as potential ravishers of their wives and daughters. Turning this idea on its head, the gynecologists claimed that they were the victims, preyed upon by wanton women patients. This suggests a complicated web of constructed sexuality. Competing images of women emerge in these debates: the Messalina type, throwing herself on the doctor, spewing lewd and lascivious language, demanding sexual gratification; the previously pure and modest woman, a "good girl," now caught in the throes of uncontrollable sexual urges; and the seductress, duplicitous, luring the unwitting physician with her downcast, languid eyes to insert a speculum or catheter in order to fulfill her perverted desires. In all of these personifications, woman is out of control, inverting the natural order because of her aggressive, powerful sexual demands.

Once again, the nature of female sexuality is at issue here. Physicians expressed surprise and shock at the violent excitement and "loss of control" they witnessed in the doctors' offices during gynecological examinations. According to one doctor, his patient took "very evident delight" in the gynecological examination (Payne 1859, 569–70). Another commented that female masturbators were easily detectable because "the clitoris will usually be found erect, and on touching it, the patient will almost invariably show her want of self control" (Chapman 1883: 457–58; see also Poovey 1987). Female orgasm was known and described in the medical literature (albeit with much debate over its nature), and many doctors recognized that the clitoris contributed to that excitement (see, e.g., Hillary 1883; McCully 1883, 845; also Martin 1987, 102–17). The ideological assumptions of the period, however, imagined that female desire was passive and latent, connected to true love, marriage, and motherhood. A woman's strong physical response to a doctor's touching the clitoris or labia, or her vaginal contractions upon insertion of a speculum, were interpreted by some as the signs of excessive sexuality, indicative of a masturbator or a nymphomaniac.

Nymphomaniacs, Lesbians, and Prostitutes

By the late nineteenth century, physicians tended to group all sexualized women together: nymphomaniacs, lesbians, and prostitutes. The descriptions of lesbians sound remarkably like those of nymphomaniacs. Two physicians in the

Journal of Nervous and Mental Disease described cases of inversion (homosexuality) by pointing to the patients' morbid excitability of sexual desire and weak, irritable nerves. They said of one thirty-five-year-old (lesbian) woman that she "grew quite passionate, threw things about and used improper language" (Shaw and Ferris, 1883, 189). According to the gynecologist Carlton Frederick, "All sorts of degenerate practices are followed by some [nymphomaniacs]. One of the most frequent is tribadism—the so-called 'Lesbian Love,' which consists in various degenerate acts between two women in order to stimulate the sexual orgasm" (1907, 810). Literature describing "lesbic love" should be kept from "young girls and neuropathic women," the British psychiatrist Daniel Hack Tuke argued, because the sensations aroused would "enslave" them and "nymphomania itself" would be established (Tuke 1892, 865; Thoinot 1911, 465–68; Chauncey 1982–83).

Nymphomaniacs were driven to prostitution to satisfy their desires; prostitutes were often lesbians.[28] According to the New York gynecologist Bernard Talmey, "It is known that Lesbianism is very prevalent among the prostitutes of Paris. . . . One-fourth of all the prostitutes in Paris serve as tribadists for the rich women who patronize public houses" ([1904] 1912, 150–51).

Just as physicians assumed that an enlarged clitoris was the sign of nymphomaniacs and prostitutes, they drew attention to hypertrophy of the lesbian's clitoris, which was used, many believed, like a penis in the "imitation of coitus." This construction vividly illustrates the physicians' inability to conceive of the sexual act in any way other than a male, heterosexual model. Indeed, it was the presumed "inversion" of gender role by the "masculine" partner in a lesbian couple that most troubled the Victorians.[29]

The late-nineteenth-century medical model of biologically based gender roles meant that women who stepped outside the norm were assumed to be diseased. These atavistic women who evidenced too much sexual desire, excessive sexual activity, or "inversion" of their assigned role severely challenged notions of innate female modesty and passivity. The Victorians believed that sexual restraint and adherence to highly differentiated gender roles were both evidence of and necessary for the continuation of the advanced level of civilization they had achieved. Lesbians, nymphomaniacs, and prostitutes—and by extension, suffragists, feminists, and the modern woman—were considered not only diseased, but dangerous as well (Lombroso and Ferrero 1897, 246; Thoinot 1911, 469–70; Chauncey 1982–83).

Physicians were particularly upset by the sexual response of very young girls or old women. Too young or too old to reproduce, little girls' and postmenopausal women's sexual desires were considered by some doctors as signs of disease.[30] In a case in 1894, Dr. A. J. Block decided that a thorough physical examination of a nine-year-old girl brought to him by her mother was needed to determine the degree of her perversion (diagnosed as masturbation tending toward nymphomania). He touched the vagina and labia minora and got no response. "As soon as I reached the clitoris," he reported, "the legs were thrown widely open, the face became pale, the breathing short and rapid, the body twitched from excitement, slight groans came from the patient" (Block 1894, 3). Block stated emphatically that the child's violent response proved that the clitoris alone was responsible for her "disease." He performed a clitoridectomy (Block 1894).

The Autobiography of a Nymphomaniac

Desperate women patients, embodying the cultural notion of what was appropriate for them to feel sexually, were described in the medical reports as begging doctors to operate on them or on their daughters because excessive sexuality had become unbearable. Unfortunately, these cases are told only from the physician's point of view. I have found only one case in the words of a woman who calls herself a nymphomaniac, and it, too, was mediated through the neurosurgeon Charles K. Mills's presentation in 1885. The twenty-nine-year-old woman in "A Case of Nymphomania . . . The Patient's History as Told by Herself" incorporated many of the prevailing Victorian notions about nymphomania. She began her story by saying, "I inherited from my mother a morbid disposition" (Mills 1885, 535). She recounted her attempts to exercise her will against the overpowering nature of the desire: "When I felt tempted, I would kneel and honestly pray to be kept from doing wrong, and then get up and do it [masturbate]; not because I wanted to, but because my life could not go on until the excitement was quieted" (Mills 1885, 537). She struggled with the feelings, "At times I felt tempted to seek the company of men to gratify my passion, but was too modest" (Mills 1885, 535). She is "treated" by having her clitoris removed, "but it grew again. . . . I tormented doctors to operate again" (Mills 1885, 535). They did. "Since removal of the ovaries I have been able to control the desire when awake, but at times in my sleep I can feel something like an orgasm taking place" (Mills 1885, 536).

Yet she was also aware of how her limited options had shaped her life. "I had not been educated as I wanted. I had earned my living by labor that occupied my hands, while my mind ambitiously dreamed of work that I would have to climb to. [In seven months in the hospital] I was not once troubled with the nymphomania [because she was studying nursing]; but when I had to give it up and go away, crushed with disappointment, with weakness and poverty . . . when I had to again spend my days in work that held no interest for me, the old morbid depression came back and with it the disease" (Mills 1885, 536). In general, however, the autobiography was permeated by her belief that her feelings of sexual desire were a sign of her disease. "[Even] while I was praying my body was so contorted with the disease that I could not get away from it even while seeking God's help" (Mills 1885, 537).

Conclusion

By the late nineteenth and early twentieth centuries, discussions about nymphomania reflected increasing concern over the "New Woman's" greater independence and potential opportunity for sexual experience. Commentators feared the "proletarianization of sexuality"—that is, that middle- and upper-class women who left the safe confines of home to work or attend school would become like working-class women, who were perceived as inordinately lustful and as sexual opportunists.[31]

The ideal of marriage itself would be transformed in the early twentieth century. Greater emphasis on female sexual pleasure as a measure of a successful marriage and a growing acceptance of the separation between reproduction and sexuality would lead to heightened concern about the nature of female sexuality. While Krafft-Ebing had argued in the 1880s that the "normal, untainted wife knows how to control herself" against the urges of unrequited love when a husband does not satisfy her, early-twentieth-century sexologists were not so sure (Krafft-Ebing [1886] 1965, 84). The ideology of companionate marriage with its assumption of mutual sexual satisfaction contained potential risks. Too much sexual desire by the wife—"semi-nymphomania" according to one physician—obviously threatened the husband in ways in which the older notion of female passivity had not (Magian 1922, 76; Freedman 1982, 210; Seidman 1991, 85).

Furthermore, some authors began to focus on the potential "masculinization" of women who stepped outside the boundaries of family and home. Career women, feminists, educated women who did not marry—a growing number at the turn of the century—were taking on male roles and potentially acquiring the "masculine" trait of aggressive sexual behavior. This concern about women's masculinization coincided with the development of new psychoanalytic theories that reasserted the essential passivity of female sexuality and underscored the notion that a mature, fulfilling sexual experience for a woman could be achieved only through vaginal orgasm in heterosexual intercourse

(Freud [1905] 1962, 86–93). Eventually, those women who did not experience vaginal orgasm but maintained their sexual focus and excitement in the clitoris would be diagnosed by psychoanalysts as "frigid."[32]

These new psychodynamic theories opened the way for an understanding of nymphomania as a symptom of a disordered psyche rather than as a biological disease. But they also allowed for a new interpretation of appropriate female sexuality, one in which the threat of a woman's being labeled "not a real woman" could be used to control women's sexual behavior, to shape it in the image of male pleasure, that is, vaginal orgasm.[33]

This shift from a physiological to a psychological explanation of nymphomania during the twentieth century, with all its ramifications, remains to be explored. In the early part of the twentieth century, the pervasive belief that female reproductive organs could cause insanity through reflex action between the brain and pelvis began to be replaced by newer physiological models based on endocrinological discoveries. In addition, late-nineteenth-century pessimistic psychological theories rooted in deterministic notions of degeneration and heredity would be superseded by more hopeful psychodynamic explanations. Biological models of nymphomania were not totally discarded, but psychological explanations that pointed to nymphomania as a personality disorder took precedence. New causes of nymphomania—such as an inadequate sense of self, repressed homosexuality, or incomplete psychosexual development—were introduced and psychotherapy recommended as the treatment.

Concerns about sexual desire itself were transformed in the twentieth century from a major focus on hypersexuality to a concentration on the syndrome called "ISD," inhibited sexual desire. Medical writers paid increasing attention to the theory that nymphomaniacs were actually frigid and did not experience orgasm, thus their "insatiability." In a future study, I plan to explore nymphomania and the relationship between twentieth-century biological and psychological theories, new constructions of the nature of women, and the changing realities of women's lives.

The medical diagnosis of nymphomania in the nineteenth century, constructed within a social and cultural context as well as within a scientific one, reflected and reproduced prevailing attitudes about appropriate behavior. Physicians, however, did not speak with a single voice: they did not agree on the nature of the disease, its extent, its treatment, or even what constituted normal female sexuality. In the overlapping and contradictory descriptions of nymphomania, in the intertwining of moral and medical explanations, these physicians reveal much about the nineteenth-century construction and understanding of female sexuality and the nature of women.

Notes

I would like to thank the following people for their criticism and support: Joan Jacobs Brumberg, Steve Curry, Elisabeth Gitter, Jacqueline Jaffe, Richard Lieberman, Ron Rainey, Jennifer Terry, Judith Zinsser, the John Jay College of Criminal Justice Seminar on Research By or About Women, the Keen's Women's Group on Sexuality, and the Columbia University Seminar on Women and Society. This research was partially supported by a grant from the Professional Staff Congress/Faculty Research Grant Award Program of the City University of New York.

This essay originally appeared in *Signs: Journal of Women in Culture and Society* 1994, vol. 19, no. 2,

1. As discussed below, these treatments reflected humoral theories intended to restore the harmony of the elements of the body by cooling that which was overheated or by bleeding the body to eliminate blockage or excess blood. Although humoral theories had been mainly discredited by the medical community, no new paradigms had emerged in their stead; thus treatment based on such theories continued into the nineteenth century.

2. Phillips does not comment on Mrs. L.'s attraction to the women's bodies in the *tableau vivant*, even though "nymphomaniacal lesbians" had been discussed in medical literature by this date.

3. I have included only those case studies in which nymphomania was part of the diagnosis. The medical journals I consulted include *Alienist and Neurologist, American Gynecological and Obstetrical Journal, American Journal of Insanity, American Journal of Medical Science, American Journal of Obstetrics and Diseases of Women and Children, American Journal of Urology and Sexology, American Practitioner, Boston Medical and Surgical Journal, British Gynaecological Journal, Journal of the American Medical Association, Journal of Nervous and Mental Disease, Journal of Psychological Medicine and Mental Pathology, Lancet, Medical and Surgical Reporter, New York Medical and Physical Journal, Transactions of the American Association of Obstetricians and Gynecologists,* and *Transactions of the American Medical Association;* I also consulted dozens of contemporary texts.

4. In a recent book, Edward Shorter states that nymphomania "usually meant chronic masturbation" (1992, 81). The cases that I have read suggest that behaviors over a much wider range were labeled nymphomania.

5. *American Journal of Medical Science* 1837; Hor and Sprague 1841; Vaille 1868; Chunn 1887; Francez 1888.

6. Of course, these diagnoses are also socially constructed. Physicians interpreted various kinds of behavior as "sexual" when they could have had other meanings. The point here is not that these are "real" nymphomaniacs but, rather, that some of them, at least, were suffering from serious mental illness. See *Lancet* 1825–26; *American Journal of Medical Science* 1837; Payne 1859; *American Journal of Insanity* 1860; Mills 1885; Chunn 1887; Polack 1897; Clouston 1898; and Goldberg 1992, 14, 57–58. For the eighteenth century, see Porter 1986.

7. While considerable differences exist in the attitudes, training, and traditions of the medical professions in various countries, enough similarity exists in their shared ideas about nymphomania to warrant these generalizations; this is evident in the numerous translations of and references to a wide variety of works from England, the United States, France, Germany, and Italy in medical journals and texts.

8. Poovey 1987; Digby 1989; Jordanova 1989; Laqueur 1990; Moscucci 1990. The pioneering work in the field is Smith-Rosenberg and Rosenberg 1973.

9. Medical theories reflect the class and racial stereotypes of the period. In Europe and

the United States, black and lower-class women in general, and in the United States African American and immigrant women in particular, were seen by the Victorian middle and upper classes as more promiscuous, animal-like, and unrestrained than white middle-class women. What appears to us as a contradiction between the Victorians' biological construction of "woman" as controlled by her reproductive organs (thus all women would presumably be the same) and the distinctions they drew between middle- and lower-class and black and white women was explained in light of evolutionary theory. The "more highly evolved," white, middle-class woman was thought to be more civilized, refined, and moral and consequently to have less sexual desire. Scientific theory was called upon to support these notions; for example, ethnographical, anatomical, phrenological, and other studies of "primitive" societies were used to support arguments about these distinctions. See Gilman 1985; Weeks 1986, 39–40; D'Emilio and Freedman 1988, 35, 86, 142; Russett 1989, 26–28, 51–54; and Oppenheim 1991, 205.

10. In discussing the development of gynecology as a medical specialization in the nineteenth century, Moscucci argues that gynecology was mainly the province of midwives up through the eighteenth century, and that before 1800 no particular group of practioners claimed women's diseases as their special province.

11. Nymphomania has received surprisingly little historical study. The most recent full-scale examination (Ellis and Sagarin 1964) is a popular, psychological work that contains a brief historical chapter. Four articles and a dissertation are of interest: Diethelm 1966; Goulemot 1980; Maaskant-Kleibrink 1980 (my thanks to Sarah Pomeroy for drawing my attention to this article); Rousseau 1982; and Goldberg 1992.

12. The entry on nymphomania (Louyer-Villermay 1819) in the *Dictionnaire des Sciences Médicales* contains an extensive review of earlier theories of the disease.

13. See Rousseau 1982 for a discussion of the relationship between the rise of erotic sensibility and the change in attitudes toward sexuality in the late eighteenth and early nineteenth centuries.

14. Rousseau (1982, 106) is mistaken in stating that William Cullen's *Nosology* (1769) is the first work to refer to nymphomania and that "as late as the seventeenth-century *fin-de-siecle,* a sexually hysterical woman is still labeled 'possessed demonically,' not called nymphomaniacal or discussed in physiological, neurological or other medical terms." Furthermore, Bienville is not the first writer to refer to a connection between morbid imagination and nymphomania, as Rousseau suggests. See also Foucault 1965, 97–101.

15. *Journal of Psychological Medicine and Mental Pathology* 1849a, 1849b; *Lancet* 1849; Acton 1862, 125; *Transactions of the American Medical Association* 1868; Tilt 1869, 96–98; Ferrier 1876, 122; Howe 1883, 116; Cushing 1887; Clouston 1898, 339–40; Shortland 1987.

16. Wells, Hegar, and Battey 1886; Cushing 1887; *American Journal of Obstetrics and Diseases of Women and Children* 1901; Bucknill and Tuke (1858) 1968, 212–13; Smith-Rosenberg and Rosenberg 1973; Shortt 1986.

17. Churchill 1857, 70–73; *American Journal of Obstetrics and Diseases of Women and Children* 1873–74; Chapman 1883; Routh 1886–87; Wylie 1901. There was no comparable body of writing about male genitalia, even though doctors were also very concerned about male masturbation. See Dwyer 1984, 40–41.

18. Barker-Benfield 1976, 80–90; Showalter 1985, 74–79; Scull 1986; Moscucci 1990: 108–9, 131–32; Perrot 1990, 600; Ripa 1990, 133; Shorter 1992, 77–80. The medical historian Nancy Tomes offers an important cautionary note to simplistic theories of male doctors' motivations. "The majority of gynecological surgery performed by neurologists," she argues, "aimed simply at repairing the common injuries of childbirth, while asylum doctors rarely did gynecological exams, much less surgery" (1990, 159–60). Gynecology was the only one of the three specialties treating women's mental illness (neurology and psychiatry were the other two) likely to advocate surgical treatment. See also Mitchinson 1982. An article in the British medical journal *Lancet,* titled "Dr. Blundell on the Genital Parts" (1828–29), which recommended the removal of the ovaries in a case of nymphomania, is the earliest that I have found.

19. Wells, Hegar, and Battey 1886, 472; Cushing 1887, 442; Church 1893, 493; Hunter and Macalpine 1963, 861; Esquirol (1845) 1965, 339.

20. Sutcliffe 1889; Scull 1986. In France, according to Perrot (1990, 496), clitoridectomies were rare. The British gynecologist Heywood Smith (Tait 1888, 315) stated that clitoridectomy was "the best, indeed, the only, cure" for those women whose "lives were a misery to them on account of excessive sexual desire" (Heywood Smith in Tait 1888, 315). In Louyer-Villermay (1819, 600), a famous case from 1676 of an operation on a woman begging to have a circumcision because of large labia is described.

21. Martin (1978, 50–62) reviewed Victorian medical opinion concerning female sexuality and concluded that most physicians recognized that women experienced both sexual desire and pleasure in intercourse, although male sexual desire was understood to be stronger. Ellis, in his monumental work *Studies in the Psychology of Sex* ([1906] 1936, 1:189–255), provided a lengthy analysis of the views of female sexuality from the ancients to the twentieth century. See also Degler 1974; Fellman and Fellman 1981.

22. Sarcastically, Wells continued, "Fancy the reflected picture of a coterie of the Marthas of the profession in conclave, promulgating the doctrine that most of the unmanageable maladies of men were to be traced to some morbid change in their genitals, founding societies for the discussion of them and hospitals for the cure of them, one of them sitting in her consultation chair, with her little stove by her side and her irons all hot, searing every man as he passed before her" (Wells, Hegar, and Battey 1886, 470–71). See also Hegan 1884; Chunn 1887, 121; Tait 1888, 315; Smith 1900; *Occidental Medical Times* 1901.

23. Duprest-Rony 1820; Hammond 1883, 552; Lydston 1889, 283; Huhner (1916) 1920, 159; Féré (1904) 1932, 86.

24. The very opposite of this pathologizing of female sexuality can be seen in a comment by the mid-nineteenth-century American physician and popular medical writer William Alcott, who wrote, "The husband, moreover, in these days, who finds himself united, for life, to a woman whose only defect or weakness is a slight nymphomania, may think himself quite fortunate" ([1866] 1972, 170).

25. Hamilton 1841; Bigelow 1859; Welch 1889; Hamilton 1903. Surgical and pharmacological methods in the treatment of excessive male sexuality, usually masturbation, were widespread in the second half of the nineteenth century, including blistering the prepuce, placing a silver ring through the prepuce, faradization of the spine, infibulation, and others. See Hare 1962, 10–11.

26. In a lengthy discussion of the character of these nymphomaniacal seductresses, the British gynecologist C. H. F. Routh said they could be identified by the following characteristics: they had a strong neurotic history, had some uterine or ovarian complication (leukorrhea, ovarian pain, etc.), and were pretty, vain, and decided liars. It was clear that they were liars, he argued, because of the extraordinary stories they told, even about their loved ones, such as brothers who entered their rooms at night and forced them to submit sexually, and fathers who demanded that they live as husband and wife (1887, 487–89). Routh's dismissal of these "lies" foreshadowed Freud's revised theory that his female patients had fantasized the sexual abuse by family members described in his case studies.

27. Chapman 1883, 595. See also Sharpe 1874; McCully 1883, 846; Lydston 1889, 282; Block 1894, 5; *Boston City Hospital* 1900; Talmey (1904) 1912, 112; Magian 1922, 78; Bloch (1908) 1928, 430.

28. Schrenk-Notzing 1895, 32; Lydston 1904, 316; Talmey (1904) 1912, 114–15; Huhner (1916) 1920, 164; Forel (1906) 1924, 252; Hirschfeld (1944) 1956, 92; Krafft-Ebing (1886) 1965, 402–3; Gilman 1989, 299.

29. One gynecologist described a case in which a "young married woman became pregnant through her married sister who committed the similacrum of the male act on her, just after copulating with her husband" (Talmey [1904] 1912, 150–51). See also Kiernan 1891, 203; Kiernan 1916; Sahli 1979, 24–27; Chauncey 1982–83, 132; Greenberg 1988, 404. For an early-

seventeenth-century discussion of penis-like clitorises, see Ferrand (1623) 1990, 231. For a study of presumed gender inversion and lesbians in the twentieth century, see Terry 1990.

30. A few physicians were beginning to recognize sexuality in children, but many would have agreed with the well-known British gynecologist Robert Lawson Tait, who reported on "four year olds who have powerful erections in whom there could be no sexual desire possibly present" (1888: 315). In the same article, C. H. F. Routh stated that, although women over seventy were not supposed to possess sexual feelings because their ovaries were atrophied, he had performed a clitoridectomy on a woman of seventy-eight years who had "experienced extreme erotic feelings on going to stool" (315). See also Sloan 1888; Parke 1908, 345.

31. For important historical examinations of female sexuality at the turn of the century not already mentioned, see Bullough and Voght 1973; Skultans 1975; Schlossman and Wallach 1978; Dykstra 1986; Freedman 1987; Lunbeck 1987.

32. Freud and his followers made a connection between frigidity, unconscious refusal of the female role, lesbianism, and the inauthenticity of the clitoral orgasm. See, e.g., Abraham 1922; Fenichel 1933; Horney 1933; Hitschmann and Bergler 1949; Young-Bruehl 1990, 22–24.

33. One intriguing possibility to explore is whether we might also find women responding to this new construction of female sexuality by a new form of "passionlessness." Did women use the theory of the vaginal orgasm to empower themselves within male-female sexual relations?

References

Abraham, Karl. 1922. "Manifestations of the Female Castration Complex." *International Journal of Psycho-Analysis* 3(1):1–29.

Acton, William. 1862. *Functions and Disorders of Reproductive Organs.* 3d ed. London: John Churchill.

Alcott, William. (1866) 1972. *The Physiology of Marriage.* New York: Arno.

American Journal of Insanity. 1860. "Cases of Hysteria and Hysteromania." *American Journal of Insanity* 17(2):126–52.

American Journal of Medical Science. 1837. "Nymphomania before Puberty." *American Journal of Medical Science* 41 (November): 228–30.

American Journal of Obstetrics and Diseases of Women and Children. 1873–74. "Case of Excessive Masturbation." *American Journal of Obstetrics and Diseases of Women and Children* 6(2):294–95.

———. 1901. "Transactions of the Woman's Hospital Society." *American Journal of Obstetrics and Diseases of Women and Children* 43 (January): 720–22.

Astruc, Jean. (1740) 1743. *A Treatise on the Diseases Incident to Women,* trans. J. R——n. London: T. Cooper.

Barker-Benfield, Graham. 1976. *The Horrors of the Half-Known Life: Male Attitudes toward Women and Sexuality in Nineteenth-Century America.* New York: Harper & Row.

Bienville, M. D. T. 1775. *Nymphomania, or a Dissertation concerning the Furor Uterinus,* trans. Edward Sloane Wilmot. London: J. Bew.

Bigelow, H. J. 1859. "Castration as a Means of Cure for Satyriasis." *Boston Medical and Surgical Journal* 61(8):165–66.
Bloch, Iwan. (1908) 1928. *The Sexual Life of Our Time.* New York: Allied.
Bloch, Ruth. 1978a. "American Feminine Ideals in Transition: The Rise of the Moral Mother, 1785–1815." *Feminist Studies* 4(3):101–26.
———. 1978b. "Untangling the Roots of Modern Sex Roles: A Survey of Four Centuries of Change." *Signs: Journal of Women and Culture in Society* 4(2):237–52.
Block, A. J. 1894. "Sexual Perversion in the Female." *New Orleans Medical and Surgical Journal.* 22(1):1–6.
Boston City Hospital. 1900. "Case of Unique Surgery." *Boston City Hospital,* 11th ser., 143.
Brown, Isaac B. 1866. *On The Curability of Certain Forms of Insanity, Epilepsy, Catalepsy and Hysteria in Females.* London: Robert Hardwicke.
Bucknill, Charles, and Daniel H. Tuke. (1858) 1968. *A Manual of Psychological Medicine.* New York: Hafner.
Bullough, Vern, and Martha Voght. 1973. "Women, Menstruation and Nineteenth-Century Medicine." *Bulletin of the History of Medicine* 47(1):66–82.
Chapman, J. Milne. 1883. "On Masturbation as an Etiological Factor in the Production of Gynic Diseases." *American Journal of Obstetrics and Diseases of Women and Children* 16(5):449–58; 16(6):578–98.
Chauncey, George. 1982–83. "From Sexual Inversion to Homosexuality: Medicine and the Changing Conceptualization of Female Deviance." *Salmagundi,* nos. 58–59 (Fall–Winter), 114–46.
Chunn, Wm. Pawson. 1887. "A Case of Nymphomania." *Maryland Medical Journal* 17 (December): 121.
Church, Archibald. 1893. "Removal of Ovaries and Tubes in the Insane and Neurotic." *American Journal of Obstetrics and Diseases of Women and Children* 28(4):491–98, 569–73.
Churchill, Fleetwood. 1857. *On the Diseases of Women.* Philadelphia: Blanchard & Lea.
Clouston, T. S. 1898. *Clinical Lectures on Mental Diseases.* 5th ed. Philadelphia: Lea Brothers.
Copley, Antony. 1989. *Sexual Moralities in France, 1780–1980: New Ideas on the Family, Divorce, and Homosexuality, an Essay on Moral Change.* London: Routledge.
Cott, Nancy. 1978. "Passionlessness: An Interpretation of Victorian Sexual Ideology." *Signs* 4(2):219–36.
Cushing, E. W. 1887. "Melancholia; Masturbation; Cured by Removal of Both Ovaries." *Journal of the American Medical Association* 8 (April): 441–42.
Degler, Carl. 1974. "What Ought to Be and What Was: Woman's Sexuality in the Nineteenth Century." *American Historical Review* 79(5):1467–90.
D'Emilio, John, and Estelle B. Freedman. 1988. *Intimate Matters: A History of Sexuality in America.* New York: Harper & Row.
Diethelm, Oscar. 1966. "La surexcitation sexuelle" (Sexual overexcitation). *L'évolution psychiatrique* (Psychiatric development) 31(2):233–46.
———. 1971. *Medical Dissertations of Psychiatric Interest before 1750.* Basel: Karger.
Digby, Anne. 1989. "Women's Biological Straitjacket." In *Sexuality and Subordination: Interdisciplinary Studies of Gender in the Nineteenth Century,* ed. Susan Mendus and Jane Rendall. London: Routledge.

Dunn, R. 1849. "Case of Apoplexy of the Cerebellum." *Lancet* 1:321.
Duprest-Rony, A. P. 1820. "Satyriasis." In *Dictionnaire des Sciences Médicales* (Dictionary of medical sciences). Vol. 50. Paris: Pancoucke.
Dwyer, Ellen. 1984. "A Historical Perspective." In *Sex Roles and Psychopathology*, ed. Cathy Widom. New York: Plenum.
Dykstra, Bram. 1986. *Idols of Perversity: Fantasies of Feminine Evil in the Fin de-Siècle Culture.* New York: Oxford University Press.
Eccles, Audrey. 1982. *Obstetrics and Gynecology in Tudor and Stuart England.* Kent, Ohio: Kent University Press.
Ellis, Albert, and Edward Sagarin. 1964. *Nymphomania: A Study of the Oversexed Woman.* New York: Gilbert.
Ellis, Havelock. (1906) 1936. *Studies in the Psychology of Sex.* 2 vols. New York: Random House.
Esquirol, J. E. D. (1845) 1965. *Mental Maladies: A Treatise on Insanity.* New York: Hafner.
Fellman, Clair, and Michael Fellman. 1981. "The Rule of Moderation in Late Nineteenth-Century American Sexual Ideology." *Journal of Sexual Research* 17(3):238–55.
Fenichel, Otto. 1933. "Outline of Clinical Psychoanalysis." *Psychoanalytic Quarterly* 2(3–4):562–91.
Féré, Charles. (1904) 1932. *The Sexual Urge: How It Grows or Wanes*, trans. Ulrich van der Horst. New York: Falstaff.
Ferrand, Jacques. (1623) 1990. *A Treatise on Lovesickness*, ed. and trans. Donald A. Beecher and Massimo Ciavolella. Syracuse, N.Y.: Syracuse University Press.
Ferrier, David. 1876. *Functions of the Brain.* London: Smith, Elder.
Forel, A. (1906) 1924. *The Sexual Question: A Scientific, Psychological, Hygienic and Sociological Study.* New York: Physicians & Surgeons.
Foucault, Michel. 1965. *Madness and Civilization: History of Insanity in the Age of Reason.* New York: Vintage.
Francez, J. P. 1888. "Malarial or Paroxysmal Nymphomania." *New Orleans Medical and Surgical Journal* 15 (February): 623–24.
Frederick, Carlton. 1907. "Nymphomania as a Cause of Excessive Venery." *American Journal of Obstetrics and Diseases of Women and Children* 56(6):742–44, 807–12.
Freedman, Estelle B. 1982. "Sexuality in Nineteenth-Century America: Behavior, Ideology and Politics." *Reviews in American History* 10 (December): 196–215.
———. 1987. " 'Uncontrolled Desires': The Response to the Sexual Psychopath, 1920–1960." *Journal of American History* 74 (June): 83–106.
Freud, Sigmund. (1905) 1962. *Three Essays on the Theory of Sexuality.* New York: Basic.
Friedli, Lynne. 1988. "Passing Women." In *Sexual Underworlds of the Enlightenment*, ed. G. S. Rousseau and Roy Porter. Chapel Hill: University of North Carolina Press.
Gilman, Sander. 1985. "Black Bodies, White Bodies: Toward an Iconography of Female Sexuality in Late Nineteenth-Century Art, Medicine and Literature." *Critical Inquiry* 12(1):204–42.
———. 1989. *Sexuality: An Illustrated History.* New York: Wiley.
Goldberg, Ann E. 1992. "A Social Analysis of Insanity in Nineteenth-Century Germany: Sexuality, Delinquency and Anti-Semitism in the Records of the Eberbach Asylum." Ph.D. dissertation, University of California, Los Angeles.

Goulemot, Jean Marie. 1980. "Il Fureurs Utérines" (Uterine fury). In "Représentations de la vie sexuelle" (Representations of the sexual life), special issue of *Dix-Huitième Siècle* (Eighteenth century), 97–111.

Greenberg, David F. 1988. *The Construction of Homosexuality.* Chicago: University of Chicago Press.

Hamilton, F. H. 1841. "Varicocele and Extirpation of the Testes, with Remarks upon the Radical Treatment of Varicocele." *Boston Medical and Surgical Journal* 25(10):153–59.

Hamilton, J. A. G. 1903. "Treatment of Nymphomania by Division of Branches of Internal Pudic and Inferior Pudendal Nerves." *Australasian Medical Gazette* 22 (May): 205–6.

Hammond, William. 1883. *A Treatise on Insanity in Its Medical Relations.* New York: Appleton.

Hare, E. H. 1962. "Masturbatory Insanity: The History of an Idea." *Journal of Mental Science* 108(452):2–25.

Hegar, Alfred. 1884. "Spaying as a Remedy for Nervous and Psychical Affections." *American Journal of Obstetrics and Diseases of Women and Children* 17(11):1199–1202.

Hillary, James Jager. 1883. "Behavior of the Uterus during Orgasm." *American Journal of Obstetrics and Diseases of Women and Children* 16(11):1170–71.

Hirschfeld, Magnus. (1944) 1956. *Sexual Anomalies: The Origins, Nature and Treatment of Sexual Disorders.* New York: Emerson.

Hitschmann, Edward, and Edmund Bergler. 1949. "Frigidity in Women: Restatement and Renewed Experiences." *Psychoanalytic Review* 36(1):45–53.

Hor and Sprague. 1841. "Case of Nymphomania." *Boston Medical and Surgical Journal* 25(4):61–62.

Horney, Karen. 1933. "Psychogenic Factors in Functional Female Disorders." *American Journal of Obstetrics and Gynecology* 25(5):694–704.

Howe, Joseph W. 1883. *Excessive Venery, Masturbation and Continence.* New York: Bermingham.

Huhner, Max. (1916) 1920. *Disorders of the Sexual Function in the Male and Female.* Philadelphia: F. A. Davis.

Hunter, Richard, and Ida Macalpine. 1963. *Three Hundred Years of Psychiatry.* London: Oxford University Press.

Jordanova, Ludmilla. 1989. *Sexual Visions: Images of Gender in Science and Medicine between the Eighteenth and Twentieth Centuries.* Madison: University of Wisconsin Press.

Journal of Psychological Medicine and Mental Pathology. 1849a. "Analytic Reviews." *Journal of Psychological Medicine and Mental Pathology* 2 (January): 19–30.

———. 1849b. "On the Pathology and Characteristics of Insanity." *Journal of Psychological Medicine and Mental Pathology* 2 (October): 534–54.

Kiernan, James G. 1891. "Psychological Aspects of the Sexual Appetite." *Alienist and Neurologist* 12(2):188–219.

———. 1916. "Sexology: Increase of American Inversion." *Urologic and Cutaneous Review* 20(1):44–46.

Krafft-Ebing, Richard von. (1886) 1965. *Psychopathia Sexualis.* New York: G. P. Putnam's Sons.

Lancet. 1825–26. "Case of Idiocy in a Female, accompanied with Nymphomania, cured by the excision of the Clitoris." *Lancet* 1:420–21.

———. 1828–29. "Dr. Blundell on the Genital Parts and Diseases of the Vulva." *Lancet* 2:642–44.

———. 1849. "Case of Apoplexy in the Cerebellum." *Lancet* 1:319–20.

Laqueur, Thomas. 1987. "Orgasm, Generation, and the Politics of Reproductive Biology." In *The Making of the Modern Body: Sexuality and Society in the Nineteenth Century*, ed. Catherine Gallagher and Thomas Laqueur. Berkeley and Los Angeles: University of California Press.

———. 1990. *Making Sex: Body and Gender from the Greeks to Freud*. Cambridge, Mass.: Harvard University Press.

Lombroso, Cesare, and William Ferrero. 1897. *The Female Offender*. New York: D. Appleton.

Louyer-Villermay, J. B. 1819. "Nymphomanie." In *Dictionnaire des Sciences Médicales*. Vol. 36. Paris: Pancoucke.

Lunbeck, Elizabeth. 1987. " 'A New Generation of Women': Progressive Psychiatrists and the Hypersexual Female." *Feminist Studies* 13(3):513–43.

Lydston, Frank. 1889. "Sexual Perversion, Satyriasis and Nymphomania." *Medical and Surgical Reporter*. 61(10):253–85.

———. 1904. *The Diseases of Society*. Philadelphia: J. B. Lippincott.

Maaskant-Kleibrink, Marianne. 1980. "Nymphomania." In *Sexual Asymmetry: Studies in Ancient Society*, ed. Josine Blok and Peter Mason. Amsterdam: J. C. Gieben.

McCully, S. E. 1883. "Masturbation in the Female." *American Journal of Obstetrics and Diseases of Women and Children* 16(8):844–46.

Magendie, M. 1836–37. "Lectures on the Physiology of the Nervous System." *Lancet* 2:463–65, 505.

Magian, A. C. 1922. *Sex Problems in Women*. London: Heinemann.

Martin, John Rutledge. 1978. "Sex and Science: Victorian and Post-Scientific Ideas in Sexuality." Ph.D. dissertation, Duke University.

Maudsley, Henry. 1867. *The Pathology of Mind*. London: Julian Friedmann.

———. 1873. *Body and Mind*. London: Macmillan.

Meigs, Charles D. (1848) 1859. *Woman: Her Diseases and Remedies*. Philadelphia: Blanchard & Lea.

Mills, Charles K. 1885. "A Case of Nymphomania with Hystero-Epilepsy and Peculiar Mental Perversions—the Results of Clitoridectomy and Oophorectomy—The Patient's History as Told by Herself." *Philadelphia Medical Times* 15 (April): 534–40.

Mitchinson, Wendy. 1982. "Gynecological Operations on Insane Women: London, Ontario, 1895–1901." *Journal of Social History* (Spring), 467–84.

Moscucci, Ornella. 1990. *The Science of Woman: Gynaecology and Gender in England, 1800–1929*. Cambridge: Cambridge University Press.

Occidental Medical Times. "Nymphomania; Phleboliths." 1901. *Occidental Medical Times* 15 (June): 213–14.

Oppenheim, Janet. 1991. *Shattered Nerves: Doctors, Patients and Depression in Victorian England*. New York: Oxford University Press.

Parke, J. Richardson. 1908. *Human Sexuality: A Medico-Literary Treatise on the Laws,*

Anomolies, and Relations of Sex with Special Reference to Contrary Sex Desires. 2d ed. Philadelphia: Professional.

Payne, R. L. 1859. "A Case of Nymphomania." *Medical Journal of North Carolina* 2(3):569–70.

Perrot, Michelle, ed. 1990. *A History of Private Life.* Vol. 4, *From the Fires of Revolution to the Great War.* Cambridge, Mass.: Belknap Press, Harvard University Press.

Phillips, L. M. 1895. "Nymphomania; Reply to Questions." *Cincinnati Medical Journal* 10(7):467–71.

Polack, John O. 1897. "A Case of Nymphomania." *Medical News: New York* 71 (September): 301–2.

Poovey, Mary. 1987. " 'Scenes of an Indelicate Character': The Medical 'Treatment' of Victorian Women." In *The Making of the Modern Body: Sexuality and Society in the Nineteenth Century,* ed. Catherine Gallagher and Thomas Laqueur. Berkeley and Los Angeles: University of California Press.

Porter, Roy. 1986. "Love, Sex and Madness in Eighteenth-Century England." *Social Research* 53(2):211–42.

Ripa, Yannick. 1990. *Women and Madness: The Incarceration of Women in Nineteenth-Century France.* Minneapolis: University of Minnesota Press.

Rohé, George H. 1892. "The Relation of Pelvic Disease and Psychical Disturbances in Women." *American Journal of Obstetrics and Diseases of Women and Children* 26(5):694–726.

Rousseau, G. S. 1982. "Nymphomania, Bienville and the Rise of Erotic Sensibility." In *Sexuality in Eighteenth-Century Britain,* ed. Paul Gabriel Bouce. Manchester: Manchester University Press.

Routh, C. H. F. 1887. "On the Etiology and Diagnosis Considered Specially from a Medico-Legal Point of View for Those Cases of Nymphomania Which Lead Women to Make False Charges against Their Medical Attendants." *British Gynaecological Journal* 2(8):485–511.

Russett, Cynthia Eagle. 1989. *Sexual Science: The Victorian Construction of Womanhood.* Cambridge, Mass.: Harvard University Press.

Sahli, Nancy. 1979. "Smashing: Women's Relationships before the Fall." *Chrysalis* 8 (Summer): 17–27.

Schiebinger, Londa. 1987. "Skeletons in the Closet: The First Illustrations of the Female Skeleton in Eighteenth-Century Anatomy." In *The Making of the Modern Body: Sexuality and Society in the Nineteenth Century,* ed. Catherine Gallagher and Thomas Laqueur. Berkeley and Los Angeles: University of California Press.

Schlossman, Steven, and Stephanie Wallach. 1978. "Precocious Sexuality: Female Juvenile Delinquency in the Progressive Era." *Harvard Educational Review* 48(1):63–87.

Schrenk-Notzing, A. von. 1895. *Therapeutic Suggestions in Psychopathia Sexualis,* trans. Charles Gilbert Chaddock. Philadelphia: F. A. Davis.

Scull, Andrew. 1986. "The Clitoridectomy Craze." *Social Research* 53(2):243–60.

Seidman, Steven. 1991. *Romantic Longings: Love in America, 1830–1980.* New York: Routledge.

Sharpe, F. S. 1874. "Hair-pin in the Female bladder." *American Journal of Medical Sciences* 68:577–78.

Shaw, J. C., and G. N. Ferris. 1883. "Perverted Sex Instinct." *Journal of Nervous and Mental Disease* 10(2):185–204.

Sherwood-Dunn, B. 1897. "Conservation of the Ovary." *Transactions of the American Association of Obstetricians and Gynecologists: Minutes of the Proceedings of the Tenth Annual Meeting, August 17–20, 1897* 10:195–234.

Shorter, Edward. 1992. *From Paralysis to Fatigue: A History of Psychosomatic Illness in the Modern Era.* New York: Free Press.

Shortland, Michael. 1987. "Courting the Cerebellum: Early Organological and Phrenological Views of Sexuality." *British Journal for the History of Science* 20:173–99.

Shortt, S. E. D. 1986. *Victorian Lunacy: Richard M. Bucke and the Practice of Late Nineteenth-Century Psychiatry.* Cambridge: Cambridge University Press.

Showalter, Elaine. 1985. *The Female Malady: Women, Madness and English Culture, 1830–1980.* New York: Pantheon.

Skultans, Vieda. 1975. *Madness and Morals: Ideas on Insanity in the Nineteenth Century.* London: Routledge & Kegan Paul.

Sloan, M. G. 1888. "Sexual Desire in Infancy." *Journal of the American Medical Association* 10(23):731–32.

Smith, A. Laptham. 1900. "A Case in Which Sex Feeling First Appeared after Removal of Both Ovaries." *American Journal of Obstetrics and Diseases of Women and Children* 42(6):839–42.

Smith-Rosenberg, Carroll. 1985. *Disorderly Conduct: Visions of Gender in Victorian America.* New York: Knopf.

Smith-Rosenberg, Carroll, and Charles Rosenberg. 1973. "The Female Animal: Medical and Biological Views of Woman and Her Role in Nineteenth-Century America." *Journal of American History* 60(2):332–56.

Stewart, William. 1889. "A Remarkable Case of Nymphomania and Its Cure." *Transactions of the American Association of Obstetricians and Gynecologists* 2:260–62.

Storer, Horatio. 1856. "Cases of Nymphomania." *American Journal of Medical Science* 32(10):378–87.

———. 1869. "Obstinate Erotomania." *American Journal of Obstetrics and Diseases of Women and Children* 1(4):423–26.

Sutcliffe, J. A. 1889. "Excision of the Clitoris in a Child for Nymphomania." *Indiana Medical Journal* 8(3):64–65.

Tait, Robert Lawson. 1888. "Note on the Influence of Removal of the Uterus and Its Appendages on the Sexual Appetite." *British Gynaecological Journal* 4(15):310–17.

Talmey, Bernard. (1904) 1912. *Woman, A Treatise on the Normal and Pathological Emotions of Feminine Love.* New York: Practitioners.

Terry, Jennifer. 1990. "Lesbians under the Medical Gaze: Scientists Search for Remarkable Differences." *Journal of Sex Research* 27(3):317–39.

Thoinot, Leon. 1911. *Medicolegal Aspects of Moral Offenses.* Philadelphia: F. A. Davis.

Tilt, Edw. J. 1869. *A Handbook of Uterine Therapeutics and of Diseases of Women.* 2d ed. New York: Appleton.

Tomes, Nancy. 1990. "Historical Perspectives on Women and Mental Illness." In *Women, Health and Medicine in America,* ed. Rima D. Apple. New York: Garland.

Transactions of the American Medical Association. 1868. "Report on Insanity." *Transac-*

tions of the American Medical Association. Minutes of the Nineteenth Annual Meeting 19:161–88.

Tuke, Daniel Hack. 1892. *A Dictionary of Psychological Medicine.* Vol. 2. Philadelphia: P. Blakeston, Son.

Vaille, H. R. 1868. "Case of Puerperal Nymphomania." *Boston Medical and Surgical Journal* 79(10):184.

Wack, Mary. 1990. *Lovesickness in the Middle Ages.* Philadelphia: University of Pennsylvania Press.

Walton, John Tompkins. 1857. "Case of Nymphomania Successfully Treated." *American Journal of Medical Science* 33(1):47–50.

Weeks, Jeffrey. 1986. *Sexuality.* London: Routledge.

Welch, George T. 1889. "Satyriasis Caused by Varicocele, and Ceasing after Successful Operation for the Latter." *Medical Record* 36(7):181.

Wells, T. Spencer, Alfred Hegar, and Robert Battey. 1886. "Castration in Nervous Diseases: A Symposium." *American Journal of Medical Science* 92(10):455–90.

Wylie, W. Gill. 1901. "Amputation of the Clitoris." *American Journal of Obstetrics and Diseases of Women and Children* 43(5):720–23.

Young-Bruehl, Elisabeth. 1990. *Freud on Women: A Reader.* New York: Norton.

9

Theatres of Madness

Susan Jahoda

Theatres of Madness is an ongoing work. Containing images and texts, it is configured and reconfigured to suit the context of its appearance. In this particular case, a series of nineteenth- and twentieth-century representations are combined to explore the conceptual interdependence of sexuality, reproduction, family life, and "female disorders." The subjects of *Theatres of Madness* are white, Anglo-European women who are diagnosed and treated for their "insanity," based on the interrelations of their class and gender. Definitions of "female disorders" are revealed discursively: described within documented case-histories, medical treatises, pharmaceutical advertising, "found" photographs (that I have sometimes manipulated), and fictional and diaristic texts.[1]

By pairing and layering these various source materials I have attempted to allow for a reading that dislocates and questions the "scientific" nature of observation. The juxtapositions also serve to address complex sets of relations between individuals and institutions, relations that overdetermine the internalization of oppression and, in turn, the degrees of complicity and resistance to that oppression.

Introduction

My childhood and early adolescence were spent in the industrial North of England. We lived at the bottom-end of Butterstile Lane in Prestwich, a village in Lancashire at the edge of Manchester. The street name derived from a stile marking the entrance to a vanished dairy farm. Through our kitchen window I could see gigantic cooling-towers. Billowing steam issued from their concrete mouths, like burned milk bubbling over the brim of giant saucepans. Our semi-detached was southeast of the Prestwich Lunatic Asylum and southwest of Strangeways Prison. These living relics of late Victorian social hygiene were the coordinates and echoes of my mental space. My imagination strained to see beyond this blackened brick, suburban horizon.

The village was panoptically contained between prisons: between the inside-and-outside of high-walled edifices—imposing and ever-present paradigms of interconnected transgression. In our various states of incarceration, we were present in each other's lives, invisibly implicated in each other's lives. Indeed, the strategic placement of both institutions served to maintain order, visible and palpable on the epidermis of daily life. The asylum and the prison were the sentry boxes at the borders of all my childhood images. Nothing escaped surveillance: nursery rhymes, lies, fantasies and fears.

Women newly released from the asylum would come knocking on our back door, searching for work. Sometimes my mother would pay them to scrub the front steps. I would look on, fixed, somewhere between horror and fascination. I used to lie awake at night, terrified that "mad" Barbara was burrowing into our house. Terrified that the earth tremors, caused by underground explosions from the local coal mines, were Barbara coming to take me away. On the school playgrounds and in the streets, children divided themselves into prisoners and wardens, the sane and the insane. We tormented the labeled. We tormented each other.

I heard my mother scream at my father that he was driving her mad. I heard my mother scream at my sisters that we were all driving her mad. I overheard fragmented conversations between my mother and other women. I saw movies on television about women: women accused of inheriting their mother's madness. Were all women potentially mad? Would my mother, my sisters—would I myself—end up behind the wall? I gradually realized, after many years of observing women enter and re-enter asylums, how fragile is the line separating the sane from the insane, how telling the stains of insanity, visible upon the surfaces of women's bodies. For a woman, it seemed as though the inside-and-outside distinction was contingent. The weave of her dress, the perception of her desires and maternal inclinations determined when, how, and if she might be put away.

A series of nineteenth-century photographs and case histories of women diagnosed as hysterical and insane confronted me twenty years later while researching at the New York Psychiatric Institute. Twisted bodies and faces, simultaneously gazing inward and outward, appeared frozen—stilled at the instant of misrecognition. Sitting and standing before the camera, coerced by the mechanical power of their own reflected images, they submitted: they were, finally, what they had become. Hysterical.

Accumulations of evidence jolt the fragile strata of memory, demanding exposure. Words and images form to name the raw, uncovered appearances of sadness and anger. I had found what was inextricably bound and connected to my own past and present. In rewriting their stories I was rewriting my own.

Enlargement of the Clitoris.

This organ is not only found much larger than usual as a congenital malformation, but it sometimes requires the care of the surgeon from hypertrophy of its natural structure or morbid deposition into its tissue. Scarcely any organ is so liable to enlargement from frequent excitation,† and this in its turn prompts to a repetition of the excitement. The examples on record are very numerous, and, in some instances, it has been found of enormous size,‡ in others more moderate, it has given rise to a doubt as to the sex of the individual. In the majority of these cases, however, it does not exceed two inches in length.

The primary *symptoms*, or those which arise from the mechanical disproportion of the parts, are trifling; in some cases, sexual intercourse has been impeded, and in most, from the situation of the part and its great sensibility, it is liable to irritation from motion, and the consequences of this susceptibility form by far the most important feature of the disease. The sexual desire naturally leads to its gratification, and this again aggravates the complaint, and impels to further excess, until the patient at length falls a victim to nymphomania.

The hypertrophy may be congenital, or the result of inflammation. This part has also been found the seat of scirrhous deposition, most frequently connected with a similar morbid condition of the uterus, and ultimately running into ulceration, with lancinating pain and foetid discharge, but giving rise to few or none of the secondary or nymphomaniacal symptoms.

Treatment. If the hypertrophy be slight and the symptoms not excessive, relief may sometimes be afforded by cooling or astringent lotions, or touching the part with caustic: but if the enlargement be so considerable as to occasion physical inconvenience or excessive sexual indulgence, amputation will be necessary.* Some blood is usually lost, but cold or caustics will always restrain the hemorrhage. Astringent lotions should be used for some time, and the patient kept in a state of absolute rest.

If, when the clitoris is enlarged from morbid deposition, we can ascertain that the uterus is free from disease, we might, under favourable circumstances, remove that organ, but there are very few cases which will be permanently cured by this proceeding, so apt is the disease to be reproduced and extended. In performing the operation, great care should be taken to excise the whole of the diseased portion.

—Fleetwood Churchill, *Outlines of the Principal Diseases of Females,* 1839

The female body is a scrutinized and supervised (anti)body. Invented in the name of science and deciphered in the name of truth, its disease is located in its gender. Its gender is its fate. A body (dis/un)covered, (de/re)sexed, (dis)guised.

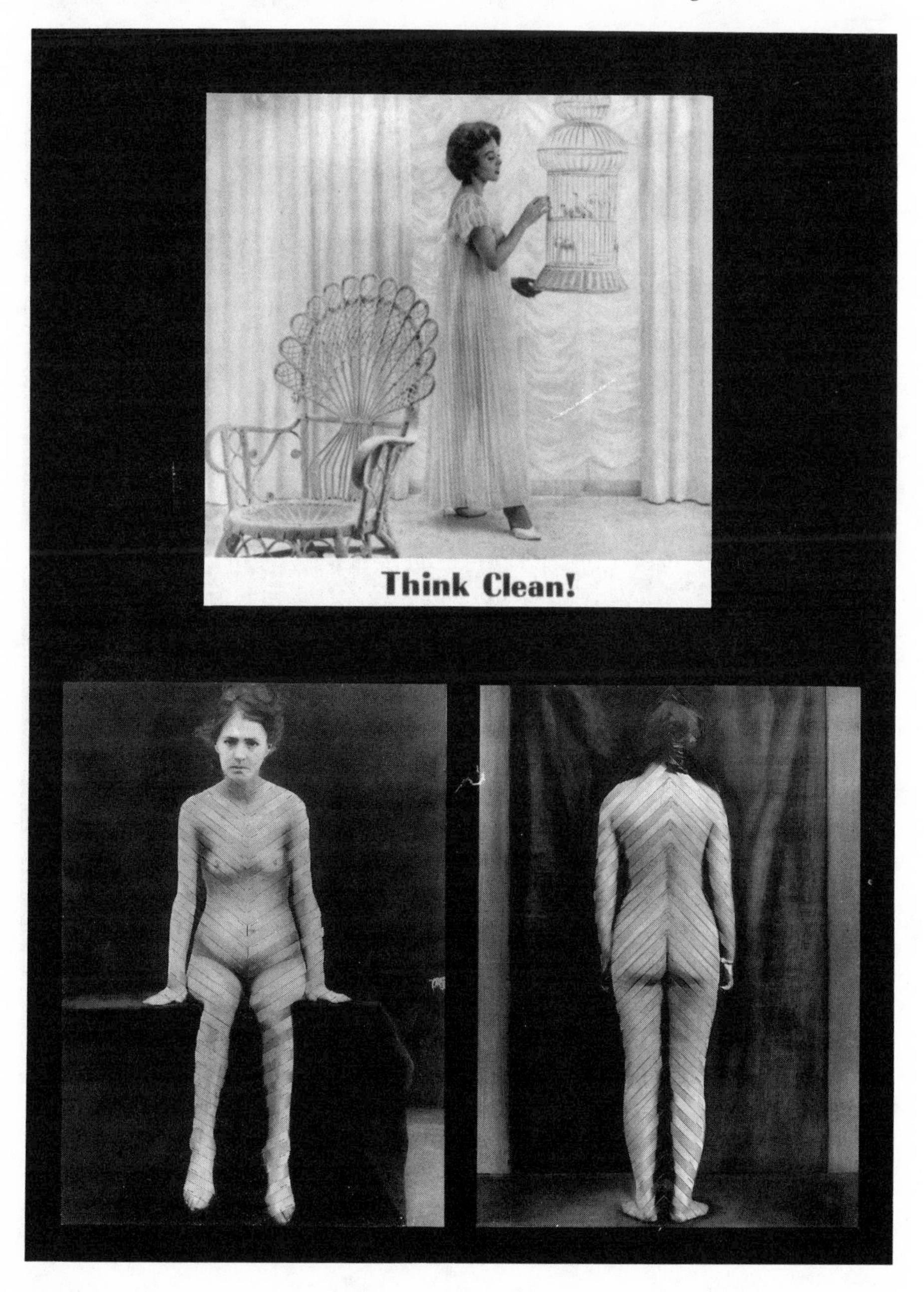

I feel ill. My nerves are raw and I have pains in my groin. I sit with my head down. The shadows in the room are creating faces, intestines, and petals. She is staring at me. An image on the wall. Pain(t)ed face—yellow, green, pink flesh.

I am aging. My body is changing shape. I crawl into myself, into my mother. If only I could sever the root. Starve the egg. Murder the connection. Imago. I saw my newly born daughter encased in a tall, transparent body. Half male, half female. She wandered out of her room, across the hallway and disappeared. I breathed a sigh of relief. A sharp pain traversed my chest. My breasts filled with salt water. I expressed it into a watering can. I fed it to a dying jade plant in the living room. My daughter reappeared and asked me for some milk. I explained that I didn't have any. I said it had turned to blood. I suggested she ask her father. I said he might be able to produce some.

We visited my parents last week. As I unpacked the children's clothing, I suddenly remembered a remark the doctor made to my husband after the birth of our daughter. "Congratulations," he said, "and oh, by the way, I put in an extra stitch for you."

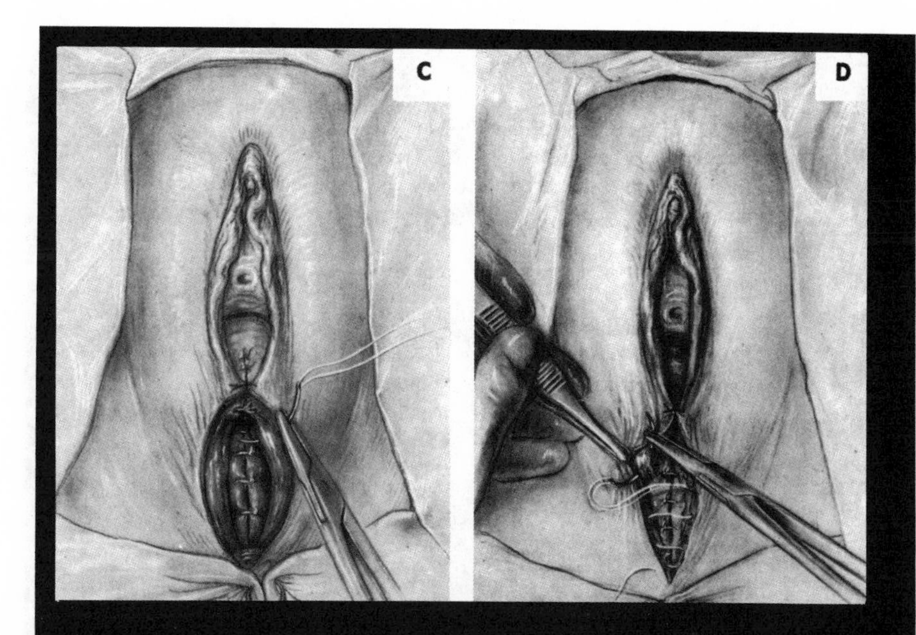

to help you transform a tense, irritable, depressed patient into a woman who is receptive to your counsel and adjusted to her environment

DEXAMYL® SPANSULE®

Case 10

M. G., a single needle-woman, aged 17; height 4 ft. 9 in. Came to Dr. Head in the London Hospital on July 16, 1897. History.—Patient's mother was a servant, and was married at 20. Her father was a dock labourer and was married at age 22. Her mother is alive. No collateral insanity. Father died of an accident, aged 46.

There were twelve children in the family as follows:—(1) Dead, five months foetus; (2) Dead, five or six month foetus; (3) Dead, six or seven month foetus; (4) Dead, seven months foetus, lived eight hours; (5) Born alive; living "very delicate;" ulcers on his legs; inflammation of his eyes; (6) Patient; (7) Girl, living, well, aged 16; (8) Boy, living, well, aged 14; (9) Boy, living well, aged 12; (10) Boy, died of convulsions; (11) Girl, died at three months, something the matter with her brain and club foot; (12) Boy, living, well, aged 7. Patient's History.—Patient is a medium-sized, somewhat squat girl of 17. There is no scarring on her body and her breasts are largely developed. Her control over her sphincters is good, and she has only once wetted the bed. She has a heavy demented look. She obeys well. She is careless in her dress and her hair is untidy. She is not destructive. No masturbation has been observed.

On October 21, 1897, patient was transferred to Claybury Asylum under the following medical certificate:—Patient is evidently of weak mind and her general condition indicates general paralysis in an early stage. The patient looks as though she has put on flesh. Responds to calls of nature. Has had no faints or fits. Is quiet and obedient.

October 20, 1898.—Patient is dull and heavy and cannot answer questions intelligently. Attention poor and memory bad. No improvement, eats and sleeps well.

November 2, 1898. Is in laundry; has been brighter lately. Dirty habits. Is very constipated.

February 21, 1899. Became much worse and stuporose. Her symptoms and behaviour, including grinding of the teeth, are those of general paralysis.

May 18.—Progressive enfeeblement. She is very nasty tempered and most destructive.

July 2.—Gradually became worse and died at 3.55 am.

Post-mortem. Body emaciated. Abrasion of skin over sacrum.

Cause of Death. General paralysis of insane; gangrene of lung and general tuberculosis.

—Frederick Walker Mott, "Twenty-Two Cases of Juvenile General Paralysis," 1899

6 August, 1652. About seven weeks after I married it pleased God to give me the blessing of conception. The first quarter I was exceedingly sick in breeding, till I was quick with child; . . . Mr Thornton had a desire that I should visit his friends at Newton. I passed down a foot a very high wall . . . Each step did very much strain me . . . This . . . killed my sweet infant in my womb . . . who lived not so long as we could get a minister to baptize it . . . after the miscarriage I fell into a terrible shaking ague. . . . The hair on my head came off, my nails of my fingers came off, my teeth did shake; and ready to come out and grew black. . . .

Alice Thornton, my second child was born near Richmond in Yorkshire the 3rd day of January, 1654.

Elizabeth Thornton, my third child was born at Hipswell the 14th of February, 1655 and died the 5th of September, 1656.

Katherine Thornton, my fourth child, was born at Hipswell . . . the 12th of June, 1656.

. . . on the delivery of my first son and fifth child at Hipswell the 10th of December, 1657 . . . the child stayed in the birth, and came crosswise with his feet first, and in this condition continued till Thursday morning . . . at which I was upon the rack in bearing my child with such exquisite torments, as if each limb were divided from the other . . . but the child was almost strangled in the birth, only living about half an hour, so died before we could get the minister to baptize him. . . .

17th of December, 1660. It was the pleasure of God . . . to bring forth my sixth child . . . a very goodly son . . . after a hard labour and hazardous. The child died two weeks later.

19th September, 1662. . . . I was delivered of Robert Thornton . . . it pleased the great God to lay upon me, his weak handmaid, an exceeding great weakness, beginning, a little after my child was born, by a most violent and terrible flux of blood, with such excessive floods all that night that . . . my dear husband, and children and friends had taken their last farewell. I was delivered and spared from that death. . . .

23rd September, 1665. Pregnant once more. I being terrified with my last extremity, could have little hopes to be preserved this . . . if my strength were not in the Almighty. . . . It pleased the Lord to make me happy with a goodly strong child, a daughter, after an exceeding sharp and perilous time the child died on January 24.

Christopher Thornton, my ninth child, was born on Monday, 11th November, 1667 . . . it pleased his Saviour . . . to deliver him out of this miserable world on 1st December, 1667.

—From "The Autobiography of Mrs Alice Thornton," 1875

Many of the women in my family were diagnosed as mentally unstable. Aunt Dora committed suicide in 1978. She took an overdose of Mellaril. Her doctor had prescribed it for "her condition." Her husband was an alchoholic. He was often unemployed. He was seldom home. Aunt Rose was convinced she had throat cancer. Her husband always completed her sentences. She found it hard to swallow. He was a marriage-guidance counselor in his spare time. He left her for a younger woman; one of his clients. Her doctor prescribed Niamid. He told her to take up knitting. Aunt Sadie disliked sexual relations with her husband. Her father had forced her to marry a man thirty years her senior. She was fifteen at the time. Her doctor prescribed Sinequan. He said it would help her to relax. Aunt Vera had insomnia. Her husband attended weekly "business" dinners Tuesday and Thursday nights. He didn't get home until the following evening. She discovered he had another "wife" in a neighboring town. He lived under a different name three days a week. Her doctor prescribed Placidyl. He said it would relieve her nervousness.

Case 5

M. B., aged 20, single; occupation, stringer. Admitted, April 9, 1895. Died, May 8, 1895.

On admission the patient was found to be a fairly-nourished girl, with brown hair. Height 5 ft. 2 in.; weight, 8 st. Mentally.—She is exceedingly noisy and restless, constantly throws herself about, and laughs and shouts. Her answers are incoherent. Before admission she is noted in the certificate as follows:—"She says she is followed about all day and night by men and women. She hears marriage bells and voices in America calling her. She sees strange people in the room at night."

Past History.—She has suffered from pains in the head and has been much worried by the want of work. She has never been insane before, and has only been noticed to be so for the last three weeks.

April 13.—Patient is suffering from mania. She is noisy and excitable and erotic. Is shamefaced and coy, will not answer questions, refuses to give particulars of herself, will suddenly shout out tirades against everyone in general; sings and behaves in an uncontrollable manner. She is in fair health and good condition, and looks more than her age.

April 25.—Patient is very restless and troublesome. She throws herself about and has erotic manners. She eats and sleeps well.

April 27.—She is quieter and more staid. Clean in her habits.

May 3.—Patient is inclined to be restless and troublesome. Dirty and of wild appearance.

May 5.—She was very noisy and restless two nights ago, and threw herself about. She was exhausted in the morning. To-day there is distension of the abdomen, and pain. The bowels are not open, and a glycerine enema was given, with no result. The bed was wet and slightly stained with dirty brown discharge. The catheter was passed but nothing came away.

May 7.—To-day patient is very collapsed, lies on her back, and the abdomen is distended. The catheter was passed, and two ounces of brown, slightly turbid urine drawn off. She takes nourishment well, and there is no pain or tenderness.

May 8.—She gradually sank, and died of syncope.—H. BOYLE.

Post-mortem.—The cause of death was found to be general paralysis. Ruptured bladder. Both ovaries were collections of cysts.

—Frederick Walker Mott, "Twenty-two Cases of Juvenile General Paralysis," 1899

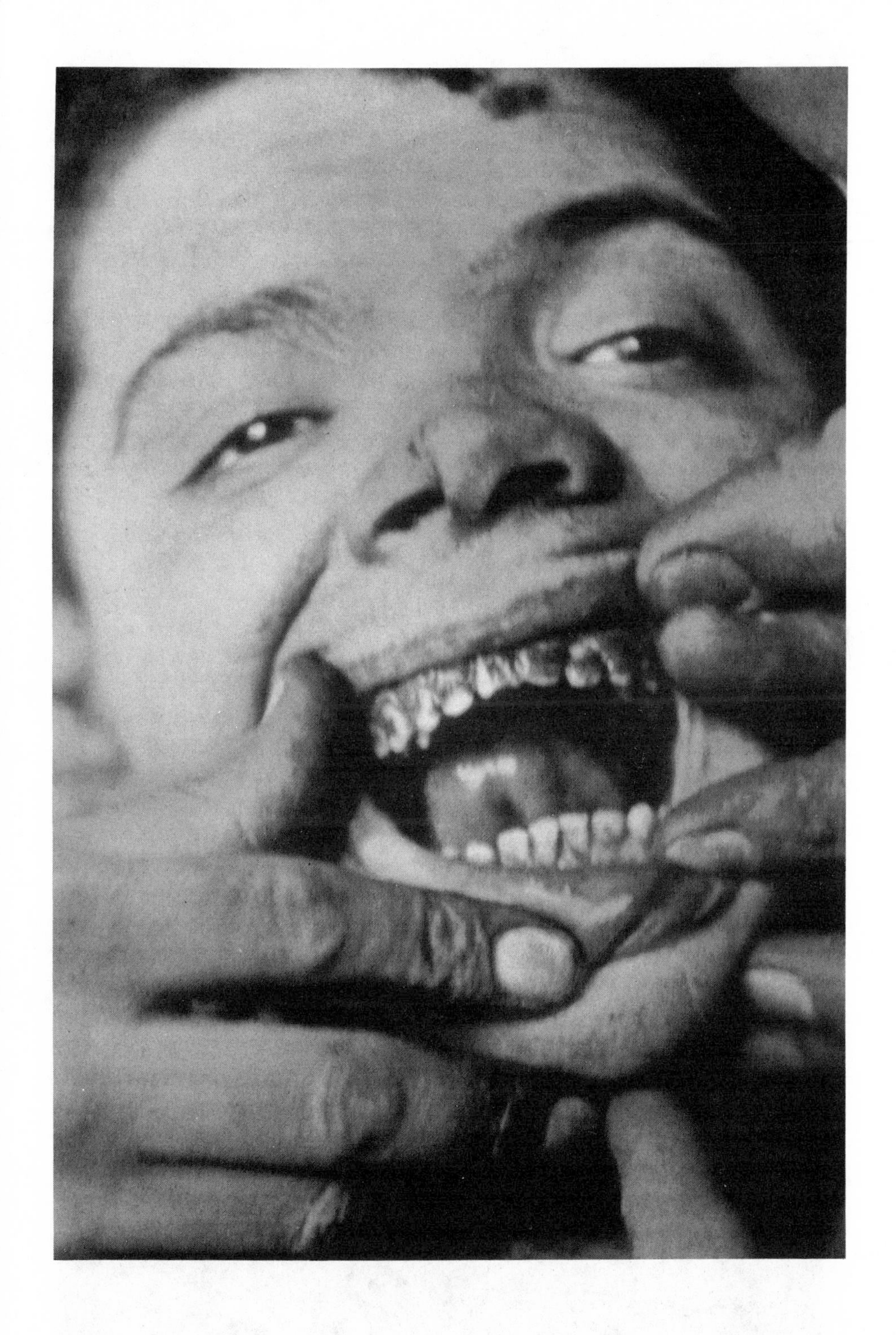

Ann Morgan's Love: A Pedestrian Poem

Her strong, bare sinewy arms and rugged hands
Blacken'd with labour; and her peasant dress
Rude, coarse in texture, yet most picturesque,
And suited to her station and her ways;
All these, transfigured by that sentiment
of lowly contrast to the man she served,

Grew dignified with beauty and herself
A noble working woman, not ashamed
Of what her work had made her
A grace, a glow of quick intelligence
And ardour, such as only Nature gives
And only gives through Man.

—A. J. Munby, 1896

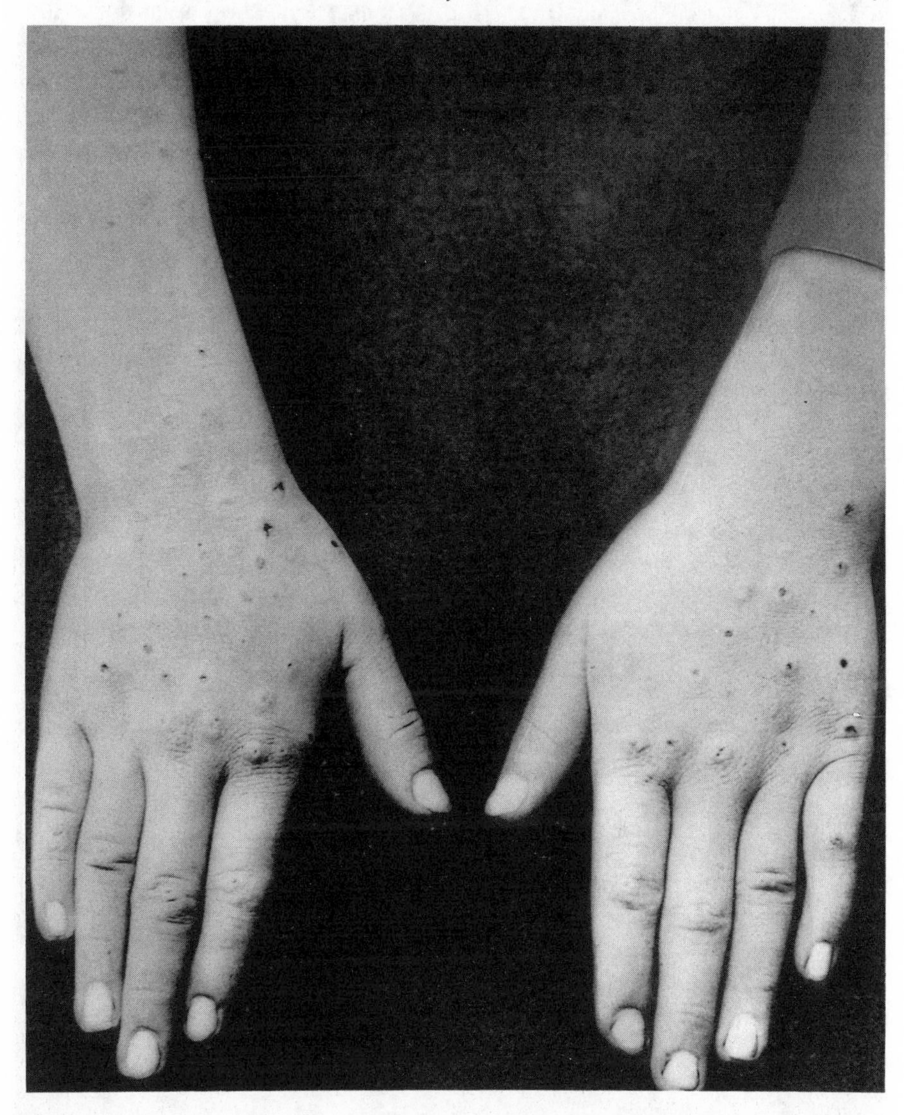

Every medical practitioner must have met with a certain class of cases which has set at defiance every effort of diagnosis, baffled every treatment, and belied every prognosis. . . .

The period when such illness attacks the patient is about the age of puberty and from that time up to almost every age the following train of symptoms may be observed, some being more or less marked than others in various cases. The patient becomes restless and excited, or melancholy and retiring; listless and indifferent to the social influences of domestic life. . . . She will always be ailing, and complaining of different affectations. . . . There will be quivering of the eyelids, and an inability to look one straight in the face. . . . Often a great disposition for novelties is exhibited, the patient desiring to escape from home, fond of becoming a nurse in hospitals or other pursuits of the like nature, according to station and opportunities.

To these symptoms in the single female will be added, in the married, distaste for marital intercourse, and very frequently either sterility or a tendency to abort in the early months of pregnancy. These physical evidences of derangement, if left unchecked, gradually lead to more serious consequences. . . .

Having ascertained the cause and nature of the disease, there are two points to be considered before operative measures are decided on. First, as to age. Although there is no doubt that patients may suffer from peripheral irritation of the pudic nerve from the earliest childhood, I never operate or sanction an operation on any patient under ten years of age, which is the earliest date of puberty.

There are again, after puberty, cases which give rise to but slight disturbance, but in which the sufferers are they who love to enlist sympathy from the charitable and will be ill, or affect to be ill, in spite of any and every treatment.

When I have decided that my patient is a fit subject for sugical treatment, I at once proceed to operate, after the ordinary preliminary measures of a warm bath and clearance of the portal circulation. The patient having been placed completely under the influence of chloroform, the clitoris is freely excised either by scissors or knife—I always prefer the scissors. The wound is then firmly plugged with graduated compresses of lint, and a pad, well secured by a T bandage.

A grain of opium is introduced per rectum, the patient placed in bed, and most carefully watched by a nurse, to prevent haemorrhage by any disturbance of the dressing. The neglect of this precaution will be frequently followed by alarming haemorrhage, and consequent injurious results. The diet must be unstimulating, and consist of milk, farinaceous food, fish and occasionally chicken; all alcoholic or fermented liquors being strictly prohibited. The strictest quiet must be enjoined, and the attention of relatives, if possible, avoided, so that the moral influence of medical attendant and nurse may be uninterruptedly maintained.

A month is generally required for perfect healing of the wound, at the end of which time it is difficult for the uninformed, or non-medical to discover the operation. The rapid improvement of the patient immediately after removal of the source of irritation is most marked; first in countenance, and soon afterwards by improved digestion and other evidences of healthy assimilation.

It cannot be too often repeated, that this improvement can only be made permanent, in many cases, by careful watching and moral training, on the part of both patients and friends. In the large majority of cases, I have administered no medicines, trusting entirely to recovery, after the removal of the source of irritation.

—Isaac Baker Brown, *On the Curability of Certain Forms of Insanity, Epilepsy, Catalepsy and Hysteria in Females*, 1866

[illegible] illness attacks the patient about the age of puberty an[illegible]
the following train of symptoms may be observed, some being more or l[illegible]
ees.

t becomes restless and excited, or melancholy and retiring, listless
of domestic life ... She will always be ailing and complaining of
be quivering of the eyelids, and an inability to look one straight
eat disposition for novelties is exhibited, the patient desiring to
nurse in hospitals or other pursuits of the like nature, according

mptoms in the single female will be added, in the married, distaste
ently either sterility or a tendency to abort in the early months of
f derangement, if left unchecked, gradually lead to more serious co[illegible]

rtained the cause and nature of the disease, there are two points t[illegible]
e decided on. First, as to age. Although there is no doubt that [illegible]
irritation of the pudic nerve from the earliest childhood, [illegible]
ent under ten years of age, which is the earliest date of pu[illegible]

gain, after puberty, cases which give rise to but slight distu[illegible]
o love to enlist sympathy from the charitable and will be ill, [illegible]
reatment.

decided that my patient is a fit subject for surgical treatment, I
rdinary preliminary measures of a warm bath and clearance of the po[illegible]
placed completely under the influence of chloroform, the clitoris
knife - I always prefer the scissors. The wound is then firmly pl[illegible]
d a [illegible] secured by a T bandage.

[illegible] is introduced per rectum, the patient placed in bed and most
aemorrhage by any disturbance of the dress[illegible] neglect of thi[illegible]
alarming haemorrhage, and consequent inju[illegible]ults. The diet
milk, farinaceous food, fish and occasion[illegible]; all alcoholi[illegible]
ohibited. The strictest quiet must be enj[illegible] the attention
that the moral influence of medical atte[illegible]se may be un-

By the middle of the nineteenth century, women constituted the majority of cases in public lunatic asylums. Uncontrolled sexuality was diagnosed as a symptom of insanity. The root of this malady was traced to the body's forbidden source of pleasure.

New fields of medical practice produced treatises on diseases peculiar to women. Guardians of morality praised and rewarded those who diagnosed desire and enforced institutionalization. The law sanctioned and protected those who called themselves experts in female disorders: those who mutilated the body by performing clitoridectomies and, in some cases, the removal of the labia, on prepubescent girls. The (dis)ease was cut out like a malignant tumor.

CASE XLIII. INCIPIENT SUICIDAL MANIA—MANY YEARS' GRADUAL ILLNESS—OPERATION—CURE.

R. T., aet. 39, single; admitted to the London Surgical Home Oct. 22, 1861.

History.—Has been ailing for many years, and given great trouble and anxiety to her friends. For some time past she has been very strange in her manner, very restless, never quiet, constantly wakeful, threatening suicide, talking to people, even perfect strangers, of her ailments and their causes, of which she is fully conscious. Was formerly modest and quiet.

On examination, she is a fine woman, of restless appearance and manner; eye wandering and unsteady; pupil dilated. The cause of her mental derangement being obvious, on October 24 the usual operation was performed. The improvement in her mental and bodily health was wonderful: she gained flesh and became cheerful and modest. She was discharged six weeks after admission. When heard of in February, 1863, this patient continued quite well.

—Isaac Baker Brown, *On the Curability of Certain Forms of Insanity, Epilepsy, Catalepsy and Hysteria in Females,* 1866

CASE XXII. NINE YEARS' ILLNESS—EPILEPTIFORM ATTACKS—THREE YEARS' DURATION—OPERATION—CURE.

G. M., single; admitted to the London Surgical Home December 18, 1860.

History.—For the last nine years has suffered greatly and regularly during the menstrual periods. Has been much worse for the last three years, during which time has, at each menstrual period, been frequently taken in a fit, dropping down suddenly and fainting right off; this state lasting for two or three hours. Being in service, this has caused her much trouble, as none of her employers would keep her. For the last six months has suffered severe pain over right ovary, increased by exercise or pressure, and at the menstrual period. Believing that the dysmenorrhoea and fits both arose from the same cause, on January 3, clitoris was cut down to the base. After this operation she never had a fit, and all untoward symptoms left her except the dysmenorrhoea; she was therefore re-admitted May 27, 1861, and there being some narrowing of the cervix, it was incised with hysterotome. June 21, catemenia came on without pain, and continued to do so regularly. In July she was well enough to return to service.

April, 1865. Her mother called at my house to say that this patient had been married some months, and was shortly expecting her confinement. She had remained quite well since the operation.

CASE XXXI. CATALEPTIC FITS—TWO YEARS' ILLNESS—OPERATION—CURE

M. N., aet. 17; admitted into the London Surgical Home September 4, 1861.

History.—Was perfectly well up to the age of fifteen, when she went to a boarding school in the West of England. In the course of three or four months she became subject to all symptoms of hysteria, and from that time gradually got worse, having fits, at first mild in character and of rare occurrence, but gradually more severe and frequent, till she became a confirmed cataleptic. For several months before admission, she had been attacked with as many as four or five fits a day, and during the whole journey from the North of England to London she was unconscious and rigidly cataleptic. She was seen immediately upon arrival, and there was no doubt that it was a genuine case of this disease. So sensitive was she, that if any one merely touched her bed, or walked across the room, she would immediately be thrown into a cataleptic state.

Before making any personal examination, Mr Brown ascertained both from her mother and herself, that she had long indulged in self-excitation of the clitoris, having first been taught by a school fellow. The commencement of her illness corresponded exactly with the origin of its cause; in fact, cause and effect were here so perfectly manifested, that it hardly wanted anything more than the history to enable one to form a correct diagnosis. All the other symptoms attending these cases were, however, well marked.

The next day after admission she was operated upon, and from that date she never had a fit. She remained in the Home for several weeks. Five weeks after operation, she walked all over Westminster Abbey, whereas for quite a year and a half before treatment, she had been incapable of the slightest exertion.

—Isaac Baker Brown, *On the Curability of Certain Forms of Insanity, Epilepsy, Catalepsy and Hysteria in Females,* 1866

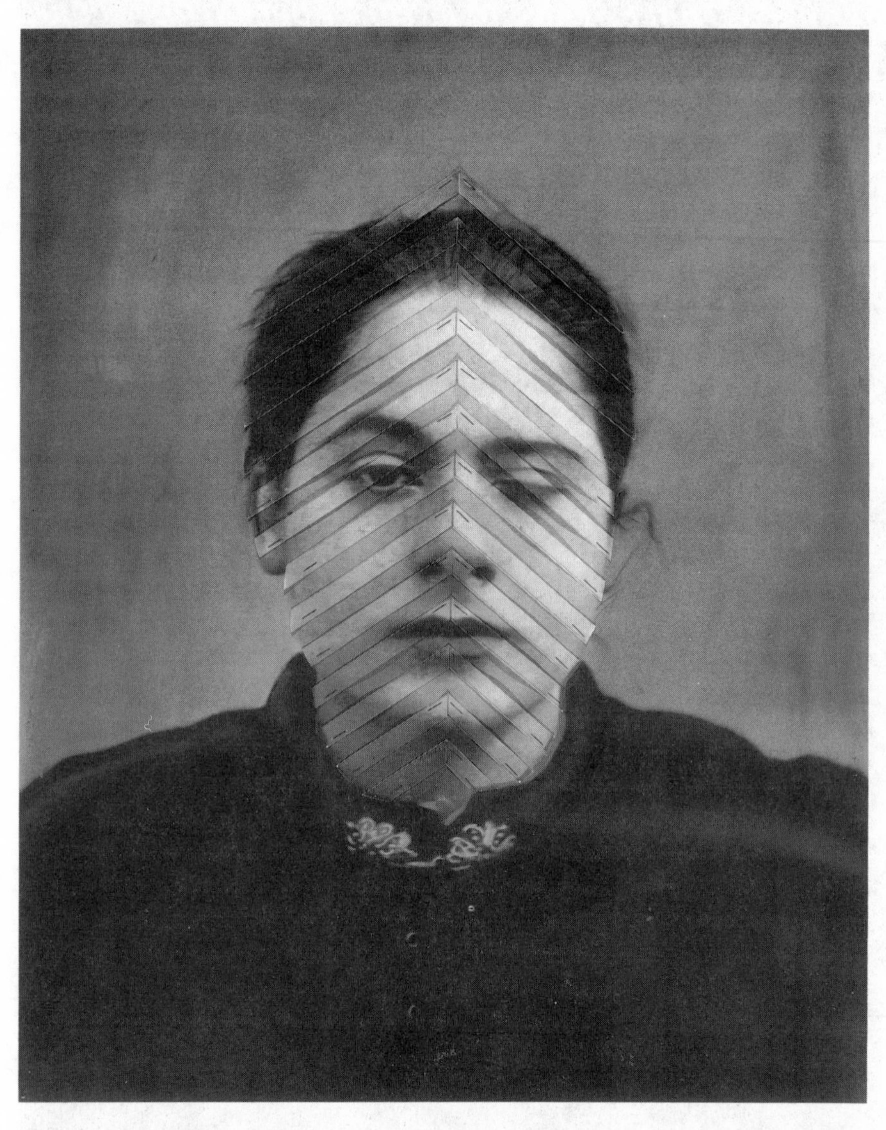

Chastity. Domesticity. Marriage. Motherhood. The middle-class body bears the burden of bourgeois ideology. It is measured, managed, and regulated. Treatments are introduced to control its reproductive cycles and system. It mirrors its hysterical construction.

I am so angry. I can hardly compose this letter. In your March issue, the article on genital mutilation made me dizzy. This anger later turned to fury. I work in a nursing home as a certified nurse's aide. Many of the patients need bed changes at least twice a shift, so I see about 90 percent of the people's genitals. They must be inspected often and cleaned thoroughly, as needed. I didn't quite understand the strange scars I saw. Then I read your article. Out of the forty residents of this home I can count five women who have had clitoridectomies. My God! Why? Who decided to deny them orgasm? Who made them go through such a procedure? I want to know. Was it fashionable? Or was it to correct a "condition"? I'd like to know what this so-called civilized country used as its criteria for such a procedure. And how widespread is it here in the United States?

—Letter to the editors, *Ms* 9; no. 1 (July 1980), p. 12.

Clitoris

Where the upper folds of the inner lips converge is a structure which resembles a very small penis (*clitoris*). It is plentifully supplied with nerves and blood vessels, becomes engorged, and enlarges and throbs under sexual excitement. It is a major source of pleasure in sexual intercourse. Since it is probably the most sensitive of the female sex organs, girls and women sometimes agitate it with their hands or with thigh pressure to gain sexual gratification. In female masturbation it is the organ most frequently used to bring on a climax, particularly by the young girl who does not wish to disturb her hymen by inserting things into her vagina or manipulating it.

Because the clitoris is so sensitive, and is sometimes the instrument of orgasm, a man preparing a woman for sexual intercourse may stimulate it by gentle massage. This and similar preparatory techniques—kissing, breastplay, caressing and fondling—are called sex foreplay. The newly married woman may be accustomed to deriving pleasure and even climax from clitoral manipulation, but not from insertion of the penis into the vagina. When full response, or orgasm, to the insertion and action of the penis is achieved, it may be found to be the more strongly pleasurable and more deeply satisfying of the two. The husband, however, may have to stimulate the clitoris considerably with his penis or hands before and during intercourse to assist his wife with the transition.

Female Orgasm

There is some question whether the female experiences two distinct types of orgasm. A woman usually reaches orgasm either by stimulation of the clitoris, or by vaginal penetration, or by some combination of both. Orgasm when reached, however, is a complex response difficult to ascribe to a single cause. At least the physiological response in orgasm for women seems to be the same no matter how the orgasm is precipitated. But women often note a qualitative difference between clitoral and vaginal orgasm, and express a distinct preference for the latter.

—Lawrence Q. Crawley, *Reproduction, Sex, and Preparation for Marriage,* 1964

It came on the youngest one's thirteenth birthday. They were eating cake. The chocolate icing was burned. There was a paraffin aftertaste from the dripping candles. The blood trickled down her thighs onto a green vinyl seat. Her mother said the cramping wouldn't last long. She offered her aspirin. Her doctor prescribed Valium. She went into the bathroom to get her a sanitary napkin. She kept them in an old toy box in the bathroom. She went to get a cigarette for herself. Those she stored in a biscuit tin on top of the refrigerator. They were born exactly five years apart. There were three of them. Girls. Her father had never wanted sons. He imagined daughters were easier to control. When it was time for the eldest child to start school, she thought about getting a job. They needed the money. He refused to let her work. He was uneasy when she left the house. He felt abandoned. He felt ashamed. She felt bereft. She became pregnant. Part of her shriveled. She became agoraphobic. When it was time for the second child to start school, she looked in the newspaper for a job. He told her she had no skills. He told her she was inefficient. She believed him. She became pregnant. She started to smoke. His words began to sound foreign. Her voice began to sound thin. The children learned to be suspicious of one another. Her mother explained how to attach a napkin to the belt. She complained it felt like a harness. She didn't know what to say. They went back to the table. Nobody wanted any more cake. The middle child said she had homework to do. Simultaneous equations. Once she'd asked her father to help her with a geometry assignment. She didn't understand his explanations. He often said she was stupid. She was afraid of her math teacher. She cleaned off the dinner table. She washed the dishes. She watched her mother creep upstairs and enter the linen closet. A string-bag hung on the inside of the closet door. It contained her daily chocolate supplies. She felt nauseous. She ate a walnut twist. She squeezed her body in between the bottom shelf and the floor. She pulled the door shut. It was her space. The youngest child started to cry. The aspirin wasn't helping. She needed somebody. The belt was rubbing against her swollen abdomen. It needed adjusting.

References

Brown, Isaac Baker
1866 *On the Curability of Certain Forms of Insanity, Epilepsy, Catalepsy and Hysteria in Females.* London: Robert Hardwicke, pp. 16–18.

Churchill, Fleetwood
1839 *Outlines of the Principal Diseases of Females.* Reprinted in *Classics of Medicine Library.* Birmingham, Ala., 1986, pp. 17, 18.

Crawley, Lawrence Q.
1964 *Reproduction, Sex, and Preparation for Marriage.* Englewood Cliffs, N.J.: Prentice-Hall.

Mott, Frederick Walker
1899 "Twenty-two Cases of Juvenile General Paralysis." In *Archives of Neurology and Psychiatry* (Pathological Laboratory of the London County Asylums) 1: 308, 315.

Ms.
1980 Letters to the editors, July, p. 12.

Munby, Arthur
1896 *Ann Morgan's Love: A Pedestrian Poem.* London: Reeves and Turner.

Thornton, Alice Wandesford
1875 *The Autobiography of Mrs Alice Thornton of East Newton, Co. York.* Durham, London, and Edinburgh: Surtees Society. Quoted in *Not in God's Image: Women in History from the Greeks to the Victorians.* Ed. Julie O'Faolain and Lauro Martines, pp. 237–39. New York: Harper and Row, 1973.

10

The Anthropometry of Barbie

Unsettling Ideals of the Feminine Body in Popular Culture

Jacqueline Urla and Alan C. Swedlund

It is no secret that thousands of healthy women in the United States perceive their bodies as defective. The signs are everywhere: from potentially lethal cosmetic surgery and drugs to the more familiar routines of dieting, curling, crimping, and aerobicizing, women seek to take control over their unruly physical selves. Every year at least 150,000 women undergo breast implant surgery (Williams 1992), while Asian women have their noses rebuilt and their eyes widened to make themselves look "less dull" (Kaw 1993). Studies show that the obsession with body size and the sense of inadequacy start frighteningly early; as many as 80 percent of 9-year-old suburban girls are concerned about dieting and their weight (Bordo 1991: 125). Reports like these, together with the dramatic rise in eating disorders among young women, are just some of the more noticeable fallout from what Naomi Wolf calls "the beauty myth." Fueled by the hugely profitable cosmetic, weight-loss, and fashion industries, the beauty myth's glamorized notions of the ideal body reverberate back upon women as "a dark vein of self hatred, physical obsessions, terror of aging, and dread of lost control" (Wolf 1991: 10).

It is this conundrum of somatic femininity, that female bodies are never feminine enough, that they must be deliberately and oftentimes painfully remade to be what "nature" intended—a condition dramatically accentuated under consumer capitalism—that motivates us to focus our inquiry into deviant bodies on images of the feminine ideal. Neither universal nor changeless, idealized notions of both masculine and feminine bodies have a long history that shifts considerably across time, racial or ethnic group, class, and culture. Body ideals in twentieth-century North America are influenced and shaped by images from classical or "high" art, the discourses of science and medicine, and increasingly via a multitude of commercial interests, ranging from mundane life insurance standards to the more high-profile fashion, fitness, and entertainment industries. Each have played contributing, and sometimes conflicting, roles in determining what will count as a desirable body in the late-twentieth-century United States. In this essay, we focus our attention on the domain of popular culture and the ideal feminine body as it is conveyed by one of pop culture's longest lasting and most illustrious icons: the Barbie doll.

Making her debut in 1959 as Mattel's new teenage fashion doll, Barbie rose quickly to become the top-selling toy in the United States. Thirty-four years and a woman's movement later, Barbie dolls remain Mattel's best-selling item, netting over one billion dollars in revenues worldwide (Adelson 1992), or roughly one Barbie sold every two seconds (Stevenson 1991). Mattel estimates that in the United States over 95 percent of girls between the ages of three and eleven own at least one Barbie, and that the average number of dolls per owner is seven (E. Shapiro 1992). Barbie is clearly a force to contend with, eliciting over the years a combination of critique, parody, and adoration. A legacy of the postwar era, she remains an incredibly resilient visual and tactile model of femininity for prepubescent girls headed straight for the twenty-first century.

It is not our intention to settle the debate over whether Barbie is a good or bad role model for little girls or whether her unrealistic body wrecks havoc on girls' self-esteem. Though that issue surrounds Barbie like a dark cloud, such debates have too often been based on literal-minded, decontextualized readings of popular culture. We want to suggest that Barbie dolls, in fact, offer a much more complex and contradictory set of possible meanings that take shape and mutate in a period marked by the growth of consumer society, intense debate over gender and racial relations, and changing notions of the body. Building on Marilyn Motz's (1983) study of the cultural significance of Barbie, and fashion designer extraordinaire BillyBoy's adoring biography, *Barbie, Her Life and Times,* we want to explore not only how it is that this popular doll has been able to survive such dramatic social changes, but also how she takes on new significance in relation to these changing contexts.

We begin by tracing Barbie's origins and some of the image makeovers she has undergone since her creation. From there we turn to an experiment in the anthropometry of Barbie to understand how she compares to standards for the "average American woman" that were emerging in the postwar period.[1] Not surprisingly, our measurements show Barbie's body to be thin—very thin—far from anything approaching the norm. Inundated as our society is with conflicting and exaggerated images of the feminine body, statistical measures can help us to see that exaggeration more clearly. But we cannot stop there. First, as our brief foray into the history of anthropometry shows, the measurement and creation of body averages have their own politically inflected and culturally biased histories. Standards for the "average" American body, male or female, have always been imbricated in histories of nationalism and race purity. Secondly, to say that Barbie is unrealistic seems to beg the issue. Barbie *is* fantasy: a fantasy whose relationship to the hyperspace of consumerist society is multiplex. What of the pleasures of Barbie bodies? What alternative meanings of power and self-fashioning might her thin body hold for women/girls? Our aim is not, then, to offer another rant against Barbie, but to clear a space where the range of her contradictory meanings and ironic uses can be contemplated: in short, to approach her body as a meaning system in itself, which, in tandem with her mutable fashion image, serves to crystallize some of the predicaments of femininity and feminine bodies in late-twentieth-century North America.

A Doll Is Born

> Parents thank us for the educational values in the world of Barbie. . . . They say that they could never get their daughters well groomed before—get them out of slacks or blue jeans and into a dress . . . get them to scrub their necks and wash their hair. Well, that's where Barbie comes in. The doll has clean hair and a clean face, and she dresses fashionably, and she wears gloves and shoes that match.
>
> —Ruth Handler, 1964, quoted in Motz

Legend has it that Barbie was the brainchild of Mattel owner Ruth Handler, who first thought of creating a three-dimensional fashion doll after seeing her daughter play with paper dolls. As an origin story, this one is touching and no doubt true. But Barbie was not the first doll of her kind, nor was she just a mother's invention. Making sense of Barbie requires that we look to the larger sociopolitical and cultural milieu that made her genesis both possible and meaningful. Based on a German prototype, the "Lili" doll, Barbie was from "birth" implicated in the ideologies of the Cold War and the research and technology exchanges of the military-industrial complex. Her finely crafted durable plastic mold was, in fact, designed by Jack Ryan, well known for his work in designing the Hawk and Sparrow missiles for the Raytheon Company. Conceived at the hands of a military-weapons-designer-turned-toy-inventor, Barbie dolls came onto the market the same year that the infamous Nixon-Krushchev "kitchen debate" took place at the American National Exhibition in Moscow. Here, in front of the cameras of the world, the leaders of the capitalist and socialist worlds faced off, not over missile counts, but over "the relative merits of American and Soviet washing machines, televisions, and electric ranges" (May 1988: 16). As Elaine Tyler May has noted in her study of the Cold War, this much-celebrated media event signaled the transformation of American-made commodities and the model suburban home into key symbols and safeguards of democracy and freedom. It was thus with fears of nuclear annihilation and sexually charged fantasies of the perfect bomb shelter running rampant in the American imaginary, that Barbie and her torpedo-like breasts emerged into popular culture as an emblem of the aspirations of prosperity, domestic containment, and rigid gender roles that were to characterize the burgeoning postwar consumer economy and its image of the American Dream.

Marketed as the first "teenage" fashion doll, Barbie's rise in popularity also coincided with, and no doubt contributed to, the postwar creation of a distinctive teenage lifestyle.[2] Teens, their tastes, and their behaviors were becoming the object of both sociologists and criminologists as well as market survey researchers intent on capturing their discretionary dollars. While J. Edgar Hoover was pronouncing "the juvenile jungle" a menace to American society, retailers, the music industry, and moviemakers declared the thirteen to nineteen-year-old age bracket "the seven golden years" (Doherty 1988:51–52).

Barbie dolls seemed to cleverly reconcile both of these concerns by personifying the good girl who was sexy, but didn't have sex, and was willing to spend, spend, spend. Amidst the palpable moral panic over juvenile delinquency and teenagers' new-found sexual freedom, Barbie was a reassuring symbol of solidly middle-class values. Popular teen magazines, advertising, television, and movies of the period painted a highly dichotomized world divided into good (i.e., middle-class) and bad (i.e., working-class) kids: the clean-cut, college-bound junior achiever versus the street-corner boy; the wholesome American Bandstander versus the uncontrollable bad seed (cf. Doherty 1988; and Frith 1981, for England). It was no mystery where Barbie stood in this thinly disguised class discourse. As Motz notes, Barbie's world bore no trace of the "greasers" and "hoods" that inhabited the many B movies about teenage vice and ruin. In the life Mattel laid out for her in storybooks and comics, Barbie, who started out looking like a somewhat vampy, slightly Bardot-esque doll, was gradually transformed into a " 'soc' or a 'frat'—affluent, well-groomed, socially conservative" (Motz 1983:130). In lieu of backseat sex and teenage angst, Barbie had pajama parties, barbecues, and her favorite pastime, shopping.

Every former Barbie owner knows that to buy a Barbie is to lust after Barbie accessories—that pair of sandals and matching handbag, canopy bedroom set, or country camper. Both conspicuous consumer and a consumable item herself, Barbie surely was as much the fantasy of U.S. retailers as she was the panacea of middle-class parents. For every "need" Barbie had, there was a deliciously miniature product to fulfill it. As Paula Rabinowitz has noted, Barbie dolls, with their focus on frills and fashion, epitomize the way that teenage girls and girl culture in general have figured as accessories in the historiography of post-war culture; that is as both essential to the burgeoning commodity culture as consumers, but seemingly irrelevant to the central narrative defining cold war existence (Rabinowitz 1993). Over the years, Mattel has kept Barbie's love of shopping alive, creating a Suburban Shopper Outfit and her own personal Mall to shop in (Motz 1983: 131). More recently, in an attempt to edge into the computer game market, we now have an electronic "Game Girl Barbie" in which (what else?) the object of the game is to take Barbie on a shopping spree. In "Game Girl Barbie," shopping takes skill, and Barbie plays to win.

Perhaps what makes Barbie such a perfect icon of late capitalist constructions of femininity is the way in which her persona pairs endless consumption with the achievement of femininity and the appearance of an appropriately gendered body. By buying for Barbie, girls practice how to be discriminating consumers knowledgeable about the cultural capital of different name brands, how to read packaging, and the overall importance of fashion and taste for social status (Motz 1987: 131–32). Being a teenage girl in the world of Barbie dolls becomes quite literally a performance of commodity display, requiring numerous and complex rehearsals. In making this argument, we want to stress that we are drawing on more than just the doll. "Barbie" is also the packaging, spin-off products, cartoons, commercials, magazines, and fan club paraphernalia, all of which contribute to creating her persona. Clearly, as we will discuss below, children may engage more or less with those products, subverting or ignoring various aspects of Barbie's "official" presentation. However, to the extent that little girls *do* participate in the prepackaged world of Barbie, they come into contact with a number of beliefs central to femininity under consumer capitalism. Little girls learn, among other things, about the crucial importance of their appearance to their personal happiness and to their ability to gain favor with their friends. Barbie's social calendar is constantly full, and the stories in her fan magazines show her frequently engaged in preparation for the rituals of heterosexual teenage life: dates, proms, and weddings. A perusal of Barbie magazines, and the product advertisements and pictorials within them, shows an overwhelming preoccupation with grooming for those events. Magazines abound with tips on the proper ways of washing hair, putting on makeup, and assembling stunning wardrobes. Through these play scenarios, little girls learn Ruth Handler's lesson about the importance of hygiene, occasion-specific clothing, knowledgeable buying, and artful display as key elements to popularity and a successful career in femininity.

Barbie exemplifies the way in which gender in the late twentieth century has become a commodity itself, "something we can buy into . . . the same way we buy into a style" (Willis 1991: 23). In her insightful analysis of the logics of consumer capitalism, cultural critic Susan Willis pays particular attention to the way in which children's toys like Barbie and the popular muscle-bound "He-Man" for boys link highly conservative and narrowed images of masculinity and femininity with commodity consumption (1991: 27). In the imaginary world of Barbie and teen advertising, observes Willis, being or becoming a teenager, having a "grown-up" body, is inextricably bound up with the acquisition of certain commodities, signaled by styles of clothing, cars, music, etc. In play groups and fan clubs (collectors are a whole world unto themselves), children exchange knowledge about the latest accessories and outfits, their relative merit, and how to find them. They become members of a community of Barbie owners whose shared identity is defined by the commodities they have or desire to have. The articulation of social ties through commodities, is, as Willis argues, at the heart of how sociality is experienced in consumer capitalism. In this way, we might say that playing with Barbie serves not only as a training ground for the production of the appropriately gendered woman, but also as an introduction to the kinds of knowledge and social relations one can expect to encounter as a citizen of a post-Fordist economy.

Barbie Is a Survivor

A field trip in 1991 to Evelyn Burkhalter's Barbie Hall of Fame, located just above her husband's eye, ear, and throat clinic in downtown Palo Alto, California, revealed a remarkable array of Barbie dolls from across the globe. With over 1,500 dolls on display, several thousand more in storage, and an encyclopedic knowledge about Barbie's history, Mrs. Burkhalter proudly concluded her tour of the dolls with an emphatic, "Barbie is a survivor!" Indeed! In the past three decades, this popular children's doll has undergone numerous changes in her fashion image and "occupations" and has acquired a panoply of ethnic "friends" and analogues that have allowed her to weather the dramatic social changes in gender and race relations that arose in the course of the sixties and seventies.

As the women's movement gained strength in the seventies, the media and popular culture felt the impact of a growing self-consciousness about sexist imagery of women. The toy industry was no exception. Barbie, the ever-beautiful bride-to-be, became a target of some criticism and concern for parents who worried about the effects such a toy would have on their daughters. Barbie buffs like BillyBoy describe the seventies as the doll's dark decade, a time when sales dipped, quality worsened as production was transferred from Japan to Taiwan, and Barbie was lampooned in the press (BillyBoy 1987). Mattel responded by trying to give Barbie a more diversified wardrobe and a more "now" image. A glance at Barbie's resumé, published in *Harper's* magazine in August 1990, while incomplete, shows Mattel's attempt to expand Barbie's career options beyond the original fashion model:

Positions Held

1959–present	Fashion model
1961–present	Ballerina
1961–64	Stewardess (American Airlines)
1964	Candy striper
1965	Teacher
1965	Fashion editor
1966	Stewardess (Pan Am)
1973–75	Flight attendant (American Airlines)
1973–present	Medical doctor
1976	Olympic athlete
1984	Aerobics instructor
1985	TV news reporter
1985	Fashion designer
1985	Corporate executive
1988	Perfume designer
1989–present	Animal rights volunteer

It is only fitting, given her origin, to note that Barbie has also had a career in the military and aeronautics space industry: she has been an astronaut, a marine, and, during the Gulf War, a Desert Storm trooper. Going from pink to green, Barbie has also acquired a social conscience, taking up the causes of UNICEF, animal rights, and environmental protection. According to Mattel, the doll's careers are chosen to "reflect the activities and professions that modern women are involved in" (quoted in *Harpers,* August 2, 1990, p. 20). Ironically, former Mattel manager of marketing Beverly Cannady noted that the doctor and astronaut uniforms never sold well. As Cannady candidly admitted to *Ms.* magazine in a 1979 interview, "Frankly, we only kept the doctor' s uniform in line as long as we did because public relations begged us to give them something they could point to as progressive" (Leavy 1979: 102). Despite their efforts to dodge criticism and present Barbie as a liberated woman, it is clear that glitz and glamour are at the heart of the Barbie doll fantasy. [3] Motz reports, for example, that in 1963 only one out of sixty-four outfits on the market was job-related. There is no doubt that Barbie has had her day as astronaut, doctor, rock star, and even presidential candidate. She can be anything she wishes to be, although it is interesting that the difference between occupation and outfit has never been entirely clear. As her publicists emphasize, Barbie's purpose is to let little girls dream. And that dream continues to be fundamentally about leisure and consumption, not production.

For anyone tracking Barbiana, it is abundantly clear that Mattel's marketing strategies are sensitive to a changing social climate. Just as Mattel has sought to present Barbie as a career woman with more than air in her vinyl head, they have also tried to diversify her otherwise lily-white suburban world. About the same time that Martin Luther King was assassinated and Detroit and Watts were burning in some of the worst race riots of the century, Barbie acquired her first black friend. "Colored Francie" appeared in 1967, failed, and was replaced the following year with Christie, who also did not do terribly well on the market. In 1980, Mattel went on to introduce Black Barbie, the first doll with Afro-style hair. She, too, appears to have suffered from a low advertising profile and low sales (Jones 1991). Nevertheless, the eighties saw a concerted effort on Mattel's part to "go multicultural," coinciding with a parallel preference in the pages of high-fashion magazines, such as *Elle* and *Vogue*, for racially diverse models. With the expansion of sales worldwide, Barbie has acquired multiple national guises (Spanish Barbie, Jamaican Barbie, Malaysian Barbie, etc.).[4] In addition, her cohort of "friends" has become increasingly ethnically diversified, as has Barbie advertising, which now regularly features Asian, Hispanic, and African American little girls playing with Barbie. Today, Barbie pals include a smattering of brown and yellow plastic friends, like Teresa, Kira, and Miko, who appear in her adventures and, very importantly, can share her clothes. This diversification has not spelled an end to reigning Anglo beauty norms and body image. Quite the reverse. When we line the dolls up together, they look virtually identical. Cultural difference is reduced to surface variations in skin tone and costumes that can be exchanged at will. Like the concomitant move toward racially diverse fashion models, "difference" is remarkably made over into sameness, as ethnicity is tamed to conform to a restricted range of feminine beauty.

Perhaps Mattel's most glamorous concession to multiculturalism is their latest creation, Shani. Billed as tomorrow's African American woman, Shani, whose name, according to Mattel, means "marvelous" in Swahili, premiered at the 1991 Toy Fair with great fanfare and media attention. Unlike her predecessors, who were essentially "brown plastic poured into blond Barbie's mold," Shani, together with her two friends, Asha and Nichelle (each a slightly different shade of brown), and boyfriend, Jamal, created in 1992, were decidely Afrocentric, with outfits in "ethnic" fabrics rather than the traditional Barbie pink (Jones 1991). The packaging also announced that these dolls' bodies and facial features were meant to be more like those of real African American women, although they too can interchange clothes with Barbie.

A realization of the growing market share of African American and Hispanic consumers has no doubt played a role in the changing face of Barbie. However, as *Village Voice* writer Lisa Jones has pointed out, there is a story other than simple economic calculus here. On the one hand, Mattel's social consciousness reflects the small but significant inroads black women have made into the company's top-level employee structure. It also underscores the growing authority of and recourse to expert knowledge, particularly psychological experts, as a way of understanding the social consequences of popular culture. As it turns out, Mattel product manager Deborah Mitchell and the principal fashion designer, Kitty Black-Perkins, are both African American (Barbie's hair designer is also non-Anglo). Both women had read clinical psychologist Dr. Darlene Powell-Hopson's *Different and Wonderful: Raising Black Children in a Race-Conscious Society* (Powell-Hopson and Hopson 1987), and, according to Jones's account, became interested in creating a doll that could help give African American girls a positive self-image. Mattel eventually hired Powell-Hopson as a consultant, signed on a public relations firm with experience in targeting black consumers, and got to work creating Shani. Now, Mattel announced, "ethnic Barbie lovers will be able to dream in their own image" (*Newsweek* 1990: 48). Multiculturalism cracked open the door to Barbie-dom, and diversity could walk in, so long as she was big-busted and slim-hipped, had long flowing hair and tiny feet, and was very, very thin.

"The icons of twentieth-century mass culture," writes Susan Willis, "are all deeply infused with the desire for change," and Barbie is no exception (1991: 37). In looking over the course of Barbie's career, it is clear that part of her resilience, appeal, and profitability stems from the fact that her identity is constructed primarily through fantasy and is consequently open to change and reinterpretation. As a fashion model, Barbie continually creates her identity anew with every costume change. In that sense, we might want to call Barbie the prototype of the "transformer dolls" that cultural critics have come to see as emblematic of the restless desire for change that permeates postmodern capitalist society (Wilson 1985: 63). Not only can she renew her image with a change of clothes, Barbie also is seemingly able to clone herself effortlessly into new identities—Malibu Barbie; Totally Hair Barbie; Teen Talk Barbie; even Afrocentric Barbie/Shani—without somehow suggesting a serious personality disorder. Furthermore, Barbie's owners are at liberty to fantasize any number of life choices for the perpetual teenager; she might be a high-powered fashion executive, or she just might marry Ken and "settle down" in her luxury condo. Her history is a barometer of changing fashions and changing gender and race relations, as well as a keen index of corporate America's anxious attempts to find new and more palatable ways of selling the beauty myth and commodity fetishism to new generations of parents and their daughters. The multiplication of Barbie and her friends translates the challenge of gender inequality and racial diversity into an ever-expanding array of costumes, a new "look" that can be easily accommodated into a harmonious and illusory pluralism that never ends up rocking the boat of WASP beauty.

What is striking, then, is that, while Barbie's identity may be mutable—one day she might be an astronaut, another a cheerleader—*her hyper-slender, big-chested body has remained fundamentally unchanged over the years*—a remarkable fact in a society that fetishizes the new and improved. Barbie did acquire flexible arms and legs, as we know, and her hair, in particular, has grown by leaps and bounds, making Superstar Barbie the epitome of a "big-hair girl." Collectors also identify three distinctive changes in Barbie's face: the original cool, pale look with arched brows, red, pursed lips, and coy sideways glance gave way in the late sixties to a more youthful, straight-haired, teenage look. This look lasted about a decade, and in 1977 Barbie acquired the exaggerated, wide-eyed, smiling look associated with Superstar Barbie that she still has today (Melosh and Simmons 1986). But her measurements, pointed toes, and proportions have not altered significantly in her thirty-five years of existence. We turn now from Barbie's "persona" to the conundrum of her body and to our class experiment in the anthropometry of feminine ideals. In so doing, our aim is deliberately subversive. We wish to use the tools of calibration and measurement—tools of normalization that have an unsavory history for women and racial or ethnic minorities—to destabilize the ideal. In this way, our project represents a strategic use of scientific measurement and the authority it commands against the powerfully normative image of the feminine body in commodity culture. We begin with a very brief historical overview of the anthropometry of women and the emergence of an "average" American female body in the postwar United States, before using our calipers on Barbie and her friends.

The Measured Body: Norms and Ideals

> The paramount objective of physical anthropology is the gradual completion, in collaboration with the anatomists, the physiologists, and the chemists, of the study of the normal white man under ordinary circumstances.
>
> —Ales Hrdlicka, 1918

As the science of measuring human bodies, anthropometry belongs to a long line of techniques of the eighteenth and nineteenth centuries concerned with measuring, comparing, and interpreting variability in different zones of the human body: craniometry, phrenology, physiognomy, and comparative anatomy. Early anthropometry shared with these an understanding and expectation that the body was a window into a host of moral, temperamental, racial, or gender characteristics. It sought to distinguish itself from its predecessors, however, by adhering to rigorously standardized methods and quantifable results that would, it was hoped, lead to the "complete elimination of personal bias" that anthropometrists believed had tainted earlier measurement techniques (Hrdlicka 1939: 12).[5] Although the head (especially cranial capacity) continued to be a source of special fascination, by the early part of this century, physical anthropologists, together with anatomists and medical doctors, were developing a precise and routine set of measurements for the entire body that would permit systematic comparison of the human body across race, nationality, and gender.[6]

Under the aegis of Earnest Hooton, Ales Hrdlicka, and Franz Boas, located respectively at Harvard University, the Smithsonian, and Columbia University, anthropometric studies within U.S. physical anthropology were utilized mainly in the pursuit of three general areas of interest: identifying racial and or national types; the measurement of adaptation and "degeneracy"; and a comparison of the sexes. Anthropometry was, in other words, believed to be a useful technique in resolving three critical border disputes: the boundaries between races or ethnic groups; the normal and the degenerate; and the border between the sexes.

As is well documented by now, women and non-Europeans did not fare well in these emerging sciences of the body (see the work of Blakey 1987; Gould 1981; Schiebinger 1989, 1993; Fee 1979; Russett 1989; also Horn and Fausto-Sterling, this volume); measurements of women's bodies, their skulls in particular, tended to place them as inferior to or less intelligent than males. In the great chain of being, women as a class were believed to share certain atavistic characteristics with both children and so-called savages. Not everything about women was regarded negatively. In some cases it was argued that women possessed physical and moral qualities that were superior to those of males. Above all, woman's body was understood through the lens of her reproductive function; her physical characteristics, whether inferior or superior to those of males, were inexorably dictated by her capacity to bear children. These ideas, none of which were new, were part of the widespread scientific wisdom that reverberated throughout the development of physical anthropology and informed a great deal of what many of its leading figures had to say about the shape and size of women's bodies. Hooton, in his classic *Up From the Ape* (1931) was to reguarly compare women, and especially non-Europeans of both sexes, to primates. Similarly, Hrdlicka's 1925 comparative study of male and female skulls went to rather extraordinary lengths to explain how it could be that women's brains (and hence, intelligence) were actually smaller than men's, even though his measurements showed females to have a cranial capacity relatively larger than that of males.[7] Boas stood alone as an exception to this trend toward evolutionary ranking and typing. Although he did not address sex differences per se, his work on European migrants early in the century served to refute existing hypotheses on the hereditary nature of perceived racial or ethnic physical differences. As such, his work pushed physical anthropology and anthropometry toward the study of human adaptability and variation rather than the construction of fixed racial, ethnic, or gender physical types.

It is striking that, aside from those studies specifically focused on the comparison of the sexes, women did not figure prominently in physical anthropology's attempt to quantify and typologize human bodies. In the studies of race and nationality, anthropometric studies of males far outnumbered those of females in the pages of the *American Journal of Physical Anthropology.* And where female bodies were measured, non-European women far outnumbered white women as the subjects of the calipers.[8] Although Hrdlicka and others considered it necessary to measure both males and females, textbooks reveal that more often than not it was the biologically male body that stood in as the generic and ideal representative of the race or of humankind. It is, in fact, somewhat unusual that, in the quote from Hrdlicka given above, he should have called attention to the fact that physical anthropology's main object of study was indeed the "white male," rather than the "human body." With males as the unspoken prototype, women's bodies were frequently described (subtly or not) as deviations from the norm: as subjects, the measurement of their bodies was occasionally risky to the male scientists,[9] and as bodies they were variations from the generic or ideal type (their body fat "excessive," their pelvises maladaptive to a bipedal [i.e., more evolved] posture, their musculature weak.) Understood primarily in terms of their reproductive capacity, women's bodies, particularly their reproductive organs, genitalia, and secondary sex characteristics, were instead more carefully scrutinized and measured within "marital adjustment" studies and in the emerging science of gynecology, whose practitioners borrowed liberally from the techniques used by physical anthropologists (see Terry, this volume).

In the United States, an attempt to elaborate a scientifically sanctioned notion of a normative "American" female body, however, was taking place in the college studies of the late nineteenth and early twentieth centuries. By the 1860s, Harvard and other universities had begun to regularly collect anthropometric data on their male student populations, and in the 1890s comparable data began to be collected from the East Coast women's colleges as well. Conducted by departments of hygiene, physical education, and home economics, as well as physical anthropology, these large-scale studies gathered data on the elite, primarily WASP youth, in order to determine the dimensions of the "normal" American male and female. The data from one of the earliest cosexual studies, carried out by Dr. Dudley Sargent, a professor of physical education at Harvard, were then used to create two life-sized statues that were exhibited at the Chicago World's Fair in 1893 and put on display at the Peabody Museum. Effectively excluded from these attempts to define the "normal" or average body, of course, were those "other" Americans—descendants of African slaves, North American Indians, and the many recent European immigrants from Ireland, southern Europe, and eastern Europe—whose bodies were the subject of racist, evolution-oriented studies concerned with "race crossing," degeneracy, and the effects of the "civilizing" process (see Blakey 1987).

Standards for the average American male and female were also being elaborated in a variety of domains outside of academia. By the early part of the twentieth century, industry began to make widespread commercial use of practical anthropometry: the demand for standardized measures of the "average" body manifested in everything from Taylorist designs for labor-efficient workstations and kitchens to standardized sizes in the ready-to-wear clothing industry (cf. Schwartz 1986). Certainly, one of the most common ways in which individuals encountered body norms was in the medical examination required for life insurance. It was not long before such companies as Metropolitan Life would rival the army, colleges, and prisons as the most reliable source of anthropometric statistics. Between 1900 and 1920, the first medicoactuarial standards of weight and height began to appear in conjunction with new theories linking weight and health. The most significant of these, the Dublin Standard Table of Heights and Weights, developed in 1908 by Louis Dublin, a student of Franz Boas and statistician for Metropolitan Life, became the authoritative reference in every doctor's office (cf. Bennett and Gurin 1982: 130–38). However, what began as a table of statistical averages soon became a means of setting ideal norms. Within a few years of its creation, the Dublin table shifted from providing a record of statistically "average" weights to becoming a guide to "desirable" weights that, interestingly enough, were notably below the average weight for most adult women. In her history of anorexia in the United States, Joan Brumberg points to the Dublin table, widely disseminated to doctors and published in popular magazines, and the invention of the personal, or bathroom, scale as the two devices most responsible for popularizing the notion that the human figure could be standardized and that abstract and often unrealistic norms could be uniformly applied (1988: 232–35).

By the 1940s the search to describe the normal American male and female bodies in anthropometric terms was being conducted on many fronts. Data on the average measurements of men and women were now available from a number of different sources, including surveys of army recruits from World War I, the longitudinal college studies, sample measurements from the Chicago World's Fair, actuarial data, and extensive data from the Bureau of Home Economics, which had amassed measurements to assist in developing standardized sizing for the garment industry. Between the two wars, nationalist interests had fueled eugenic interests and provoked a deepening concern about the physical fitness of the American people. Did Americans constitute a distinctive physical "type"; were they puny and weak as some Europeans had alleged, or were they physically bigger and stronger than their European ancestors? Could they defend themselves in time of war? And who did this category of "Americans" include? Questions such as these fed into an already long-standing preoccupation with defining a specifically American national character and, in 1945, led to the creation of one of the most celebrated and widely publicized anthropometric models of the century: Norm and Norma, the average American male and female. Based on the composite measurements of thousands of young people, described only as "native white Americans," across the United States, the statues of Norm and Norma were the product of a collaboration between obstetrician-gynecologist Robert Latou Dickinson, well known for his studies of human reproductive anatomy, and Abram Belskie, the prize student of Malvina Hoffman, who had sculpted the Races of Mankind series.[10] Of the two, Norma received the greatest media attention when the Cleveland Health Museum, which had purchased the pair, decided to sponsor, with the help of a local newspaper, the YWCA, and several other health and educational organizations, a contest to find the woman in Ohio whose body most closely matched the dimensions of Norma. Under the catchy headline, "Are You Norma, Typical Woman?" the publicity surrounding this contest instructed women in how to measure themselves at the same time that it extolled the virtues of Norma's body compared to those of her "grandmother," Dudley Sargent's composite of the 1890s woman.[11] Within ten days, 3,863 women had sent in their measurements to compete for the $100 prize in U.S. War Bonds that would go to the woman who most resembled the average American girl.

Although anthropometric studies such as these were ostensibly descriptive rather than prescriptive, the normal or average and the ideal were routinely conflated. Nowhere is this more evident than in the discussions surrounding the Norma contest. Described in the press as the "ideal" young woman, Norma was said to be everything an American woman should be in a time of war: she was fit, strong-bodied, and at the peak of her reproductive potential. Commentators waxed eloquent about the model character traits—maturity, modesty, and virtuousity—that this perfectly average body suggested. Curiously, although

Norma was based on the measurements of living women, only about one percent of the contestants came close to her proportions. Harry Shapiro, curator of physical anthropology at the American Museum of Natural History, explained in the pages of *Natural History* why it was so rare to find a living, breathing Norma. Both Norma and Norman, he pointed out:

> . . . exhibit a harmony of proportion that seems far indeed from the usual or the average. One might well look at a multitude of young men and women before finding an approximation to these normal standards. We have to do here then with apparent paradoxes. Let us state it this way: the average American figure approaches a kind of perfection of bodily form and proportion; the average is excessively rare. (Shapiro 1945: 51)

Besides bolstering the circulation of the *Cleveland Plain Dealer,* the idea behind the contest was to promote interest in physical fitness. Newspaper articles emphasized that women had a national responsibility to be fit, if America was to continue to excel after the war. Commenting on the search for Norma, Dr. Bruno Gebhard, director of the Cleveland Health Museum, was quoted in one of the many newspaper articles as saying that "if a national inventory of the female population of this country were taken there would be as many '4Fs' among the women as were revealed among the men in the draft" (Robertson 1945:4). The contest provided the occasion for many health reformers to voice their concern about the need for eugenic marital selection and breeding. Beside weakening the "American stock," Gebhard claimed, "the unfit are both bad producers and bad consumers. One of the outstanding needs in this country is more emphasis everywhere on physical fitness" (ibid). Norma was presented to the public as a reminder to women of their duty to the nation, and, not incidentally, Norma could also serve as a hypothetical standard in women's colleges for the detection of faulty posture and "so that students who need to lose or gain in spots or generally may have a mark to shoot at" (ibid).

Norma and Norman were thus more than statistical composites, they were ideals. It is striking how thoroughly racial and ethnic differences were erased from these scientific representations of the American male and female. Based on the measurements of white Americans, eighteen to twenty-five years old, Norm and Norma emerged carved out of white alabaster, with the facial features and appearance of Anglo-Saxon gods. Here, as in the college studies that preceded them, the "average American" of the postwar period was to be visualized only as a youthful white body.

However, they were not the only ideal. The health reformers, educators, and doctors who approved and promoted Norma as an ideal for American women were well aware that her sensible, strong, thick-waisted body differed signifi-

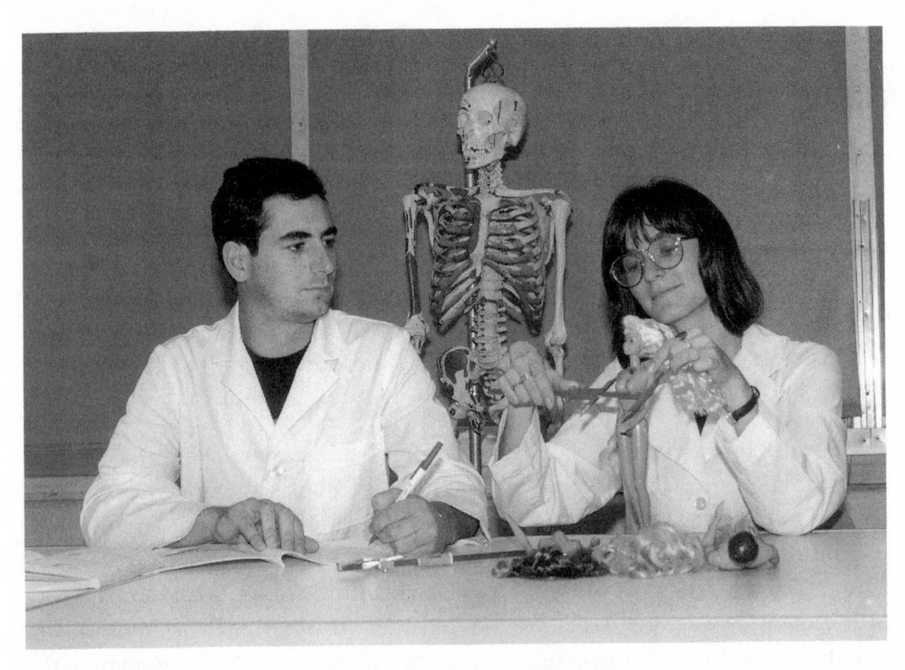

Figure 10.1. Measuring Barbie. (Photo by Ann Marie Mires)

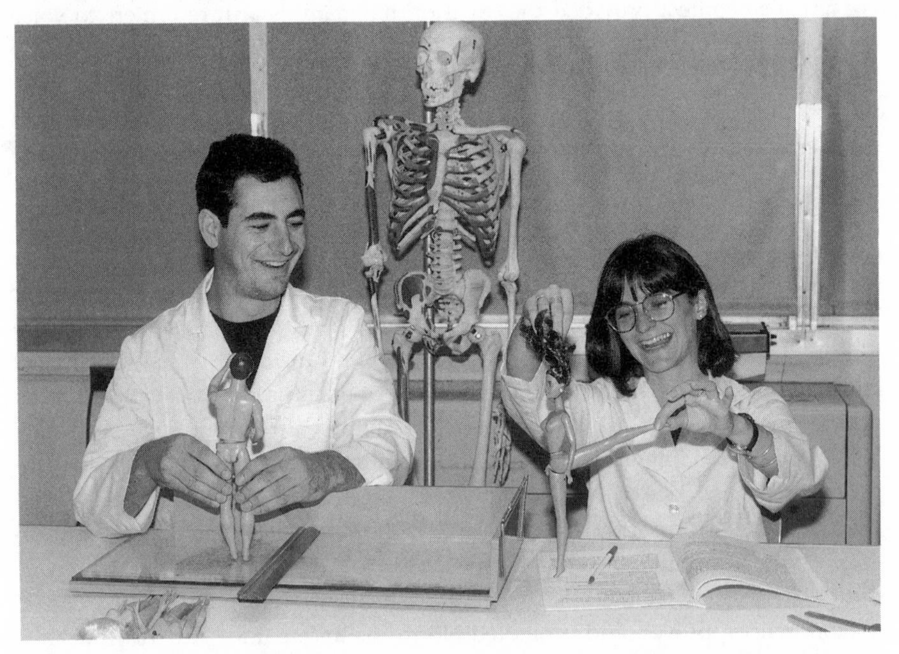

Figure 10.2. Anthropometry Fun. (Photo by Ann Marie Mires)

cantly from the tall, slim-hipped bodies of fashion models in vogue at the time.[12] Gebhard and others tried through a variety of means to encourage women to ignore the temptations of "vanity" and fashion, but they were ill equipped to compete with the persuasive powers of a rapidly expanding mass media that marketed a very different kind of female body. As the postwar period advanced, Norma would continue to be trotted out in home economics and health education classes. But in the iconography of desirable female bodies, she would be overshadowed by the array of images of fashion models and pinup girls put out by advertisers, the entertainment industry, and a burgeoning consumer culture. These idealized images were becoming, as we will see below, increasingly thin in the sixties and seventies while the "average" woman' s body was in fact getting heavier. With the the thinning of the American feminine ideal, Norma and subsequent representations of the statistically average woman would become increasingly aberrant, as slenderness and sex appeal—not physical fitness—became the premier concern of postwar femininity.

The Anthropometry of Barbie: Turning the Tables

As the preceding discussion makes abundantly clear, the anthropometrically measured "normal" body has been anything but value-free. Formulated in the context of a race-, class-, and gender-stratified society, there is no doubt that quantitatively defined ideal types or standards have been both biased and oppressive. Incorporated into weight tables, put on display in museums and world's fairs, and reprinted in popular magazines, these scientifically endorsed standards produce what Foucault calls "normalizing effects," shaping, in not altogether healthy ways, how individuals understand themselves and their bodies. Nevertheless, in the contemporary cultural context, where an impossibly thin image of women's bodies has become the most popular children's toy ever sold, it strikes us that recourse to the "normal" body might just be the power tool we need for destabilizing a fashion fantasy spun out of control. It was with this in mind that we asked students in one of our social biology classes to measure Barbie to see how her body compared to the average measurements of young American women of the same period. Besides estimating Barbie's dimensions if she were life-sized, we see the experiment as an occasion to turn the anthropometric tables from disciplining the bodies of living women to measuring the ideals by which we have come to judge ourselves and others. We also see it as an opportunity for students who have grown up under the regimes of normalizing science—students who no doubt have been measured, weighed, and compared to standards since birth—to use those very tools to unsettle a highly popular cultural ideal.

Initially, this foray into the anthropometry of Barbie was motivated by an exercise in a course entitled Issues in Social Biology. Since one objective of the course was to learn about human variation, our first task in understanding more about Barbie was to consider the fact that Barbie's friends and family do represent some variation, limited though it may be. Through colleagues and donations from students or (in one case) their children we assembled seventeen dolls for analysis. The sample included:

1 early '60s Barbie
4 mid-'70s-to-contemporary Barbies, including a Canadian Barbie
3 Kens
2 Skippers
1 Scooter
Assorted Barbie's friends, including Christie, Barbie's "black" friend
Assorted Ken's friends

To this sample we subsequently added the most current versions of Barbie and Ken (from the "Glitter Beach" collection) and also Jamal, Nichelle, and Shani, Barbie's more recent African American friends. As already noted, Mattel introduced these dolls (Shani, Asha, and Nichelle) as having a more authentic African American appearance, including a "rounder and more athletic" body. Noteworthy also are the skin color variations between the African American dolls, ranging from dark to light, whereas Barbie and her white friends tend to be uniformly pink or uniformly suntanned.

Normally, of course, before undertaking the somewhat invasive techniques of measuring people's bodies, we would have written a proposal to the Human Subjects Committees of the department and the university; submitted a written informed consent form detailing the measurements to be taken and seeking permission to use the data, by name, in subsequent reports; and, finally, discussed the procedures and the importance of the research with each of the subjects. However, since our subjects were unresponsive, these protocols had to be waived.

Before beginning the actual measurements, we discussed the kinds of data we thought would be most appropriate. Student interest centered on height and chest, waist, and hip circumference. Members of the class also pointed out the apparently small size of the feet and the general leanness of Barbie. As a result, we added a series of additional standardized measurements, including upper arm and thigh circumference, in order to obtain an estimate of body fat and general size.

After practicing with the calipers, discussing potential observational errors,

TABLE 1 Measurements of Glitter Beach Barbie, African American Shani, and the average measurements of the 1988 U.S. Army women recruits.[a]

Meas.	Barbie	Shani	U.S. Army "Norma"
Height	5′10″	5é10″	5é4″
Chest circum.	35″	35Æ	35.7″
Waist circum.	20″	20Æ	31″
Hip circum.[b]	32.50″	31.25Æ	38.10″
Hip breadth	11.6″	11.0″	13.49″
Thigh circum.	19.25″	20.00″	22.85Æ

[a]"Norma" is based on 2,208 army recruits, 1,140 of whom were white, 922 of whom were black.
[b]Hip circumference is referred to as "buttock circumference" in anthropometric parlance.

TABLE 2 Measurements of Glitter Beach Ken, African American Ken, and the average measurements of the 1988 U.S. Army male recruits.[a]

Meas.	Ken	Jamal	U.S. Army "Norm"
Height	6′0″	6′0″	5′9″
Chest circum.	38.4″	38.4″	39.0″
Waist circum.	28.8″	28.8″	33.1″
Hip Circum.[b]	36.0″	36.0″	38.7″
Hip breadth	12.2″	12.2″	13.46″
Thigh circum	20.4″	20.04″	23.48″

[a]"Norm" is based on 1,774 males, 1,172 of whom were white and 458 of whom were black.
[b]Hip circumference is referred to as "buttock circumference" in anthropometric parlance.

and performing repeated trial runs, we began to record. All the measurements were taken in the Physical Anthropology Laboratory at the University of Massachusetts under clean, well-lit conditions. We felt our almost entirely female group of investigators would no doubt have pleased Hrdlicka, since he believed women anthropometrists to be more skilled at the precise, small-scale measurements our experiment required.[13] In scaling Barbie to be life-sized, the students decided to translate her measurements using two standards: (a) if Barbie were a fashion model (5′10″) and (b) if she were of average height for women in the United States (5′4″). We also decided to measure Ken, using both an average male stature, which we designated as 5′8″, and the more "idealized" stature for men, 6′.

For the purposes of this chapter, we took measurements of dolls in the cur-

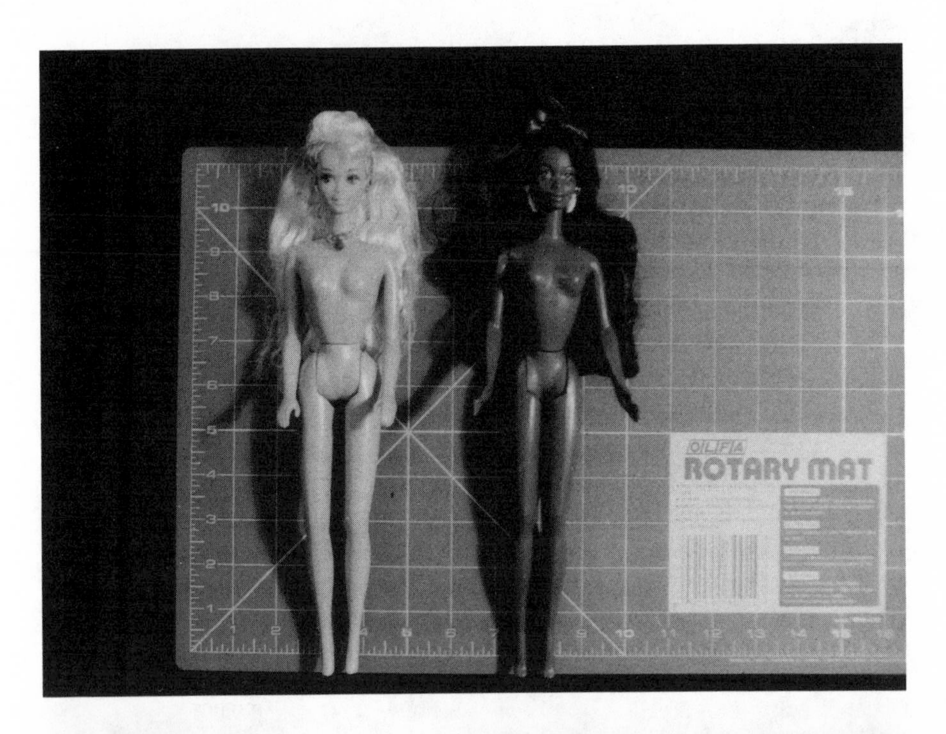

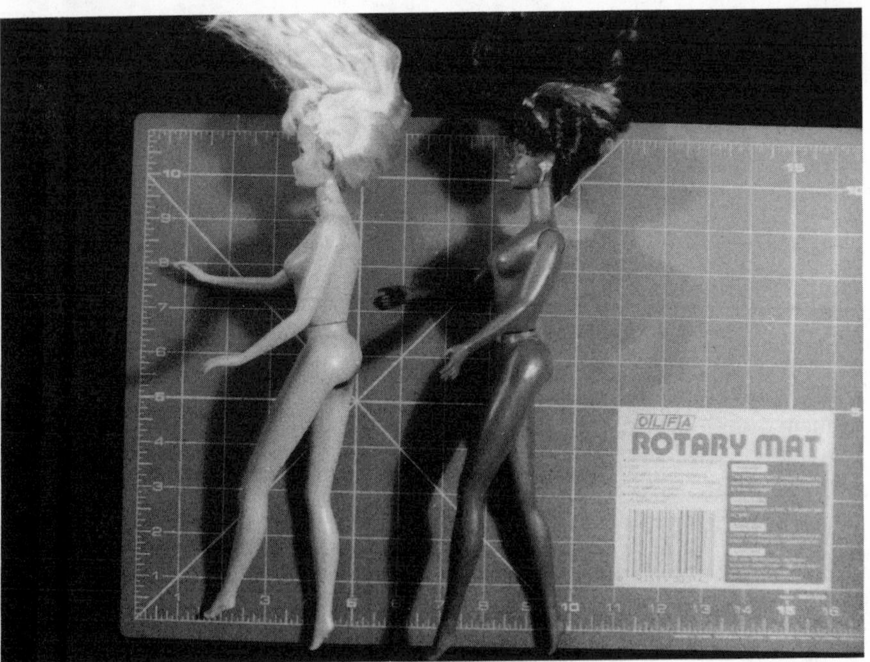

Figures 10.3. & 10.4. (Photos by Ann Marie Mires)

rent Glitter Beach and Shani collection that were not available for our original classroom experiment, and all measurements were retaken to confirm estimates. We report here only the highlights of the measurements taken on the newer Barbie and newer Ken, Jamal, and Shani, scaled at their ideal fashion-model height. For purposes of comparison, we include data on average body measurements from the standardized published tables of the 1988 Anthropometric Survey of Army Personnel. We have dubbed these composites for the female and male recruits Army "Norma" and Army "Norm," respectively.

Barbie and Shani's measurements reveal interesting similarities and subtle differences. First, considering that they are six inches taller than "Army Norma," note that their measurements tend to be considerably less *at all points.* "Army Norma" is a composite of the fit woman soldier; Barbie and Shani, as high-fashion ideals, reflect the extreme thinness expected of the runway model. To dramatize this, had we scaled Barbie to 5′4″, her chest, waist, and hip measurements would have been 32″–17″–28″, clinically anorectic to say the least. There are only subtle differences in size, which we presume intend to facilitate the exchange of costumes among the different dolls. We were curious to see the degree to which Mattel had physically changed the Barbie mold in making Shani. Most of the differences we could find appeared to be in the face. The nose of Shani is broader and her lips are ever so slightly larger. However, our measurements also showed that Barbie's hip circumference is actually larger than Shani's, and so is her hip breadth. If anything, Shani might have thinner legs than Barbie, but her back is arched in such a way that it tilts her buttocks up. This makes them appear to protrude more posteriorly, even though the hip depth measurements of both dolls are virtually the same (7.1″). Hence, the tilting of the lumbar dorsal region and the extension of the sacral pelvic area produce the visual illusion of a higher, rounder butt (see Figures 10.3 and 10.4). This is, we presume, what Mattel was referring to in claiming that Shani has a realistic, or ethnically correct, body (Jones 1991).

One of our interests in the male dolls was to ascertain whether they represent a form closer to average male values than Barbie does to average female values. Ken and Jamal provide interesting contrasts to "Army Norm," but certainly not to each other. Their postcranial bodies are identical in all respects. They, in turn, represent a somewhat slimmer, trimmer male than the so-called fit soldier of today. Visually, the newer Ken and Jamal appear very tight and muscular and "bulked out" in impressive ways. The U.S. Army males tend to carry slightly more fat, judging from the photographs and data presented in the 1988 study.[14]

Indeed, it would appear that Barbie and virtually all her friends characterize a somewhat extreme ideal of the human figure, but in Barbie and Shani, the female cases, the degree to which they vary from "normal" is much greater than in the male cases, bordering on the impossible. Barbie truly is the unobtainable representation of an imaginary femaleness. But she is certainly not unique in the realm of female ideals. Studies tracking the body measurements of *Playboy* magazine centerfolds and Miss America contestants show that between 1959 and 1978 the average weight and hip size for women in both of these groups have decreased steadily (Wiseman et al. 1992). Comparing their data to actuarial data for the same time period, researchers found that the thinning of feminine body ideals was occurring at the same time that the average weight of American women was actually increasing. A follow-up study for the years 1979–88 found this trend continuing into the eighties: approximately sixty-nine percent of *Playboy* centerfolds and sixty percent of Miss America contestants were weighing in at fifteen percent or more below their expected age and height category. In short, the majority of women presented to us in the media as having desirable feminine bodies were, like Barbie, well on their way to qualifying for anorexia nervosa.

Our Barbies, Our Selves

> I feel like Barbie; everyone calls me Barbie; I love Barbie. The main difference is she's plastic and I'm real. There isn't really any other difference.
>
> —Hayley Spicer, winner of Great Britain's Barbie-Look-Alike competition

On the surface, at least, Barbie's strikingly thin body and the repression and self-discipline that it signifies would appear to contrast with her seemingly endless desire for consumption and self-transformation. And yet, as Susan Bordo has argued in regard to anorexia, these two phenomena—hyper-thin bodies and hyper-consumption—are very much linked in advanced capitalist economies that depend upon commodity excess. Regulating desire under such circumstances is a constant, ongoing problem that plays itself out on the body. As Bordo argues:

> [In a society where we are] conditioned to lose control at the very sight of desirable products, we can only master our desires through a rigid defense against them. The slender body codes the tantalizing ideal of a well-managed self in which all is "in order" despite the contradictions of consumer culture. (1990:97)

The imperative to manage the body and "be all that you can be"—in fact, the idea that you can *choose* the body that you want to have—is a pervasive feature of consumer culture. Keeping control of one's body, not getting too fat or flabby—in other words, conforming to gendered norms of fitness and weight—

Gal has plastic surgery 18 times — to become a real-life Barbie

Plain-Jane farm girl Cindy Jackson dreamed of being beautiful and glamorous like her Barbie doll — so when she grew up she had surgery that transformed her into a real-life Barbie!

Cindy, 34, has spent $55,000 on 18 operations in the past four years to make herself gorgeous.

"You can't buy love, but you can buy lovely. And I used my money to make my childhood dream come true!" Cindy told The ENQUIRER.

By CHARLES MONTGOMERY

"People stop me on the street and tell me I look just like Barbie with my turned-up nose and pouted lips.

"That makes me feel great!"

Cindy was born and raised on a farm near the small town of Fremont, Ohio. She wore secondhand clothes and lived a dull, dreary life.

"I'd dream of wearing pretty clothes and going to Paris, Rome, London," Cindy recalled. "I wanted an exciting, glamorous life."

After high school Cindy became a commercial photographer, saved her money and moved to London on the adventure of her life.

"London was exciting and glamorous — and that's when I decided I wanted the looks that went with my dream.

"That may sound superficial, but it's what I wanted.

"I thought I was ugly. My nose was too big, my lips too thin and my eyes too small and hooded. I had a mean-looking mouth and a double chin. My chest was flat and I had a bulging stomach. I had saddlebags on my hips and thighs and even my knees were fat.

"I wasn't pretty or sensuous. Then my father died and left me some money — and with that and my savings, I started transforming myself into my dream girl.

"I started by having my eyes lifted at the corners to give me a more doll-like appearance."

Cindy followed that with liposuction, upper and lower face-lifts, chemical peels, tummy tucks, breast implants, two nose jobs and other cosmetic surgery. She had some procedures more than once.

"I even had a few moles removed from my body."

Now Cindy, who still lives in London, loves the attention she gets as a living doll.

"I've had a number of marriage proposals, but I'm not quite ready yet to say 'yes' to my own 'Ken,'" she said.

She's still a photographer but also manages and sings in her own rock band, Cindy Jackson and the Hellcats. And she operates The Cosmetic Surgery Network in London, England, a hotline that dispenses information to people considering plastic surgery.

She's also writing a book about her experiences.

Says Cindy: "I'm proud of myself.

"I was just a hick from the sticks and look at me now. I had a dream and made it come true!"

Snapshot of Cindy the way she was.

A living, breathing Barbie doll.

LOVELY CINDY now has the looks she'd only dreamed about before.

TWO NOSE JOBS: At far left, Cindy before she had surgery. At center, her first nose job. At right, today.

Did you go from Plain Jane to glamor gal? Tell us your story

If you've gone from Plain Jane to plain gorgeous, tell us about it — and maybe we'll tell your story, too!

Write your account in 50 words or less and send it to: Transform, NATIONAL ENQUIRER, Lantana, Fla. 33464.

Include "before" and "after" photos to show how you've changed. Please don't send pictures that must be returned.

NATIONAL ENQUIRER 3

Figure 10.5. *National Enquirer,* January 12, 1993, vol. 67, no. 25, page 3. Reproduced with permission.

are signs of an individual's social and moral worth. But, as feminists Bordo, Sandra Bartky, and others have been quick to point out, not all bodies are subject to the same degree of scrutiny or the same repercussions if they fail. It is women's bodies and desires in particular where the structural contradictions—the simultaneous incitement to consume and social condemnation for overindulgence—appear to be most acutely manifested in bodily regimes of intense self-monitoring and discipline. "The woman who checks her make-up half a dozen times a day to see if her foundation has caked or her mascara run, who worries that the wind or rain may spoil her hairdo has become just as surely as the inmate of the Panopticon, a self-policing subject, a self committed to a relentless self surveillance" (Bartky 1990: 80). Just as it is women's appearance that is subject to greater social scrutiny, so it is that women's desires, hungers, and appetites are seen as most threatening and in need of control in a patriarchal society.

This cultural context is relevant to making sense of Barbie and the meaning her body holds in late consumer capitalism. In dressing and undressing Barbie, combing her hair, bathing her, turning and twisting her limbs in imaginary scenarios, children acquire a very tactile and intimate sense of Barbie's body. Barbie is presented in packaging and advertising as a role model, a best friend or older sister to little girls. Television jingles use the refrain, "I want to be just like you," while look-alike clothes and look-alike contests make it possible for girls to live out the fantasy of being Barbie. And, finally, in the pages of the *National Enquirer*, where cultural fantasies have a way of becoming nightmare reality, we find the literalization of becoming Barbie, thanks to the wonders of modern medical technology (see Figure 10.5). In short, there is no reason to believe that girls (or adult women) separate Barbie's body shape from her popularity and glamour.[15]

This is exactly what worries many feminists. As our measurements show, Barbie's body differs wildly from anything approximating "average" female body weight and proportions. Over the years her wasp-waisted body has evoked a steady stream of critique for having a negative impact on little girls' sense of self-esteem.[16] While her large breasts have always been a focus of commentary, it is interesting to note that, as eating disorders are on the rise, her weight has increasingly become the target of criticism. For example, the 1992 release of a Barbie aerobics workout video for girls was met with the following angry letter from an expert in the field of eating disorders:

> I had hoped these plastic dolls with impossible proportions would have faded away in this current health-conscious period; not at all. . . . Move over Jane Fonda. Welcome again, ever smiling, breast-thrusting Barbie with your stick legs and sweat-free aerobic routines. I'm concerned about

> the role model message she is giving our young. Surely it's hard to accept a little cellulite when the culture tells you unrelentingly how to strive for thinness and the perfect body. (Warner 1992)[17]

There is no doubt that Barbie's body contributes to what Kim Chernin (1981) has called "the tyranny of slenderness." But is repression all her hyper-thin body conveys? Looking once again to Susan Bordo's work on anorexia, we find an alternative reading of the slender body—one that emerges from taking seriously the way anorectic women see themselves and make sense of their experience:

> For them, anorectics, [the slender ideal] may have a very different meaning; it may symbolize not so much the containment of female desire, as its liberation from a domestic, reproductive destiny. The fact that the slender female body can carry both these (seemingly contradictory) meanings is one reason, I would suggest, for its compelling attraction in periods of gender change. (Bordo 1990: 103)

Similar observations have been made about cosmetic surgery: women often explain their experience as one of empowerment, taking charge of their bodies and lives (Balsamo 1993; Davis 1991). What does this mean for making sense of Barbie? We would suggest that a subtext of agency and independence, even transgression, accompanies this pencil-thin icon of femininity. One could argue that, like the anorectic body she resembles, Barbie's body displays conformity to dominant cultural imperatives for a disciplined body and contained feminine desires. As a woman, however, her excessive slenderness also signifies a rebellious manifestation of willpower, a visual denial of the maternal ideal symbolized by pendulous breasts, rounded stomach and hips. Hers is a body of hard edges, distinct borders, self-control. It is literally impenetrable. Unlike the anorectic, whose self-denial renders her gradually more androgynous in appearance, in the realm of plastic fantasy Barbie is able to remain powerfully sexualized, with her large, gravity-defying breasts, even while she is distinctly nonreproductive. Like the "hard bodies" in fitness advertising, Barbie's body may signify for women the pleasure of control and mastery, both of which are highly valued traits in American society and predominantly associated with masculinity (Bordo 1990: 105). Putting these elements together with her apparent independent wealth can make for a very different reading of Barbie than the one we often find in the popular press. To paraphrase one Barbie-doll owner: she owns a Ferrari and doesn't have a husband—she must be doing something right![18]

Invoking the testimonies and experiences of women caught up in the beauty myth is not meant to suggest that playing with Barbie, or becoming like

Barbie, is a means of empowerment for little girls. But it is meant to signal the complex and contradictory meanings that her body has in contemporary American society. Barbie functions as an ideological sign for commodity fetishism and a rather rigid gender ideology. But neither children nor adult consumers of popular culture are simply passive victims of dominant ideology. It is sensible to assume that the children who play with Barbie are themselves creative users, who respond variously to the messages about femininity encoded in her fashions and appearance. Not only do many children make their own clothes for their Barbie dolls,[19] but anecdotes abound of the imaginative uses of Barbie. In the hands of their owners, Barbies have been known to occupy roles and engage in activities anathema to the good-girl image Mattel has carefully constructed. In the course of our research in the past year, we have heard or read about Barbies that have been tattooed, decapitated, and had their flowing locks shorn into mohawks. Knowing only the limits of a child's imagination, Barbies have become amazon cave warriors, dutiful mommies, evil axe-murderers, and *Playboy* models. As *Mondo Barbie,* a recent collection of Barbie-inspired fiction and poetry, makes clear, the possibilities are endless, and sexual transgression is always just around the corner (Ebersole and Peabody 1993; see also Rand, 1994).

It is clear that a next step we would want to take in the cultural interpretation of Barbie is an ethnographic study of Barbie-doll owners.[20] In the meanwhile, we can know something about these alternative appropriations by looking to various forms of popular culture and the art world. Barbie has become a somewhat celebrated figure among avant-garde and pop artists, giving rise to a whole genre of Barbie satire, known as "Barbie Noire" (Kahn 1991). According to Peter Galassi, curator of *Pleasures and Terrors of Domestic Comfort,* an exhibit at the Museum of Modern Art, in New York "Barbie isn't just a doll. She suggests a type of behavior—something a lot of artists, especially women, have wanted to question" (quoted in Kahn 1991: 25). Perhaps the most notable sardonic use of Barbie dolls to date is the 1987 film *Superstar: The Karen Carpenter Story,* by Todd Haynes and Cynthia Schneider. In this deeply ironic exploration into the seventies, suburbia, and middle-class hypocrisy, Barbie and Ken dolls are used to tell the tragic story of Karen Carpenter's battle with anorexia and expose the perverse underbelly of the popular singing duo's candy-coated image of happy, apolitical teens. It is hard to imagine a better casting choice to tell this tale of femininity gone astray than the ever-thin, ever-plastic, ever-wholesome Barbie.

For Barbiana collectors it should come as no surprise that Barbie's excessive femininity also makes her a favorite persona of female impersonators, alongside Judy, Marilyn, Marlene, and Zsa Zsa. Appropriations of Barbie in gay camp culture have tended to favor the early, vampier Barbie look: with the arched eyebrows, heavy black eyeliner, and coy sideways look—the later superstar version

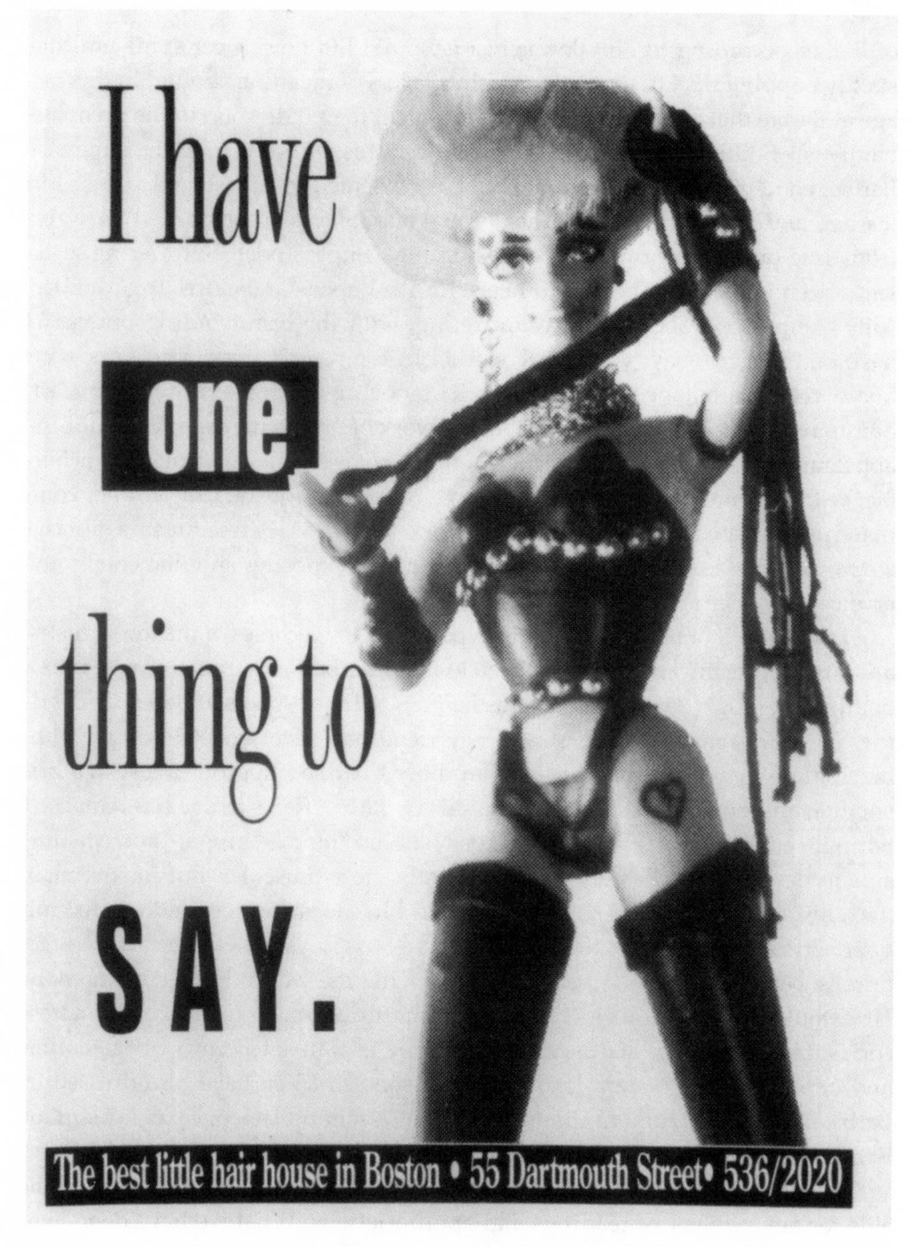

Figure 10.6. "You Better Work!" Advertisement, Larry Baron and Linda Frisco, proprietors, The Best Little HairHouse in Boston. 1993. Reproduced with permission.

of Barbie, according to BillyBoy, is just *too* pink. But new queer spins are constantly popping up. For example, multiple layers of meaning abound in this image of Barbie that appeared in an advertisement for a hair salon in the predominantly gay South End neighborhood of Boston (see Figure 10.6). Here, superstar Barbie, good-girl teenage fashion model, is presented to us tattooed, dressed in leather, and painted over with heavy makeup. She is accompanied by lyrics from drag-queen performer Ru Paul's hit song, "Supermodel." A towering seven feet of gender-bending beauty in heels, Ru Paul opens and closes this wonderfully campy song about runway modeling, with the commanding phrase: "I have one thing to say: you better work!" In the song, "work" or "work your body," refers simultaneously to the work of moving down the runway with "savoir-faire" and to the work of illusion, the work of producing a perfect feminine appearance, a "million-dollar derrière." In the advertisement, of course, it is Barbie, with her molded-by-Mattel body, who stands in for the drag queen, commanding the spectator, whip in hand, to *work her body.* Barbie, in this fantasyscape, becomes the mistress of body discipline, exposing simultaneously the artifice of gender and the feminine body.

In the world of Barbie Noire, the hyper-rigid gender roles of the toy industry are targeted for inversion and subversion. While Barbie is transformed into a dominatrix drag queen, Ken, too, has had his share of spoofs and gender bending. Barbie's somewhat dull steady boyfriend has never been developed into much more than a reliable escort and proof of Barbie's appropriate sexual orientation and popularity. In contrast to that of Barbie, Ken's image has remained boringly constant over the years. He has had his "mod," "hippie" and Malibu-suntan days, and he has gotten significantly more muscular. But for the most part, his clothing line is less diversified, and he lacks an independent fan club or advertising campaign.[21] In a world where boys' toys are G.I. Joe-style action figures, bent on alternately saving or destroying the world, Ken is an anomaly. Few would doubt that his identity was primarily another one of Barbie's accessories. His secondary status vis-à-vis Barbie is translated into emasculation and/or a secret gay identity: cartoons and spoofs of Ken have him dressed in Barbie clothes, and rumors abound that Ken's seeming lack of sexual desire for Barbie is only a cover for his real love for his boyfriends, Alan, Steve, and Dave.

Inscrutable with her blank stare and unchanging smile, Barbie is thus available for any number of readings and appropriations. What we have done here is examine some of the ways she resonates with the complex and contradictory cultural meanings of femininity in postwar consumer society and a changing politics of the body. Barbie, as we, and many other critics, have observed, is an impossible ideal, but she is an ideal that has become curiously normalized. In a youth-obsessed society like our own, she is an ideal not just for young women, but for all women who feel that being beautiful means looking like a skinny, buxom, white twenty-year-old. It is this cultural imperative to remain ageless and lean that leads women to have skewed perceptions of their bodies, undergo painful surgeries, and punish themselves with outrageous diets. Barbie, in short, is an ideal that constructs women's bodies as hopelessly imperfect. It has been our intention to unsettle this ideal and, at the same time, to be sensitive to other possible readings, other ways in which this ideal body figures and reconfigures the female body. For example, implicit in the various strategies of technologically mediated body-sculpting and surveillance that women engage in to meet these ideals is not only a conception of the female body that is inherently pathological, but an increasingly imaginary body of malleable, replaceable parts. Fueled by an ideology demanding limitless improvement and an increasingly popular cyborgian science fiction, the modern paradigm of "body as machine," says Susan Bordo, is giving way to an understanding of the body as "cultural plastic." The explosion in technologically assisted modifications through cosmetic surgery, piercing, aerobics, and nautilus all point to a conception of the body as raw material to be fragmented into parts, molded, and reshaped into a more perfect form. Lacking any essential truth, the body has become, like Barbie, all surface, a ground for staging cultural identities.

What to make of this apparent denaturalizing of the feminine body is not clear. Feminists call attention to the way women use these techniques to take control over their bodies, while others are hesitant to join with current trends in cultural studies that would celebrate these as empowered acts of resistance. Our concern is not to decry the corruption of a fictitious "natural" body, but to underscore how these acts of self-re-creation are inflected by power and desire. "Fashion surgery," as Balsamo calls it, liberates one from the body one is born with, and as Nan Goldin's 1993 photo essays of transvestites and transsexuals make apparent, advances in this medical technology have made possible new permutations of the gendered body. What they do *not* do is erase the larger cultural matrix and power relations that propel women to undertake certain kinds of body transformations instead of others. The different matrices of power in which individuals are located make it such that, while all body transformations in some way treat the gender and the body as cultural plastic, they do not have similar meanings. Further, the potential to surgically alter bodies may challenge the naturalness of gender and the determining power of biological sex itself, but it has not unsettled the notion that gender is fundamentally located in the physical body, rather than in language, gesture, or other performative displays.[22] Indeed, a variety of social and cultural forces conspire to make body modification so normal, so necessary, that "electing *not* to have cosmetic surgery is sometimes interpreted as a failure to deploy all available resources to maintain a youthful, and therefore socially acceptable and attractive, body appearance" (Balsamo 1993: 216).[23] It is the complexity of this terrain that leads artists such as Barbara Kruger to describe the body, particularly the body of the socially disempowered, as a *battleground,* a terrain of multiple sites of conflict and resistances, where histories of racial, national, and gender inequalities come to bear upon the "choices" individuals make with regard to their bodies.

We have explored some of the battleground upon which the serious play of Barbie unfolds. If Barbie has taught us anything about gender, it is that femininity in consumer culture is a question of carefully performed display, of paradoxical fixity and malleability. One outfit, one occupation, one identity can be substituted for another, while Barbie's body has remained ageless, changeless, untouched by the ravages of age or cellulite. She is always a perfect fit, always able to consume and be consumed. Mattel has skillfully managed to turn the challenges of feminist protest, ethnic diversity, and a troubled multiculturalism to a new array of outfits and skin tones, annexing these to a singular anorectic body ideal. Cultural icon that she is, Barbie nevertheless cannot be permanently located in any singular cultural space. Her meaning is mobile as she is appropriated and relocated into different cultural contexts, some of which, as we have seen, make fun of many of the very notions of femininity and consumerism she personifies. As we consider Barbie's many meanings, we should remember that Barbie is not only a denizen of subcultures in the United States, she is also world traveler. A product of the global assembly line, Barbie dolls owe their existence to the internationalization of the labor market and global flows of capital and commodities that today characterize the toy industry, as well as other industries in the postwar era. Designed in Los Angeles, manufactured in Taiwan or Malaysia, distributed worldwide, Barbie ™ is American-made in name only. Speeding her way into an expanding global market, Barbie brings with her some of the North American cultural subtext we have outlined in this analysis. How this teenage survivor then gets interpolated into the cultural landscapes of Mayan villages, Bombay high-rises, and Malagasy towns is a rich topic that begs to be explored.

Notes

Acknowledgments. Friends and colleagues too numerous to mention have contributed ideas and a steady flow of clippings that have inspired us, shaped our analysis, and kept us up to date on our "Barbiana." We are especially grateful to: Jennifer Terry, for bringing our attention to the Norm and Norma statues and for generously sharing with us her documentation on the Dickinson collection, George Armelagos, Carole Vance, Laurie Godfrey, Oriol Pi Sunyer, Susan DiGiacomo, Donna Penn, Alice Echols, Joe Urla, Lisa Handwerker, Alan Berube, Arturo Escobar, Paul Mullins, Nancy Muller, Sue Hyatt, Mary Orgel, Liz Chilton, Uzi Baram, and Barry Benton. We would also like to thank the participants in Alan Swedlund's Social Biology class (Fall 1989) at the University of Massachusetts, who carried out the anthropometric study of Barbie.

1. At the time of this writing, there was no definitive history of Barbie and the molds that have been created for her body. However, Barbie studies are booming, and we expect new work in press, including M. G. Lord's *Forever Barbie: The Unauthorized Biography of a Real Doll* (1994), to provide greater insight into Barbie's history and the debates surrounding her body within Mattel and the press.

2. While the concept of adolescence as a distinct developmental stage between puberty and adulthood was not new to the fifties, Thomas Doherty (1988) notes that it wasn't until the end of World War II that the term "teenager" gained standard usage in the American language.

3. The most recent upset, for example, surrounded the release of "Teen Talk" Barbie in the summer of 1992. A formal complaint from the American Association of University Women over one of the doll's preprogrammed phrases, "Math class is tough," resulted in Mattel's apology and discontinuation of the potentially offensive comment. It also generated a flurry of jokes, cartoons, and commentaries that seized the opportunity to alternately bash and embrace feminist critiques of traditional gender stereotypes. As with the many other controversies that have enveloped the doll since the seventies, the "Barbie-hates-math" brouhaha revealed Barbie to be a kind of litmus test of gender ideology in American society: a vehicle through which competing social constituencies air their differing views on appropriate gender roles and the status of feminism in a time of flux.

4. Recent work by Ann duCille promises to offer an incisive cultural critique of the "ethnification" of Barbie and its relationship to controversies in the United States over multiculturalism and political correctness (duCille 1995). More work, however, needs to be done on how Barbie dolls are adapted to appeal to various markets outside the U.S. For example, Barbie dolls manufactured in Japan for Japanese consumption have noticeable larger, rounder eyes than those marketed in the United States (see BillyBoy 1987). For some suggestive thoughts on the cultural implications of the transnational flow of toys like Barbie dolls, TransFormers, and He-Man, see Carol Breckenridge's (1990) brief but intriguing editorial comment to *Public Culture.*

5. Closely aligned with the emergence of statistics, it was Hrdlicka's hope that the two would be joined, and that one day the state would be "enlightened" enough to incorporate regular measurements of the population with the various other tabulations of the periodic census, in order to "ascertain whether and how its human stock is progressing or regressing" (1939: 12).

6. Though measurements of skulls, noses, and facial angles for scientific comparison had been going on throughout the nineteenth century, it wasn't until the 1890s that any serious attempts were made to standardize anthropometric measurements of the living body. This culminated in the Monaco Agreement of 1906, one of the first international meetings to standardize anthropometric measurement. For a brief review of the attempts to use and systematize photography in anthropometry, see Spencer (1992).

7. Hrdlicka argued, rather fantastically, that there was more space between the brain and the cranium in women than in men. Michael Blakey refers to this as Hrdlicka's "air head" hypothesis (1987: 16).

8. In her study of eighteenth-century physical sciences, Schiebinger (1993) remarks that male bodies (skulls in particular) were routinely assumed to embody the prototype of their race in the various typologies devised by comparative anatomists of the period. "When anthropologists did compare women across cultures, their interest centered on sexual traits—feminine beauty, redness of lips, length and style of hair, size and shape of breasts or clitorises, degree

of sexual desire, fertility, and above all the size, shape, and position of the pelvis" (1993: 156). In this way, the male body "remained the touchstone of human anatomy," while females were regarded as a sexual subset of the species (1993: 159–60).

9. In *Practical Anthropometry* Hrdlicka goes to some trouble to instruct field-workers (presumably male) working among "uncivilized groups" about the steps they need to take not to offend, and thereby put themselves at risk, when measuring women (1939: 57–59).

10. Norma is described in the press reports as being based on the measurements of 15,000 "real American girls." Although we cannot be sure, it is likely this data come from the Bureau of Home Economics, which conducted extensive measurements of students "to provide more accurate dimensions and proportions for sizing women's ready-made garments" (Shapiro 1945). For further information on the Dickinson collection and Dickinson's methods of observation, see Terry (1992).

11. Norma was described thus by a reporter in the *Cleveland Plain Dealer:* "She is taller, heavier, and more athletic than her grandmother of 1890. Her legs are longer, her waist is thicker, since it is no longer fashionable to have one so small that a man can girdle it with his two hands; her hips are slightly heavier and her bosom is fuller. She is not so voluptuous as the Greek ideal, Aphrodite of Cyrene, but she seems headed that way. Anthropologically she is considered an improvement on her grandmother. Whether she is an esthetic improvement is a matter of taste" (Robertson 1945: 1).

12. Historians have noted a long-standing conflict between the physical culture movement, eugenicists and health reformers, on the one hand, and the fashion industry, on the other, that gave rise in American society to competing ideals of the fit and the fashionably fragile woman (e.g., Banner 1983; Cogan 1989).

13. In *Practical Anthropometry* (1939), Hrdlicka states that males generally make better anthropometrists because they are more adept at handling the bigger instruments, but he believes women can be more precise and useful for measuring other women and children. We take the liberty of extending this practice to dolls.

14. One aspect of the current undertaking that is clearly missing is the possible variation that exists *within* individual groups of dolls that would result from mold variation and casting processes. Determining this variation would require a much larger doll collection than we had at our disposal. We are considering a grant proposal, but not seriously.

15. This process of identification becomes mimesis, not only in Barbie look-alike contests, but also in the recent Barbie workout video. In her fascinating analysis of the semiotics of workout videos, Margaret Morse (1987) has shown how these videos structure the gaze in such a way as to establish identification between the exercise leader's body and the participant-viewer. Surrounded by mirrors, the viewer is asked to exactly model her movements on those of the leader, literally mimicking the gestures and posture of the "star" body she wishes to become. In Barbie's video, producers use animation to make it possible for Barbie to occasionally appear on the screen as the exercise leader/cheerleader—the star whose body the little girls mimic.

16. In response to this anxiety, Cathy Meredig, an enterprising computer software designer, created the "Happy to Be Me" doll. Described as a healthy alternative for little girls, "Happy to Be Me" has a shorter neck, shorter legs, wider waist, larger feet, and a lot fewer clothes—designed to make her look more like the average woman ("She's No Barbie, nor Does She Care to Be." *New York Times,* August 15, 1991, C-11).

17. When confronted with these kinds of accusations, Mattel and the fashion industry protest that Barbie dolls, like fashion models, are about fantasy, not reality. "We're not telling you to be that girl" writes Elizabeth Saltzman, fashion editor for *Vogue.* "We're trying to show you fashion" (France 1992: 22).

18. "Dolls in Playland." 1992. Colleen Toomey, producer. BBC.

19. Who makes clothes for Barbie and what kinds of outfits are made is a fascinating subject for further study. Melosh and Simmons (1986) report that a survey at a Barbie doll exhibit revealed two-thirds of all doll owners made at least some clothes for their Barbie.

20. While not exactly ethnographic, Hohmann's 1985 study offers a sociopsychological view of how children experiment with social relations during play with Barbies.

21. Signs of a Ken makeover, however, have begun to appear. In 1991, a Ken with "real" hair that can be styled was introduced and, most dramatically, in 1993, he had his hair streaked and acquired an earring in his left ear. This was presented as a "big breakthrough" by Mattel and was received by the media as a sign of a broader trend in the toy industry to break down rigid gender stereotyping in children's toys (see Lawson 1993). It doesn't appear, however, that Ken is any closer to getting a "realistic" body than Barbie. Ruth Handler notes that when Mattel was planning the Ken doll, she had wanted him to have genitals—or at least a bump, and claims the men in the marketing group vetoed her suggestion. Ken did later acquire his bump (see "Dolls in Playland," Colleen Toomey, producer. BBC. 1993).

22. Thanks to Judith Halberstam for pointing this out. Her work on female masculinity and transsexualism points to the irony that, while improvements in cosmetic surgery underscore the "inventedness of sex," we do not see a corresponding disembodying of gender. Halberstam writes, "One might expect . . . in these postmodern times that as we posit the artificiality of gender and sex with increasing awareness of how and why our bodies have been policed into gender identities, there might be a decrease in the incidence of such things as sex change operations. On the contrary . . . there has been, as I suggested, a rise in the discussions of, depictions of and requests for f to m sex change operations" (1994: 215).

23. Professional organizations and various cosmetic industries stand a lot to gain by redefining such body "adjustments" as medical problems. The Women's Health Network, which monitors and reports on health-care issues and abuses, reproduced the following excerpt from a statement by their American Society for Plastic and Reconstructive Surgery in a letter addressed to the FDA regarding the classification of breast implants:

> There is a common misconception that the enlargement of the female breast is not necessary for maintenance of health or treatment of disease. There is a substantial and enlarging [*sic*] body of medical information and opinion, however, to the effect that these deformities [small breasts] are really a disease which in most patients results in feelings of inadequacy, lack of self confidence, distortion of body image and a total lack of well-being due to a lack of self-perceived femininity. The enlargement of the under-developed female breast is, therefore, often very necessary to insure an improved quality of life for the patient. (Sprague Zones 1989: 4)

For an interesting analysis of the language of cosmetic surgery, see Balsamo (1992).

References

Adelson, Andrea
1992 "And Now, Barbie Looks Like a Billion." *New York Times*, November 26, sec. D, p. 3.

Balsamo, Anne
1992 "On the Cutting Edge: Cosmetic Surgery and the Technological Production of the Gendered Body." *Camera Obscura* 28:207–38.

Banner, Lois W.
1983 *American Beauty*. New York: Knopf.

Bartky, Sandra Lee
1990 "Foucault, Femininity, and the Modernization of Patriarchal Power." In *Femininity and Domination: Studies in the Phenomenology of Oppression,* pp. 63–82. New York: Routledge.

Bennett, William, and Joel Gurin
1982 *The Dieter's Dilemma: Eating Less and Weighing More.* New York: Basic Books.

BillyBoy
1987 *Barbie, Her Life and Times, and the New Theater of Fashion.* New York: Crown.

Blakey, Michael L.
1987 "Skull Doctors: Intrinsic Social and Political Bias in the History of American Physical Anthropology." *Critique of Anthropology* 7(2): 7–35.

Bordo, Susan R.
1990 "Reading the Slender Body." In *Body/Politics: Women and the Discourses of Science.* Ed. Mary Jacobus, Evelyn Fox Keller, and Sally Shuttleworth, pp. 83–112. New York: Routledge.
1991 "Material Girl: The Effacements of Postmodern Culture." In *The Female Body: Figures, Styles, Speculations.* Ed. Laurence Goldstein, pp. 106–30. Ann Arbor, Mich.: The University of Michigan Press.

Breckenridge, Carol A.
1990 "Editor's Comment: On Toying with Terror." *Public Culture* 2(2): i–iii.

Brumberg, Joan Jacobs
1988 *Fasting Girls: The History of Anorexia Nervosa.* Cambridge: Harvard University Press. Reprint, New York: New American Library.

Chernin, Kim
1981 *The Obsession: Reflections on the Tyranny of Slenderness.* New York: Harper and Row.

Cogan, Frances B.
1989 *All-American Girl: The Ideal of Real Woman-hood in Mid-Nineteenth-Century America.* Athens and London: University of Georgia Press.

Davis, Kathy
1991 "Remaking the She-Devil: A Critical Look at Feminist Approaches to Beauty." *Hypatia* 6(2): 21–43.

Doherty, Thomas
1988 *Teenagers and Teenpics: The Juvenilization of American Movies in the 1950s.* Boston: Unwin Hyman.

duCille, Ann
1995 "Toy Theory: Blackface Barbie and the Deep Play of Difference." In *The Skin Trade: Essays on Race, Gender, and the Merchandising of Difference.* Cambridge: Harvard University Press.

Ebersole, Lucinda, and Richard Peabody, eds.
1993 *Mondo Barbie.* New York: St. Martin's Press.

Fee, Elizabeth
1979 "Nineteenth-Century Craniology: The Study of the Female Skull." *Bulletin of the History of Medicine* 53: 415–33.

France, Kim
1992 "Tits 'R' Us." *Village Voice*, March 17, p. 22.

Frith, Simon
1981 *Sound Effects. Youth, Leisure, and the Politics of Rock'n'Roll.* New York: Pantheon.

Goldin, Nan
1993 *The Other Side.* New York: Scalo.

Gould, Stephen Jay
1981 *The Mismeasure of Man.* New York: Norton.

Halberstam, Judith
1994 "F2M: The Making of Female Masculinity." In *The Lesbian Postmodern*, ed. Laura Doan, pp. 210–28. New York: Columbia University Press.

Hohmann, Delf Maria
1985 "Jennifer and Her Barbies: A Contextual Analysis of a Child Playing Barbie Dolls." *Canadian Folklore Canadien* 7(1–2): 111–20.

Hrdlicka, Ales
1925 "Relation of the Size of the Head and Skull to Capacity in the Two Sexes." *American Journal of Physical Anthropology* 8: 249–50.
1939 *Practical Anthropometry.* Philadelphia: Wistar Institute of Anatomy and Biology.

Jones, Lisa
1991 "Skin Trade: A Doll Is Born." *Village Voice*, March 26, p. 36.

Kahn, Alice
1991 "A Onetime Bimbo Becomes a Muse." *New York Times*, September 29.

Kaw, Eugenia
1993 "Medicalization of Racial Features: Asian American Women and Cosmetic Surgery." *Medical Anthropology Quarterly* 7(1): 74–89.

Lawson, Carol
1993 "Toys Will Be Toys: The Stereotypes Unravel." *New York Times.* February 11, sec. C, pp. 1, 8.

Leavy, Jane
1979 "Is There a Barbie Doll in Your Past?" *Ms.* September, p. 102.

Lord, M. G.
1994 *Forever Barbie: The Unauthorized Biography of a Real Doll.* New York: William Morrow.

May, Elaine Tyler
1988 *Homeward Bound: American Families in the Cold War Era.* New York: Basic Books.

Melosh, Barbara, and Christina Simmons
1986 "Exhibiting Women's History." In *Presenting the Past: Essays on History and the Public.* Ed. Susan Porter Benson, Stephen Brier, and Roy Rosenzweig, pp. 203–21. Philadelphia: Temple University Press.

Morse, Margaret
1987 "Artemis Aging: Exercise and the Female Body on Video." *Discourse* 10 (1987/88): 20–53.

Motz, Marilyn Ferris
1983 "I Want to Be a Barbie Doll When I Grow Up: The Cultural Significance of the Barbie Doll." In *The Popular Culture Reader,* 3d ed. Ed Christopher D. Geist and Jack Nachbar, pp. 122–36. Bowling Green: Bowling Green University Popular Press.

Newsweek
1990 "Finally. Barbie Dolls Go Ethnic." *Newsweek,* August 13, p. 48.

Rabinowitz, Paula
1993 Accessorizing History: Girls and Popular Culture. Discussant Comments, Panel #150: Engendering Post-war Popular Culture in Britain and America. Ninth Berkshire Conference on the History of Women. Vassar College, June 11–13, 1993.

Rand, Erica
1994 "We Girls Can Do Anything, Right Barbie? Lesbian Consumption in Postmodern Circulation." In *Lesbian Postmodern,* ed. Laura Doan, pp. 189–209. New York: Columbia University Press.

Robertson, Josephine
1945 "Theatre Cashier, 23, Wins Title of 'Norma'." *Cleveland Plain Dealer,* September 21, pp. 1, 4.

Russett, Cynthia Eagle
1989 *Sexual Science: The Victorian Construction of Womanhood.* Cambridge: Harvard University Press.

Schiebinger, Londa
1989 *The Mind Has No Sex?: Women in the Origins of Modern Science.* Cambridge: Harvard University Press.
1993 *Nature's Body: Gender in the Making of Modern Science.* Boston: Beacon.

Schwartz, Hillel
1986 *Never Satisfied: A Cultural History of Diets, Fantasies and Fat.* New York: Free Press.

Shapiro, Eben
1992 " 'Totally Hot, Totally Cool.' Long-Haired Barbie Is a Hit." *New York Times.* June 22, sec. D, p. 9.

Shapiro, Harry L.
1945 *Americans: Yesterday, Today, Tomorrow.* Man and Nature Publications. (Science Guide No. 126). New York: The American Museum of Natural History.

Spencer, Frank
1992 "Some Notes on the Attempt to Apply Photography to Anthropome-

try during the Second Half of the Nineteenth Century." In *Anthropology and Photography, 1860–1920*. Ed. Elizabeth Edwards, pp. 99–107. New Haven: Yale University Press.

Sprague Zones, Jane
1989 "The Dangers of Breast Augmentation." *The Network News* (July/August), pp. 1, 4, 6, 8. Washington, D.C.: National Women's Health Network.

Stevenson, Richard
1991 "Mattel Thrives as Barbie Grows." *New York Times.* December 2.

Terry, Jennifer C.
1992 Siting Homosexuality: A History of Surveillance and the Production of Deviant Subjects (1935–1950). Ph.D. diss., University of California at Santa Cruz.

Warner, Patricia Rosalind
1992 Letter to the editor. *Boston Globe,* June 28.

Williams, Lena
1992 "Woman's Image in a Mirror: Who Defines What She Sees?" *New York Times,* February 6, sec. A, p. 1, sec. B, p. 7.

Willis, Susan
1991 *A Primer for Daily Life.* London and New York: Routledge.

Wilson, Elizabeth
1985 *Adorned in Dreams: Fashion and Modernity.* London: Virago.

Wiseman, C., J. Gray, J. Mosimann, and A. Ahrens
1992 "Cultural Expectations of Thinness in Women: An Update." *International Journal of Eating Disorders* 11 (1): 85–89.

Wolf, Naomi
1991 *The Beauty Myth: How Images of Beauty Are Used against Women.* New York: William Morrow.

11

Regulated Passions

The Invention of Inhibited Sexual Desire and Sexual Addiction

Janice M. Irvine

IN RECENT YEARS, several different headlines bannered the covers of *Cosmopolitan* magazine. In November, 1988, "When You're Not Interested in Sex, He's Not Interested: How to Reawaken Your Desire." In July, 1989, "Girls Who are Addicted to Sex: Why They Can't Stop." And finally, in November, 1989, the plaintive question, "How Much Sex Is Enough?" The *Cosmo* girl was embroiled in the labyrinthine contemporary debate about sexuality and sexual desire. The latest questions about the nature and limits of desire issue from the invention in the mid-1970s of two distinct diagnostic constructs: inhibited sexual desire (ISD) and sexual addiction. Clearly these are recognizable concerns—that one might have too little desire for sex, or conversely, experience oneself as sexually insatiable. The medicalization of these two conditions, however, with elaborate systems of diagnostic categories and treatment interventions, charts a sexual condition or "sick role"[1] and, in the case of sexual addiction, an entire identity constructed around a specific sexual pattern.

In our culture, disease constructs frequently exceed the bounds of a simple biological entity with clear and objective organic etiology, but rather serve as expanded paradigms imbued with diverse meanings. Diseases are artifacts with social history and social practice. In the area of sexuality, the discursive practices of medicine since the nineteenth century have spawned what Foucault terms a "proliferation of sexualities,"[2] most of which carry the stamp of perversion reborn as disease. Thus, the invention of sexual addiction and inhibited sexual desire can be understood in light of two related historical factors of the late nineteenth century. First, a range of socioeconomic changes prompted a commercialized sexuality in which sex is increasingly privileged as fundamental to individual identity and happiness.[3] Second, the medical profession usurped moral and religious authority in the area of sexuality, generated new and highly visible discourses, and promulgated the diversification of new sexual identities. As Foucault suggests, "Sex was driven out of hiding and forced to lead a discursive existence."[4] Inhibited sexual desire and sexual addiction are two of the most recent medical constructions of sexual disease and disorder.

It would be a mistake, however, to impute sole and uncontested power to the medical profession in the invention of the new disorders. Rather, new diseases emerge within the triangulation of medical imperatives, the demands and experiences of individuals, and cultural traditions and anxieties. This chapter will examine these three axes of influence. First, it will analyze the history of professional intervention in problematic behavior subsequently defined as desire dysfunctions. Second, it will explore the complexities of definition and treatment, and the implications for the afflicted individual. Finally, it emphasizes the ways in which disease reflects the cultural style of a period.[5] It suggests that, in the late twentieth century, these new diseases chart the medically legitimated boundaries of acceptable contemporary sexual experience and serve as signifiers for powerful cultural anxieties about sexuality and desire.

Disease Narratives

The construction of disease categories entails a complex set of negotiations among professionals, the general public, and afflicted individuals, which is always mediated by broader cultural ideologies. The particular configuration of very different circumstances for the emergence of ideas about inhibited sexual desire and sexual addiction in the mid-1970s offers clues about their powerful individual and social valence.

Although modern clinicians have anecdotally noted cases of low sexual desire as early as 1972, ISD was first identified in the medical literature in 1977 by two sexologists working independently: Harold Leif and Helen Singer Kaplan.[6] Both are well-known sex therapists, who reported the increasing prevalence of complaints about low libido in their clinical practices. This was a noteworthy departure from the presenting problems of most patients during this heyday in sex therapy. With the publication of Masters and Johnson's *Human Sexual Inadequacy* in 1970, sex therapy had grown throughout the decade to become the most visible, lucrative, and widespread enterprise of sexology.[7] On the basis of their research and clinical work, Masters and Johnson had identified several major categories of sexual problems, which were eventually adopted by the third edition of the *Diagnostic and Statistical Manual of Mental Disorders* (DSM-III). For men, the basic sexual dysfunctions included premature ejaculation, primary and secondary impotence, and ejaculatory incompetence (a rare condition in which a man cannot ejaculate intravaginally). Female sexual dysfunctions included dyspareunia (painful intercourse), vaginismus (a tightening of the vaginal muscles that prohibits penile penetration), and several types of orgasmic dysfunctions, broken down into primary or secondary and coital or masturbatory categories.[8] These dysfunctions encompassed the range of technical difficulties to which a couple might be vulnerable, and undeniably, Masters and Johnson's sex therapy program helped scores of people improve their sex lives. Their brief, symptomatic treatment seemed perfect for those with little experience or information about sex.

The mere discovery of the program's existence was enough to instill hope and confidence in some couples. By the late '70s, sexologists, with their appeal to scientific legitimacy and medical authority, were riding a wave of popularity in a vast market eager for a new approach to sexual problems. Rumors of dramatic, near-miraculous success rates for interventions with sexual problems were so persuasive that clients began reporting cures merely from sitting in the waiting room. By the end of the decade, however, sex therapists voiced a common lament about the disappearance of the "easy cases"—specifically, those problems which were essentially the result of ignorance or misinformation, and which responded well to the simple behavioral methods of Masters and Johnson. The new difficulties were reported in different ways: sexual boredom, low libido, sexual malaise, and even sexual aversion and sexual phobia. Harold Leif recommended that the diagnosis of inhibited sexual desire be applied to those patients who chronically failed to initiate or respond to sexual stimuli. The dysfunction is now routinely referred to as either inhibited sexual desire or hypoactive sexual desire. The American Psychological Association estimates that 20 percent of the population has low or absent sexual desire. Among sex therapists, ISD is now reported as the most common presenting problem, constituting half of all diagnoses, and it is considered to be the most difficult sexual problem to treat.[9] More women than men are diagnosed with inhibited sexual desire, although many therapists report that the rate among males is rising.[10]

The concept of sexual addiction had a quite different beginning, springing to life independently in several cities almost simultaneously. Not surprisingly, the idea of being addicted to sex emerged in the addiction movement among those who were in recovery from substance use. Its first manifestations were in the establishment of 12-Step groups to contain what their members describe as "sexual unmanageability." Sex and Love Addicts Anonymous was the first such group, started by a musician in Boston in 1977.[11] He had been a member of Alcoholics Anonymous for years, had a wife, a mistress, engaged in other sexual affairs, and masturbated several times a day. His perception that his sexuality was out of control led him to find others with a similar problem so that they could "get sober." Initially they met in private homes, but the growth of the group led them to seek public meeting space. A local pastor was sympathetic, but skeptical that his parishioners would support a group of sex addicts, so he suggested a name change. Sex and Love Addicts Anonymous is now also know as the Augustine Fellowship, since one of the members had been reading Augustine's *Confessions,* and claimed, "He's obviously one of us." There are now seven different nationwide fellowships for sex addicts and co-addicts, with such names as Sex Addicts Anonymous, Sexaholics Anonymous, and Sexual Compulsives Anonymous. All were founded under similar circumstances as the Augustine Fellowship.

Every week close to 2,000 meetings for sex addicts are held across the country, and the groups are said to be growing at an annual rate of 30 percent.[12] Unlike with ISD, a thriving grass-roots movement of individuals who claimed to suffer from the disorder was already in place by the time professionals engaged with the issue. Now, however, experts and clients work in tandem, since sexual addiction has spawned a robust treatment industry. The diagnosis has attracted two types of professionals: the dominant group are "addictionologists" (as they now call themselves), who are joined by a smaller cohort of clinicians who treat sex offenders. Professional awareness of sex addiction was fostered by the "opponent-process" theory of addiction introduced in the early '70s, which suggested that a substance was no longer requisite for addiction.[13] This proposal that any behavioral excess could lead to dependence fit nicely with the popular and widespread generalization of ideas about addiction represented by such figures as the workaholic, shopaholic, and compulsive gambler. Proponents of the syndrome view sexual addiction within this expanded paradigm of addiction disorders. There are now scores of texts on sex addiction, and treatment programs dot the landscape. The first inpatient program for sexual addicts was begun in 1985 at Golden Valley Health Center's Sexual Dependency Unit in Minneapolis. Addictionologists claim that 6 percent of the population are sex addicts, and approximately 30 percent of these are women. Patrick Carnes, one of the foremost popularizers of the sexual addiction concept, claims that 1 in 12 people in the U.S. is a sex addict.[14]

The Professional Divide

The construction of new medical definitions does not simply mirror a perception of illness or problematic conditions. Rather, the discursive elaboration of disease is shaped by myriad and complex factors, including the ideological and economic imperatives of the defining professionals. It is noteworthy that the new diseases of ISD and sexual addiction have each been fashioned by highly different professional cohorts, so that the medical discourses have progressed on parallel and quite distinct trajectories. There has been little connection or communication between addictionologists and sexologists and little overlap in specialization. There has been some veiled hostility between the groups, however, deriving from the clash of underlying ideologies and overt treatment goals. These conflicts highlight the constructed nature of the new diseases and reveal the nature of the loyalities and interests of each profession.

The field of addictionology has been marked by rapid professional expansion, particularly since Solomon's opponent-process theory of addiction provided theoretical legitimacy for the identification of almost anything as an addictive agent. Although the subsequent proliferation of addictions widened the professional domain of addictionologists into new areas, such as gambling and sexuality, their medical gaze remains one of vigilance about excess and admonitions for control and management. This sexual ideology of temperance and abstinence directly opposes the vision of sexual expansion and freedom so implicit in sexology. Addictionologists have criticized many of the sex-enhancing technologies, ideologies and practices that comprise standard sex therapy models. For example, some believe that penile implant surgery, a lucrative procedure that is quite acceptable among sexologists, signifies and reinforces sex addiction.[15] And addiction experts challenge sexologists on their unbridled enthusiasm for the unrestricted use of fantasy and pornography, for their encouragement of masturbation, and for their celebration of virtually any sexual activity between consenting adults. Sex addicts, it is thought, may need to practice celibacy and eliminate fantasy and sexually explicit material "in order to attain and maintain sobriety," and sex therapists may simply re-traumatize them or facilitate a relapse by the espousal of sexual freedom.[16] While addictionologists may accept the dominance of sexologists in many areas of sexuality, they are staking out their turf and becoming more visibly critical of sex therapists for an alleged lack of effectiveness in the treatment of sex addicts.[17]

Sexologists, on the other hand, have struggled for professional legitimacy and a viable commercial market for over a century.[18] They pride themselves on scientific rigor in their work and on fairly unqualified acceptance and support of all sexual expression. The very concept of sexual addiction—that there can be too much sex—threatens the foundations of the profession. Until the early '90s, sexologists have largely responded to the emerging diagnosis of sexual addiction with sarcasm, disavowal, or attacks on its scientific credibility. In a 1988 anthology on sexual desire disorders, sexologists Sandra Leiblum and Raymond Rosen noted that sexual addiction was beginning to receive considerable attention. They described the affected individuals as "sexual enthusiasts," and noted that they "tend to be admired or envied rather than diagnosed."[19] Helen Singer

Kaplan claimed that sexual addiction is exceedingly rare[20] and is "a media term that doesn't have any scientific validity or meaning."[21] Possibly because of his work treating sex offenders with Depo-Provera,[22] noted sexologist John Money was initially a supporter of the new diagnosis, telling the *New York Times* in 1984 that he had seen many patients who were sex addicts. "Their hypersexuality often exceeds normal capacities," he noted.[23] By 1989, however, Money vehemently reversed his position, charging that "the pathologizing of sex by inventing a hitherto unknown disease, sexual addiction," constituted a strategy of "the sexual counterreformation" that has exercised a destructive effect on the advancement of the science of sexology.[24]

Professional antagonisms have slowly begun to dissipate into mutual ambivalence, largely because of the impact of the AIDS epidemic; sexologists, who have been unable to avoid widespread social anxiety about freer sexual mores, have begun fashioning their practices accordingly. Workshops and interventions, such as one entitled "Falling in Love Again," abound for the sexually and relationally bored who are terrorized into monogamy by fear of HIV infection. It is not uncommon for sexologists to chart their professional course through the changing currents of cultural ideologies about sex and gender. Sexology has never presented an uncomplicated vision of sexual liberation. Rather, the field has historically managed the contradictions of a progressive sexual message and the need for conservatism and scientific credibility in order to achieve cultural legitimacy and economic viability. Despite these antinomic imperatives, sexologists generally advance a sexual value system of greater freedom and participation. Yet the cultural sex panic exacerbated by AIDS has foregrounded the addictionologists' message of sexual chaos, as well as the terror of sexual excess. The sexologists' exhortations to sexual pleasure and experimentation look increasingly unwise and unhealthy, and sexologists have begun integrating some of the ideas, if not the wholesale diagnosis, of the sexual addiction field.

The historical narrative, then, of the construction of inhibited sexual desire and sexual addiction reveals a clear bifurcation between two professional cohorts, marked by ideological tensions and distinct border anxiety. As experts laboring at opposite poles of the same continuum of sexual anxiety and control, however, these professionals share a possibly insoluble conundrum: the task of precise clarification and definition of their disease.

Defining and Treating the Disorders of Desire

The assertion of exact definitional and diagnostic criteria poses an enormous challenge when "disease" is a generalized set of signifiers of cultural chaos and social control. The fabric of overdetermined diseases such as sexual addiction and inhibited sexual desire is woven from the diverse threads of professional expertise and ideologies, cultural beliefs about sex, and the attempts of individuals to make sense out of their own sexual experiences. Nevertheless, the medical legacy of the doctrine of specific etiology[25] has inspired each field to generate myriad hypotheses concerning the individual causes of the disorders. These etiological theories of inhibited sexual desire and sexual addiction are not totalizing discourses but rather an amalgam of diverse and sometimes nebulous perspectives. There is conflict within sexology and addictionology over the origins and nature of their respective disorders, and it would be incorrect to imply a simple unity or consensus. Despite few rigorous research studies, there has been much speculation on the basis of clinical samples and case studies. There are debates over the influences of environment, family, individual personality, and such biological factors as neurochemistry.

Since the nineteenth century, however, professionals have assiduously tracked the etiology of sexual conditions within a biomedical tradition that quantifies desire and locates "this search for the primeval urge in the subject itself."[26] It is this impulse to map desire and its varied disorders in the body itself that represents historical congruence among sexuality professionals and establishes a common theoretical terrain for both inhibited sexual desire and sexual addiction. A pervasive and largely assumed underpinning of both sexual addiction and ISD is the representation of sexual desire as a biological drive or surging energy that is either flooding uncontrollably or woefully diminished. There is the intuitive belief that sex, and specifically sexual desire, resides in the body. This essentialist assumption is not surprising, since it infuses mainstream cultural norms about sexuality with the theoretical foundation of sexual science.[27] As historian Jeffrey Weeks noted about nineteenth-century sexology's search for the origins of sexual behavior, "biology became the privileged road into the mysteries of nature. . . . "[28] More than one hundred years later, although few theorists would unequivocally advance a strict biological determinism in the etiology of ISD or sexual addiction, strong essentialist themes resonate throughout the discourses of the desire disorders.

The literature of both professional cohorts reflects a striking emphasis on the brain as the site of sexual desire and the source of its myriad manifestations. Advances in neurochemistry converged with the technological revolution in computers to produce a cyborgian vision of sexuality and desire characterized by images invoked by such phrases as "hardwiring," "circuitry," "programmed into the brain," and "fixed into the system." In this representation, the bedrock of desire and its concomitant sexual possibilities reside within regulatory mechanisms of the brain that are alternately perceived as impervious to change or quite vulnerable to disruption.

The biological basis of sexual desire is most advanced in the work of Helen Singer Kaplan, who pioneered the notion of inhibited sexual desire and who is likely one of the most unreconstructed essentialists within sexology. In her 1979 landmark text on ISD, *Disorders of Sexual Desire,* Kaplan defines sexual desire as an "appetite or drive which is produced by the activation of a specific neural system in the brain."[29] In sociobiological terms, Kaplan describes the importance of sexual desire:

> Sexual desire is a drive that serves the biologic function of species survival. It instills a strong erotic hunger that prods us to engage in species specific behavior that leads to reproduction. It moves us to find a mate, to court, to seduce, to excite, to impregnate, to be impregnated.[30]

For Kaplan, desire is experienced when a specific neural system in the brain is activated, prompting genital sensations and an openness to and interest in sex. When this system is inactive or inhibited, the person "loses his appetite" or "the brain has 'decided' that it is too 'dangerous' to have sex."[31]

Addictionologists tend to discuss sexual desire very little, other than to implicitly regard it as an inherent physiological drive that has spiraled out of control. This discourse harks back to the eighteenth-century perspective described by Weeks that "desire was a dangerous force which pre-existed the individual, wracking his (usually his) feeble body with fantasies and distractions which threatened his individuality and sanity."[32] Significantly, within the sex addiction literature, the moniker "desire" is generally superseded by its moral ancestor, "lust."

Even so, brain-centered sexual theories are even more prolific among addictionologists than sexologists. The most dominant theme proposes that a finite number of polymorphous sexual possibilities are locked into the brain in early childhood, and subsequent behavior is virtually predetermined. The "lovemap" theory of sexologist John Money has been enthusiastically deployed by addictionologists, who find the notion of behavioral options programmed into the brain in childhood a compelling explanatory concept. Money describes the lovemap as "a developmental representation or template, synchronously functional in the mind and the brain, depicting the idealized lover, the idealized love affair, and the idealized program of sexuoerotic activity with that lover, projected in imagery and ideation, or in actual performance."[33] The lovemap allegedly incorporates into the brain a range of social inputs transmitted through sensory mechanisms. In addition, scientific breakthroughs in neurochemistry have informed the development of an essentialist sexual-addiction model, since, as Patrick Carnes notes, "studies generated greater scientific awareness that addiction could exist within the body's own chemistry."[34] Sexual desire and addiction are thus viewed as coterminus physiological events inside the body.

These models enable addictionologists to explain the intransigence of repetitive problematic sexual behavior. It has been encoded into the hardwiring of the brain. It is not uncommon to hear addictionologists suggest, for example, that our brains spontaneously move into preset programs of activities;[35] that the linchpin of codependency is the inability to change behavior because it has been programmed in as a child;[36] or that male and female brains are crucially different, so that the identical early-childhood trauma can affect a female differently than a male.[37] As we will see, these theoretical perspectives on sexual desire disorders are crucial in that they shape treatment strategies.

Although they occupy considerable space, the brain-centered theories of sexuality are not hegemonic. More recently, sexologists have been divided on the centrality of a biologically based model of desire. Many still adhere to a solidly essentialist theory, and scores of studies claim the androgens as the "libido" hormone.[38] Other sexologists have posited multidimensional models that privilege psychological and cognitive factors as shaping desire.[39] The medical literature frequently describes marital difficulties, fear, and anger as underpinnings of inhibited sexual desire. As Helen Singer Kaplan notes about women who experience ISD in the context of ongoing relational discord, "it is not possible for most people to feel sexual desire for 'the enemy.' "[40] And many addictionologists link sexual addiction and codependency to trauma, child sexual abuse, and a breakdown in spirituality.[41] On infrequent occasions, clinicians will point to the role of broader cultural messages that result in complex and often contradictory imperatives about sex.

Even when these professionals invoke more expansive hypotheses to explain the desire disorders, however, they are largely theoretically located in what anthropologist Carole Vance terms the "cultural influence model."[42] In this view, sexuality, although influenced by culture, is thought to encompass universal forms of expression driven by an inner force or impulse. Vance notes, "Although capable of being shaped, the drive is conceived of as powerful, moving toward expression after its awakening in puberty, sometimes exceeding social regulation, and taking a distinctively different form in men and women."[43] While the cultural-influence model as it appears in professional theories of desire disorders is a marked improvement over rigidly essentialist frameworks, it retains determinist assumptions. And, unlike social construction theory, it leaves unexamined the radical mediation of sexuality by history and culture. Only on rare occasions, for example, do sexologists and addictionologists acknowledge the role of societal norms of sex in shaping desire.

Aside from theoretical congruence concerning the nature and origins of desire, however, both professional cohorts face the challenge of precisely elucidating the parameters of their new diseases. The experts constructing sexual addiction and inhibited sexual desire share a common conceptual and practical problem of definition: the decisive question of how much is too much and how little is too little. This indeterminacy is a familiar dilemma in describing sexual disorders. Kinsey was famous for anecdotally defining the promiscuous individual as "someone who's getting more than you are." Similarly, Masters and Johnson struggled for a reasonable definition of premature ejaculation. For example, given the range of partners a man might have, he might be premature with one partner and not with another.

For ISD and sexual addiction, professionals have given diagnostic weight to outside referents. For sex addicts, repeated criminal offenses can serve as a surrogate marker for the subjective experience of being out of control sexually. An angry and dissatisfied partner is often the impetus for someone to seek professional treatment for ISD. Yet again, the fundamental subjectivity is inescapable, and calls to mind the exchange between Woody Allen's and Diane Keaton's characters in *Annie Hall* when their therapists asked how often they had sex. Woody said, "Hardly ever. Maybe three times a week," while Diane replied, "Constantly. I'd say three times a week." Clinicians admit that, especially with ISD, the concept of desire discrepancy is inevitably relational, so that individuals can easily shift diagnoses, depending on their partner. Ultimately, both ISD and sexual addiction rely heavily on self-diagnosis and serve as beacons for the individual who feels a sense of inadequacy or incongruence with cultural or interpersonal sexual norms. Yet despite its inevitably subjective character, professionals have tried to establish quantifiable frames for their diseases.

The nomenclature committee of the American Psychiatric Association recognized inhibited sexual desire as a clinical entity in 1980 and it was included in the *Diagnostic and Statistical Manual of Mental Disorders* (DSM-III), thus making it an official mental illness. The revised third edition (DSM-IIIR) elaborated this classification further by dividing the disease into two categories: Hypoactive Sexual Desire Disorder and Sexual Aversion Disorder.[44] The definition of HSD is vague but implies that the person must be distressed or that there must be an inherent disadvantage to low sexual interest (anger of a spouse, for example). Helen Singer Kaplan describes HSD as either primary—a rare, lifelong history of asexuality—or secondary, in which there is a loss of sex drive after a history of "normal sexual development." Kaplan describes the typically situational HSD woman as the one who:

> feels very erotic during the many years of her precoital experiences. She felt desire and erotic pleasure during "petting," but she loses sexual inter-

est after she has engaged in coitus, or after marriage, or after childbirth, i.e., in situations which on a symbolic and unconscious level represent danger.[45]

The diagnosis of ISD remains controversial among sexologists, with little consensus regarding operational criteria. Some sexologists have even suggested that the disorder is so vague and the diagnostic boundaries so blurred that ISD, as a "catch-all" diagnosis, represents the schizophrenia of sex therapy.[46] Many negotiate these difficulties with the strategy summarized by sex therapists Sandra Leiblum and Ray Rosen as "you know it when you see it."[47]

Proponents of the sexual addiction diagnosis suffer similar definitional quandaries, resulting in a myriad of checklists and screening questionnaires to determine one's vulnerability. Many subscribe to the AA maxim: If you think you've got a problem, you probably do. The Sexual Dependency Unit at Golden Valley defines sexual addiction as "engaging in obsessive/compulsive sexual behavior which causes severe stress to addicted individuals and their families."[48] Sex becomes the organizing principle of the addict's life, for which anything will be sacrificed. In addition, sexual addiction can include the following behaviors when they have "taken control of addicts' lives and become unmanageable: compulsive masturbation, compulsive heterosexual and homosexual relationships, pornography, prostitution, exhibitionism, voyeurism, indecent phone calls, child molesting, incest, rape and violence."[49] Among the several 12-Step groups, definitions of sexual addiction vary, as do the concepts of what constitutes "sobriety." Yet among both professionals and recovering sex addicts, two themes are consistent. First, whatever the behavior, it is practiced compulsively. And second, the common enemy is lust, which is thought to drive the sexual addiction cycle. In a manner reminiscent of the social-purity movements of earlier centuries, lust is thought to lead the victim into uncontrollable and destructive behavior. Lust, therefore, must be eliminated. Sexaholics Anonymous is perhaps the most restrictive in this sense, in that freedom from lust occupies center stage in the definition of sobriety. The literature states:

> Any form of sex with one's self or with partners other than the spouse is progressively addictive and destructive. Thus, for the married sexaholic, sexual sobriety means having sex only with the spouse, including no form of sex with one's self. For the unmarried . . . freedom from sex of any kind. For all . . . progressive victory over lust.[50]

Again, as with the social-purity movements, addictionologists view sex as simply one of the falling dominoes in a downward-spiraling cycle of destruction that may include, among other elements, gambling, eating disorders, drugs, and alcohol. One professional, for example, described as vulnerable "a particular kind of woman very involved with fantasy who is a compulsive masturbator, compulsive overeater, and reads romance novels."[51] Addictions are described as multiple and often interchangeable. Ann is another case described in the treatment literature.

> Ann spent almost every night in cocktail lounges searching for men. After several years of emotionally empty one night stands, she came to the realization that she and the men she seduced were simply using each other sexually. In desperation she swore off the bar scene and joined her friend Judy in small stakes bingo and card games. Within one month, Ann had identified several high stakes poker and bingo games and had become totally absorbed in her new-found gambling compulsion.[52]

To discourage such symptom substitution, professionals are warned that treatment must include all the addictions. Self-help groups thus cast an ever-widening net to include such compulsions as excessive masturbation, gambling, bingo, and romance novels.

In her landmark, bestselling text, *Women, Sex, and Addiction,* Charlotte Kasl expands the system of addiction even further by intertwining sexual codependency with sexual addiction in women. Codependent (or co-addict) was a term originally created to describe the partner of an alcoholic. Kasl has broadened the definition to refer to a "devastating disease" in which a woman has sex when she doesn't want to, in order to maintain a relationship or placate a partner. Codependency, Kasl notes, is "women's basic programming" and is only a slight exaggeration of the culturally prescribed norm for women.[53]

Despite internal disagreement and confusion among sexologists and addictionologists over the etiology, definitions, and operational criteria of their diseases, some consistent treatment strategies have evolved. These are shaped by the biomedical infrastructure central to the construction of the desire disorders; for, regardless of documented social correlatives such as abuse, power differences in heterosexual relationships, or cultural pressures as possible etiological factors, the disease model of sexual desire disorders retains prominence. Sexual addiction is considered dangerous, and by some, such as Anne Schaef, "a progressive, fatal disease."[54] Likewise, sexologists view inhibited sexual desire as a serious and intractable disease. Unlike the other sexual dysfunctions, it has a poor prognosis with available treatment. Treatment strategies for both sexual addiction and inhibited sexual desire remain steadfastly fixed on the individual (or often in the case of ISD, the couple) with the goal of management, adjustment, or regulation of sexual desire and sexual behavior. The most common interventions are individual or couples therapy, sometimes supplemented with pharmaceuticals. For sexual addiction, 12-Step groups are an essential complement to either inpatient or outpatient treatment.

The professional reliance on organic, neurochemical explanations for both ISD and sexual addiction has predictably led to the search for a "magic bullet," as experts in both fields look hopefully and confidently to the future of neurochemistry for unlocking the determinants of their diseases. Meanwhile, drug treatment is used as an adjunct to treatment for both dysfunctions. The antidepressant drug Wellbtruin was once the great treatment hope for ISD, until it was found to trigger seizures. Now, despite a lack of evidence of efficacy, testosterone is being prescribed for low sexual desire in premenopausal women.[55] Prozac is often prescribed for sex addiction, although there is much controversy and suspicion among sex addicts about using a drug to treat an addiction.

Medicalizing Desire

In our culture, both disease and desire are medical events, individual experiences, and social signifiers. There is no linear relationship between medical ideology and individual behavior. We are not passively shaped by broader medical ideas; yet neither does our medical discourse directly reflect an internal, universal experience of individuals. The content of medical diagnoses is shaped by social, economic, and political factors. And a both specifically medical and broader cultural ideology operates in the construction of individual experiences of sexual desire. Not simply a biological urge, sexual desire is a culturally constructed composite. It is imperative, therefore, to analyze the contemporary medicalization of sexual desire along these three dimensions.

The nineteenth century marked a shift to scientific investigation of sexual matters. Hence, sexuality has represented a site of expansion and control by the medical profession, with its interests in delineating the nature of sexual impulses and constructing new psychological categories of behavior. The themes in ISD and sexual addiction of sexual conflict, chaos, and disorder are familiar legacies from more than a century of a medical gaze on sexual expression. The invert, the sexual psychopath, the hypersexual female, and the onanist are but some of the historically demonized characters who step from the text of a medical discourse of definition and regulation.[56] It is not surprising, then, that professionals in the late twentieth century would conceptualize concerns regarding sexual desire as major medical problems, since physicians have historicaly played a significant role not only in the management of sexual behavior but in defining the existence, appropriateness, and ideal object of sexual desire or passion.

Broader societal constructs of desire have largely been based not on the felt experiences of individuals, but on ideological beliefs about sexuality and gender. For example, permission for any individual woman to experience desire, discuss sexuality, initiate a sexual encounter, or present herself as passionate varies historically and culturally. Carl Degler has documented the variability in nineteenth-century medical-advice literature regarding desire in middle-class women. One theme speaks to the strength of women's passion; another articulates the stereotypic Victorian view that women approach sex "with shrinking, or even with horror, rather than with desire."[57] Further, Nancy Cott has related variations in the dominant ideology about women's passion through the eighteenth and nineteenth centuries not to changes in individual and interpersonal sexual experiences, but to cultural shifts in metaphoric systems about the nature of women. Passionlessness, she argues, transformed women's image in the nineteenth century to one of spirituality, away from the eighteenth-century view of women as lustful creatures prone to sexual excess.[58] The most recent medical constructions of desire disorders reinscribe historically familiar themes of morality, regulation, the ambivalence of pleasure, and the ruin of excess and depravity.

The power of medical ideology in the construction of sexual desire derives from its expansion, its authoritative voice. There must be cultural recognition that desire problems are diseases, with a subsequent adoption of the language and concepts of dysfunction. This process is facilitated by popular representation, and by the early 1990s both ISD and sexual addiction had achieved a certain currency within popular culture. For sufferers of low sexual desire, articles abound on DINS (dual-income, no sex), casting ISD as the latest malady for yuppies too tired from an active day on Wall Street to have sex. And at least one popular self-help manual has appeared; in *Not Tonight, Dear,* the author promises that "the mental nature of desire makes it particularly amenable to improvement through reading."[59] Given that our cultural balance consistently tilts away from pleasure and toward prohibition, the idea of sexual addiction has more thoroughly captured the popular imagination. In addition to the thousands of 12-step groups, there is a National Sexual Addiction Hotline; the *National Enquirer* reported that Rob Lowe had entered a sexual addiction clinic;[60] and Arnie Becker on *L. A. Law* began to describe his sexual exploits as "satyriasis." The shift in the sexual spirit of our times is perhaps best captured by the new book by Erica Jong, *Any Woman's Blues,* in which the central female character is a sex addict who joins a 12-Step group by the end.[61] The *New York Times* ad blazoned, "In the seventies, Erica Jong taught women how to fly . . . now she shows them how to land."[62] Through their widespread dissemination of the concepts of inhibited sexual desire and sexual addiction, these popularizations continually reassert and legitimate the idea that cultural ideologies about appropriate sexual expression are valid medical conditions responsive to individual intervention and cure.

The existence of inhibited sexual desire and sexual addiction as medical diagnoses ensures that proposed solutions will be individual and not structural and cultural. In part ideological, this therapeutic trajectory is also driven by a financial motor, and, clearly, economic incentives are central to medical expansion. Treatment for sexual desire problems is a vast and lucrative commercial venture. The revenues of Sex and Love Addicts Anonymous, for example, soared to over $100,000 after Hazelden took over distribution of their central text in 1988.[63] Golden Valley Health Center employs an international public relations firm to manage the scores of daily calls received from around the world about the Sexual Dependency Unit. And sexologists report that at least half of the clients coming for sex therapy present with claims of low sexual desire.[64] The large numbers of individuals engaged in treatment for desire problems speaks to the widespread acceptance of medical constructs and the availability of professionals who offer medical diagnosis and treatment. But perhaps most importantly, it indicates the pain and confusion experienced by so many people concerning their sexual desire and behavior.

In this respect, then, it is important to evaluate the medicalization of desire by its therapeutic impact—how the medical creation of the sexual desire disorders operates in the lives of individuals. How does the existence of ISD and sexual addiction as disease entities shape individual experiences of sexuality? Has the creation of these diseases either limited or expanded other options for thinking about sexual desire? What does it mean to someone to take on the identity of a sex addict? If one feels little sexual desire, is it helpful to define that absence as a disease? Does it matter that it is clinicians who will offer the range of answers to the *Cosmo* girl who wonders "how much sex is enough?" While there are anecdotal or clinical reports, the desire disorders are too new for the emergence of a nuanced ethnographic and phenomenological literature on the meanings of these diagnoses for men and women. Some speculation about the broader cultural implications and the limits of individual impact is possible, given our knowledge about the nature of medicalization and of the particular theoretical contours of both inhibited sexual desire and sexual addiction.

The imposition of a biomedical paradigm over social events or problems may suggest potential advantages. These include the increased recognition it promotes and the conceptual framework it offers for worried individuals. Further, a medical diagnosis confers legitimacy on a particular set of difficulties. The seemingly neutral and scientific language of disease may offer palpable relief to those who secretly worry that their sexuality is inadequate or out of control. Especially when the definitional options are those of morality or personal failure, a medical diagnosis may sound more dispassionate and, significantly, admits one to a high-tech arena of research and psychotechnology.[65] According to one clinician, "Lack of desire is like a fever. Something is going on."[66] As with contracting the flu, feeling too much or too little desire is nobody's fault. All that remains is the breakthrough treatment discovery.

Ultimately, however, medical diagnosis offers a false neutrality, for, "as illnesses are social judgments, they are negative judgments."[67] Disease designations connote discomfort, deviance, treatment, and cure. Sexologists and addictionologists, for example, have reified the desire disorders into static and simplistic categories. Diagnostic profiles and checklists are purposely vague so as to be inclusive of a wide range of behaviors. One profile for sexually addicted women includes and indicts behavior as diverse as "multiple and serial relationships; affairs; one night stands; cruising bars, health clubs, etc.; personal columns; masturbation; fantasy; preparing and dramatizing; s/m; exposing; dangerous situations; self abuse; suicidal and homicidal; relationships with sexual compulsive men."[68] The ideal model presented for sexually addicted women is a social purity vision of a spiritually based, monogamous sexuality that is always relationally oriented.[69] Any variation from this is pathologized, and within the sexual addiction field, retro-purity terms have reemerged, such as "promiscuity," "nymphomania," and "womanizer." Accepting the disease model of inhibited sexual desire and sexual addiction in exchange for a moral framework proves, then, to be a bad bargain, for the taint of stigma and deviance inheres in the expansive diagnostic categories.

Reliance on individual treatment solutions remains a major shortcoming of the medical model. In the case of ISD and sexual addiction, the obvious limitation is that, in the absence of social and historical insight, the problem is located within the individual chemistry or psyche and is presumed amenable to medical intervention. The inadequacy of a biomedical approach to treatment of the desire disorders is glaring even when one looks at etiology as defined by the professionals themselves. Despite the preoccupation with lovemaps and brain circuitry, professional literature suggests a broader range of social correlatives. ISD is frequently related to fear, anger, and marital problems; some studies suggest that power struggles and lack of respect are major dynamics for ISD in women.[70] Sexual addiction is linked to childhood sexual trauma.[71] Given this data, a sociohistorical approach to treatment would suggest the need for a more encompassing strategy for change. Yet clinicians articulate no social vision to end sexual abuse, challenge the primacy of the nuclear family, end the double standard, improve sex education, or expose destructive and coercive sexual ideologies. Significantly, there are no treatment outcome studies for sexual addiction, and ISD is widely considered the most difficult sexual problem to treat.

Medicalizing desire, then, cannot really be said to eliminate moral stigma or enhance "cure." Other potential effects are difficult to discern clearly. There is some concern that the message from the proliferation of sexual diseases privileges certain styles of sexual expression and marginalizes others. At least one sex therapist has been critical of the broader therapeutic milieu of sex therapy, whose emphasis on sexual enhancement techniques increases the "pressure we all are under to 'always say yes.' "[72] It is widely recognized that the discourse of sickness can readily become coercive, and there is evidence that this is increasingly true for the sexual dysfunctions.[73] One client in therapy for ISD voiced precisely this complaint about her husband's appropriation of the disease frame:

> But he's got this hang-up about my having to have sex the way he does. "Doesn't it feel good to you?" he asks. So, I rub my earlobe and say, "yeah, and this feels good, too, but I never think about rubbing my earlobe and if I do I don't say, 'Wow, I can't wait to do it again.' " I like Chinese food, but if I had to go a year without any, I wouldn't be miserable. He says I'm inhibited and don't know it and that I need therapy. When he gets angry he calls me an uptight, frigid bitch, and says I'm sick.[74]

Yet individuals internalize disease models in highly variable ways, and it is important to acknowledge that, despite coercive potential, scores of people report relief and validation from the desire diagnoses. Individuals also negotiate these diagnostic systems idiosyncratically. With sexual addiction, for example, there are clearly individuals who instrumentally select from the menu of treatment options, attending recovery groups for the structure, support, and community, but eschewing the adoption of a full-blown identity. Countless others, however, opt for wholesale acceptance of the addiction ideology as an explanatory device for their fears, and they find solace in their "sobriety" from the disease. The AIDS epidemic has been the perfect impetus for many to define as out of control behavior that once would have been perfectly acceptable to them, and in 1986 the national gay newsmagazine *The Advocate* reported that thousands of gay men were reporting that they suffered from the disease of sexual addiction.[75] The out-of-control behavior defined by the men themselves ranged from masturbation once or twice a month (by a devout Catholic) to relentless cruising of peep shows. All claimed to experience great relief through their "sexual recovery plans."

This underscores the importance of individual needs and cultural anxieties in the construction of disease categories. Inhibited sexual desire and sexual addiction serve as contemporary disease categories that help people create meaning out of their sexual experiences. The diagnoses offer the hope of achieving "normalcy" to those who experience their sexual desire as either inadequate or out of control. They are bipolar constructs that map the contradictory cultural landscape of the negotiation and management of appropriate sexuality. These disorders emerged in the mid-1970s and flourish during an era of distinct and palpable tensions regarding sexual norms. They are informed by the dichotomous contemporary ideology in which sex is simultaneously heralded as the linchpin of individual fulfillment and denigrated as the source of chaos, exploitation, and death.

Desire, too, is a cultural trope for both pleasurable satisfaction and dangerous, possibly alien, hunger. Historian Joan Jacobs Brumberg and philosopher Susan Bordo speak to this ambivalence and fear in their analyses of anorexia nervosa.[76] Women are terrified and repelled by visions of themselves as voracious, needy, yearning, and hungering without restraint. "Appetite," Brumberg writes, "is an important voice in female identity."[77] Yet appetite, whether for food or for sex, carries with it the hopes of satisfaction and the fears of wanting too much or of needing and not getting. Desire is not neutral. Cultural attitudes toward high levels of sexual desire reflect this pleasure/danger dichotomy.[78] We are assured by experts on ISD that "an increase in sexual desire is invariably beneficial," since high levels of sexual desire inspire people to exercise, watch their weight, dress with flair, groom themselves carefully, and otherwise operate as healthy, attractive individuals.[79] For sex addicts, however, it is "the athlete's foot of the mind. It never goes away. It always is asking to be scratched, promising relief."[80] Desire, then, will either make you a better person or ruin your life.

These bifurcations, so dramatically visible in this era of sexual epidemics such as AIDS, were apparent in the sexual ethos of the 1970s as well. The glut of media information about sex during that time reflected both a growing openness and an increasing sexological expertise. Further, the public challenges of the feminist and lesbian and gay liberation movements to hetero/sexist imperatives created new sexual space. Many women were empowered not merely to avoid exploitive sex, but to seek out fantasy, orgasms, and thrills. Feminist consciousness-raising groups facilitated both a critique of existing sexual relationships and the exploration of new sexual terrain. A study of married couples in the early '70s revealed greater sexual experimentation among white couples of all classes. Mainstream books such as *The Joy of Sex, The Sensuous Woman,* and *My Secret Garden* spoke to a new sexual spirit. By the late '70s, women had become increasingly active partners, and couples were enthusiastically proclaiming the importance of sex to a good relationship.[81]

Yet this sexual enthusiasm was striated with oppositional impulses. The persistence of the double standard thwarted many women pursuing sexual freedom. Feminist organizing drew greater public attention to sexual violence as a mechanism for the social control of women. And the plethora of sexual options touted by sexologists and the media were experienced by many merely as increased pressure. The glaring disjuncture between expectations of an easy sexual pleasure and the realities of failed sex helped create a cultural basis for the successful development of clinical programs of sex therapy. The growing New Right launched challenges to sex education, legalized abortion, and gay liberation, reinscribing notions of abstinence, morality, and sexual self-control on the collective psyche. The calls for sexual restraint became, of course, even more widespead and entrenched throughout the '80s with the emergence of the AIDS epidemic. On parallel tracks, then, inhibited sexual desire and sexual addiction mark these contradictory themes of sexual freedom contrasted with growing sexual fear and prohibition. Together they constitute a set of regulatory discourses and serve as social signifiers that shape individual experience.

However, medical diagnoses function differently for individuals and may operate fluidly and unpredictably in the culture. While constructed diseases like ISD and sexual addiction may play a central role in the creation and reinforcement of the traditional sociosexual order, the diagnoses also contain the seeds of disruption and opposition. The diagnostic binarisms of inhibition and excess easily suggest gendered sexual norms, and, in fact, early on, the demographics revealed more women diagnosed with ISD, while men largely filled the ranks of sex addicts. The disorders therefore reified a normative system of sex/gender relations. For women, ISD was simply a reformulation of historical diagnoses of frigidity. It implied withholding responsiveness. Conversely, the male sex addict merely occupied a position a few degrees further on the continuum of male sexual energy and aggression.

Currently, the more equivalent gender ratios suggest how, in a cultural moment of instability and ambivalence, the diagnoses of ISD and sexual addiction may signify the manner in which sex/gender boundaries are also being eroded. ISD, as it is recently constructed, draws on feminist assertions of the importance of pleasure and desire for women. Despite its many shortcomings, the diagnosis of ISD can serve as a cultural protest by women, a demand for satisfaction in sex and a refusal to settle for less. Similarly, the construct of sexual addiction is sometimes formulated as a complaint against sexual accommodation and exploitation of women. Feminists in the field claim that addiction often represents women's escape from "the powerless feelings of codependency." Codependency is described as a "disease of inequality," in which oppressed people must understand and accommodate those in power.[82] For these women, then, the struggle against sex addiction is a fundamental challenge to restrictive gender roles. Similarly, some men who identify with either ISD or sex addiction have criticized traditional male sexual expectations. It is too soon to tell whether the male who identifies with ISD will simply be silently ridiculed and despised, while the female sex addict will remain an anomaly destined for *Oprah*. But the new diagnoses clearly allow for more than the simple recuperation of normative roles.

These deconstructions simply suggest that, like the nineteenth-century proliferation of sexualities, the invention of contemporary medical categories is not one-dimensional in effect. Discourse, as Foucault notes, produces and reinforces power but also exposes and destabilizes it.[83] The creation of new sexual disorders reinscribes traditional sex/gender relations, while providing a site for resistance. Central to this resistance, however, is a consistent and sharp awareness of how these new diseases, as signifiers of social relations and anxieties, are generally supportive of dominant political interests and social structures. This is especially true in an era when, as medical experts are asserting guidelines about "safer" and hence "appropriate" sex, many individuals feel more vulnerable and therefore susceptible to medical definition, intervention, and control.

Continual challenge of medical definitions is essential, particularly as the new desire disorders become widespread. For despite the potential for some regrounding of sex/gender relations, the tendencies of medicalization are such that ISD and sex addiction can easily become social practices inimical to the goals of feminism and the lesbian/gay movement. We must remember that in the earlier social-purity movements, feminist themes resounded through movements that were otherwise conservative and anti-sex.[84] The social-purity themes of lust, degradation, and loss of control inherent in the sexual addiction construct should give us pause, particularly, for example, as the model is being suggested as salient to the area of sexual abuse and sex offenders.[85] After decades of scholarship suggesting that power inequities and gender oppression underpin most sexual violence, feminists should be wary of models that suggest that rapists and sexual abusers suffer instead from individual dysfunctions. And despite enthusiastic identification by scores of lesbians and gay men, sex addiction has gotten little, and decidedly negative, attention in the gay press. The one major article in *The Advocate*, entitled "Reinventing the Sex Maniac," rightfully worried that sex addiction was simply a new expression of homophobia and self-hatred.[86]

Inhibited sexual desire and sexual addiction are not demon diagnoses; they have offered validation and community to many. But since the biomedical model is a severely limited paradigm for understanding sexuality on either a social or personal level, it is clearly time for an alternative popular and accessible frame for people to understand their experiences or engage in collective discussion and support for sexual concerns. Progressive movements currently articulate a public and oppositional discourse that inserts the elements of history, cultural ideologies, and power relations into any analysis of sexuality. It is the next challenge for them to create the space for individuals to determine how the personal might be political in their sex lives. Otherwise, the new desire disorders stand as uncontested models in which sexual anxieties, discomfort, and problems inhere in the individual body or psyche, rather than in the body politic.

Notes

1. Talcott Parsons, "The Sick Role and the Role of the Physician Reconsidered," *Health Society,* no. 53 (1975): 257–78.

2. Michel Foucault, *The History of Sexuality. Vol. 1: An Introduction,* trans. Robert Hurley (New York: Pantheon, 1978), p. 48.

3. John D'Emilio and Estelle B. Freedman, *Intimate Matters: A History of Sexuality in America* (New York: Harper and Row, 1988).

4. Foucault, *The History of Sexuality,* p. 33.

5. Elizabeth Fee, "Henry E. Sigerist: From the Social Production of Disease to Medical Management and Scientific Socialism," *The Milbank Quarterly* 67, suppl. 1 (1989).

6. See *Sexual Desire Disorders,* ed. Sandra R. Leiblum and Raymond C. Rosen (New York: Guilford, 1988), p. vii.

7. See Janice M. Irvine, *Disorders of Desire: Sex and Gender in Modern American Sexology* (Philadelphia: Temple University Press, 1990), for this discussion of sexology's history.

8. William H. Masters and Virginia E. Johnson, *Human Sexual Inadequacy* (New York: Bantam Books, 1980).

9. Leiblum and Rosen, *Sexual Desire Disorders,* introduction.

10. Anthony Pietropinto and Jacqueline Simenauer, *Not Tonight, Dear: How to Reawaken Your Sexual Desire* (New York: Doubleday, 1990).

11. This history is from Richard F. Salmon, "A History of the 12-Step Fellowships for Sexual Addicts and Co-Addicts," National Conference on Sexual Compulsivity/Addiction, Minneapolis, Minnesota, May 21, 1990.

12. Ibid.

13. Richard Solomon, "The Opponent-Process Theory of Acquired Motivation," *American Psychologist* 35 (1980): 691–712.

14. Daniel Goleman, "Some Sexual Behavior Viewed as an Addiction," *New York Times,* October 16, 1984.

15. Audience discussion during Carole G. Anderson, "Assessment and Treatment of the

Sexual Dependency, Eating Disorders, Sexual Trauma Complex," National Conference on Sexual Compulsivity/Addiction, Minneapolis, Minnesota, May 20, 1990.

16. Ginger Manley, "Sexual Health Recovery in Sex Addiction: Implications for Sex Therapists," *American Journal of Preventive Psychiatry and Neurology* 3, no. 1 (Spring 1991).

17. Mark Schwartz, "Four Paraphilias: Victim to Victimizer Triumph over Tragedy," National Conference on Sexual Compulsivity/Addiction, Minneapolis, Minnesota, May 20, 1991; and Manley "Sexual Health Recovery."

18. See Irvine for a discussion of sexology's strategies to achieve professional legitimacy.

19. Leiblum and Rosen, *Sexual Desire Disorders,* introduction.

20. Goleman, "Some Sexual Behavior."

21. Craig Rowland, "Reinventing the Sex Maniac," *The Advocate,* January 21, 1986, p. 45.

22. Judy Foreman, "Drugs May Help Sex Offenders," *Boston Globe,* March 5, 1984.

23. Goleman, "Some Sexual Behavior."

24. John Money and Margaret Lamacz, *Vandalized Lovemaps: Paraphilic Outcome of Seven Cases in Pediatric Sexology* (Buffalo, N.Y: Prometheus Books, 1989).

25. Rene J. Dubos, *Mirage of Health* (New York: Harper, 1959).

26. Jeffrey Weeks, *Against Nature: Essays on History, Sexuality and Identity* (London: Rivers Oram, 1991), p. 70.

27. See Jeffrey Weeks, *Sex, Politics, and Society: The Regulation of Sexuality since 1800* (New York: Longman, 1981); Weeks, *Against Nature;* and Irvine, *Disorders of Desire.*

28. Weeks, *Against Nature,* p. 70.

29. Helen Singer Kaplan, *Disorders of Sexual Desire and Other New Concepts and Techniques in Sex Therapy* (New York: Brunner/Mazel, 1979), p. 9.

30. Ibid., p. 78.

31. Ibid., p. 25.

32. Weeks, *Against Nature,* p. 70.

33. Money and Lamacz, *Vandalized Lovemaps,* p. 43.

34. Patrick J. Carnes, "Sexual Addiction: Progress, Criticism, Challenges," *American Journal of Preventive Psychiatry and Neurology* 2, no. 3 (May 1990): 1.

35. Ian Forster, "Co-dependency: A New Description and Theory—A Correlation between Co-dependency and the Development of Addictive Disease," National Conference on Sexual Compulsivity/Addiction, Minneapolis, Minnesota, May 21, 1991.

36. Ibid.

37. Schwartz, "Four Paraphilias."

38. Leiblum and Rosen, *Sexual Desire Disorders,* introduction.

39. See Leiblum and Rosen for the parameters of this debate.

40. Kaplan, *Disorders of Sexual Desire,* p. 90.

41. See, for example, Patrick Carnes, *Out of the Shadows: Understanding Sexual Addiction* (Minnesota: CompCare Publications, 1983); and Charlotte Kasl, *Women, Sex, and Addiction: A Search for Love and Power* (New York: Ticknor and Fields, 1989).

42. Carole S. Vance, "Anthropology Rediscovers Sexuality: A Theoretical Comment," *Social Science and Medicine* 33, no. 8 (1991): 875–84.

43. Ibid., p. 878.

44. Leiblum and Rosen, *Sexual Desire Disorders,* introduction.

45. Kaplan, *Disorders of Sexual Desire,* pp. 63–64.

46. Clearing-Sky and Thornton, cited in *Sexual Desire Disorders,* ed. Leiblum and Rosen, p. vii.

47. Leiblum and Rosen, *Sexual Desire Disorders,* p. 8.

48. "Sexual Addiction," brochure of the Golden Valley Health Center.

49. Ibid.

50. Quoted in Richard Salmon, "Twelve-Step Resources for Sexual Addicts and Co-Addicts," National Association on Sexual Addiction Problems of Colorado, 1989.

51. Keziah Hinchen and Anne McBean, "Sexually Compulsive or Addicted Women," National Conference on Sexual Compulsivity/Addiction, Minneapolis, Minnesota, May 21, 1990.

52. Marvin A. Steinberg, "Sexual Addiction and Compulsive Gambling," *American Journal of Preventive Psychiatry and Neurology* 2, no. 3 (May 1990): 40.

53. Kasl, *Women, Sex, and Addiction.*

54. Anne Wilson Schaef, *Escape from Intimacy: The Pseudo-Relationship Addictions* (San Francisco: Harper and Row, 1989), p. 34.

55. See Irvine for a discussion of ISD and drug treatment.

56. Gayle Rubin, "Thinking Sex: Notes for a Radical Theory of the Politics of Sexuality," in *Pleasure and Danger: Exploring Female Sexuality,* ed. Carole S. Vance (Boston: Routledge and Kegan Paul, 1984), pp. 267–318.

57. Carl Degler, "What Ought to Be and What Was: Women's Sexuality in the Nineteenth Century," in *Women and Health in America: Historical Readings,* ed. Judith Walzer Leavitt (Madison: University of Wisconsin Press, 1984), pp. 40–56.

58. Nancy Cott, "Passionlessness: An Interpretation of Victorian Sexual Ideology, 1790–1850," in *Women and Health in America,* ed. Leavitt, pp. 57–69.

59. Pietropinto and Simenauer, *Not Tonight, Dear,* p. 6.

60. "Rob Lowe in Sex Addiction Clinic," *National Enquirer,* June 5, 1990.

61. Erica Jong, *Any Woman's Blues* (New York: Harper and Row, 1990).

62. *New York Times Book Review,* February 4, 1990.

63. Salmon, "A History."

64. See Leiblum and Rosen, *Sexual Desire Disorders;* and Pietropinto and Simenauer, *Not Tonight, Dear.*

65. S. Chorover, "Big Brother and Psychotechnology," *Psychology Today* (October 1973) pp. 43–54.

66. Pietropinto and Simenauer, *Not Tonight, Dear,* p. 4.

67. Peter Conrad and Joseph W. Schneider, *Deviance and Medicalization: From Badness to Sickness* (St. Louis: Mosby, 1980), p. 31.

68. Handout from Keziah Hinchen and Anne McBean, "Sexually Compulsive or Addicted Women."

69. See Kasl, *Women, Sex, and Addiction.*

70. See Irvine for an expansion of this argument.

71. See Kasl, *Women, Sex, and Addiction,* and press release from Golden Valley Health Center, 1990.

72. Bernard Apfelbaum, "An Ego-Analytic Perspective on Desire Disorders," in *Sexual Desire Disorders,* ed. Leiblum and Rosen, p.78.

73. See Irvine for a broader examination of these issues.

74. Pietropinto and Simenauer, *Not Tonight, Dear,* p. 20.

75. Rowland, "Reinventing the Sex Maniac."

76. Joan Jacobs Brumberg, *Fasting Girls: The History of Anorexia Nervosa* (New York: New American Library, 1988); and Susan Bordo, "Anorexia Nervosa: Psychopathology as the Crystallization of Culture," in *Feminism and Foucault: Reflections on Resistance,* ed. Irene Diamond and Lee Quinby (Boston: Northeastern University Press, 1988), pp. 87–118.

77. Brumberg, *Fasting Girls,* p. 265.

78. Carole S. Vance, *Pleasure and Danger: Exploring Female Sexuality* (Boston: Routledge and Kegan Paul, 1984).

79. Pietropinto and Simenauer, *Not Tonight, Dear,* pp. 15–16.

80. Carnes, *Out of the Shadows,* p. vii.

81. See Irvine for a discussion of these cultural patterns.

82. Kasl, *Women, Sex, and Addiction,* p. 31.

83. Foucault, *The History of Sexuality,* vol. 1.

84. See Ellen Carol DuBois and Linda Gordon, "Seeking Ecstasy on the Battlefield: Danger and Pleasure in Nineteenth-Century Feminist Sexual Thought," in Vance, *Pleasure and Danger,* pp. 31–49; and Margaret Hunt, "The De-Eroticization of Women's Liberation: Social Purity Movements and the Revolutionary Feminism of Sheila Jeffries," *Feminist Review,* no. 34 (Spring 1990): 23–46.

85. Judith Lewis Herman, "Considering Sex Offenders: A Model of Addiction," *Signs: Journal of Women in Culture and Society* 13, no. 4 (1988): 695–724.

86. Rowland, "Reinventing the Sex Maniac."

12

Between Innocence and Safety

Epidemiologic and Popular Constructions of Young People's Need for Safe Sex

Cindy Patton

Although only a small proportion of teenagers are at high risk for contracting AIDS, a majority—like my sons—will probably have at least passing acquaintance with a person who contracts or dies from the disease.

—Dr. David Elkind, 1990

BY 1990, OVER 10,000 people under the age of twenty-five had been diagnosed with AIDS, with probably ten times that number infected with HIV (Boyer and Kegeles 1991: 12). Medical experts now believe that it takes an average of ten years from date of infection until serious symptoms occur; thus, most of these some 110,000 HIV-infected young people contracted HIV as teenagers. In the crucial first decade of the epidemic, some attention was devoted to instilling in young people a sense of tolerance toward people living with AIDS. Tragically, moralistic attitudes about sex combined with racism and homophobia to delay identification of young people as significantly at *risk* and desperately in need of risk-reduction education.

From 1985 to 1990, popular accounts of young people's risk for contracting HIV converted some, but not all, young people into deviant bodies at risk of contracting HIV. While the media appear to use the terms *teenager, adolescent,* and *young adult* interchangeably, each has a different connotation tacitly underwritten by the two most common views of adolescence: the raging hormones (or stress-and-storm) theory and the youth-as-subculture theory. Each theory has a different notion of the young person's relation to knowledge about "adult" matters, but both propose a stage between a natural and innocent childhood and an accomplished (worked-for) adulthood.

The mid-to-late-1980s media coverage of young people's relationship to HIV mirrored the early public health establishment perceptions of who was at risk. Dominant-class youth were described as the "adolescents" of the stress-and-storm theory, while gay teenagers tended to be treated as members of a subculture anxiously linked to both heterosexual peers and adult homosexuals. Young people of color were treated through contextual rhetorics that linked the *environment* of the ghetto and the atavistic black body. Understood as premodern and therefore outside either model of adolescence, youth of color were outside the discourse of innocence that veiled and protected their white, heterosexual age-peers. Youth of color did not evoke the protective concerns that innocence mobilized, nor were they considered capable of participating as full "adults" in "modern" society.

The basic logic for separating the young deviant from the "normally abnormal" adolescent (thereby determining who needs to know about safe sex) depended on the construction of normal adolescence as a passage from a precultural body (the innocent child) through a civilizing process (the disembodied adolescent with desires unfulfilled by practices) to sexually responsible adulthood (i.e., monogamous, married, procreative, white, heterosexual). Two deviations from this developmental tale were envisioned: black bodies never passed through the cultural, having neither adolescence nor civilized adulthood. Similarly, homosexual bodies passed into adolescence, but emerged into a distinct subculture in which real (heterosexual) adulthood was foreclosed. Once deviant sexual practice had occurred—homosexual sex of any kind, heterosexual intercourse in the ghetto—the homosexual and black bodies could no longer be entered into the mainstream, into culture.

Once these three categories solidified, any gender distinction was subordinated to class, race, and sexual orientation. The sharpest gender distinctions occur in the discourse about the most socially marginal class—poor youth of color. Males are seen as walking time bombs of violence and AIDS, and females are cast as irresponsible *women*, infectors of white men, and the primary source of AIDS babies, their own and those of white women infected through the white men. Gender is only barely visible in representations of dominant-class youth: the actual body of the adolescent is avoided in the frenzy to prevent carnal knowledge. Where a distinction exists, males are the subjects and females the objects of raging hormones. Young lesbians do not appear at all.

A New Disease: Risk and Deviance

In 1981, a new medical syndrome was identified. Because the initial cases were among self-identified gay men, GRID (Gay Related Immune Deficiency), renamed AIDS (Acquired Immune Deficiency Syndrome), became inextricably linked with sexual deviancy in popular discourse and in the unconscious of scientific, clinical, and pedagogic practice. By the early 1980s, the decade-old gay liberation movement had scored a number of victories: increased awareness among health-care providers created the possibility that, at least in large urban gay ghettos, openly gay people could come to the attention of openly gay or sympathetic health-care professionals.

Had the epidemic occurred a decade or two earlier, it is unlikely that the common behavioral factor in the cases—specific sexual practices—would have been as quickly identified. Even if they had, the social stereotype that homosexuals were intrinsically pathological might well have seemed sufficient to explain the syndrome, and an etiologic agent might not have been so actively sought.

In fact, a recorded epidemic of pneumonia deaths among New York City drug injectors went largely uninterrogated in 1978; it seemed unremarkable to researchers that junkies should die, even if the exact cause could not be determined. Indeed, it was the partial depathologization of homosexuality that enabled the recognition of a new syndrome and provided the crucial early clinical definition of AIDS—appearance of opportunistic infections *in previously healthy adults*.

As I have shown elsewhere (1990, 1992a, 1993), expectations about health, and the use of Western, white, middle-class, male health as the definitional norm against which clinical AIDS is defined, had fatal consequences for global AIDS policy. Stoking fears of global vectors while denying education and treatment to developing countries, "Third World beliefs" and poverty, rather than international policies of malevolent neglect, were blamed for steeply rising seroprevalence curves. The central trope of normal, white, middle-class, adult health also affected policy for gay teens and youth of color because neither was thought capable of achieving "normal" adulthood. Thus, from the first definition of GRID/AIDS through successive phases of scientific development, there was an unstable coexistence of social deviance and medical health within the body subject to HIV.

Early AIDS Education: For Deviants Only

By 1982, the emerging syndrome was linked to sexual practices (especially anal intercourse) associated with homosexuality. AIDS epidemiology and educational efforts supervised by the public health service targeted specific risk-reduction advice to groups perceived to be at risk. This strategy arose from and reinforced a broad societal perception that behavior and identity exactly correspond, so that certain *types* of people were at risk.

By the mid-1980s, it was evident that the "general public" had problematic misperceptions about AIDS, especially regarding casual transmission. A second wave of education aimed at the "general public" culminated in the controversial 1988 surgeon general's report. The report, which never imagined that the general public was actually at risk, largely assumed that, while knowledge about risk reduction was nice, the public really only needed information about the impossibility of contracting HIV through casual or social contact.

These two forms of AIDS education created the conditions for a system of policing by enunciating one educational subject as needing to protect herself/himself from HIV and the other as needing to protect herself/himself from the deviants at inherent risk for HIV. Although only military recruits were forced to be tested, this economy of knowledges engendered in the general public a belief in its "right to know" who was infected. Similarly, although few courts ruled in favor of "innocent victims" infected with HIV during consensual sex, the split encouraged individuals to hold those they perceived to be "at risk" to an obligation to ascertain and disclose their serostatus. The same split underwrote education for youth, although in a kinder, gentler form, as we see in Dr. Elkind's *Parents* column above, which continues:

> For teenagers, then, there must be two types of AIDS education. One provides young people with the factual information regarding the nature of the disease, its means of transmission, and the precautions one should take to avoid contracting it. The second type deals with the attitudes teenagers should adopt toward people with AIDS. (Elkind 1990: 192)

In reality, the first kind of education was virtually impossible or rendered incomprehensible by the restrictions that prevented discussion of condoms and nonintercourse forms of sex and demanded "teaching" abstinence, a paradox in itself—"no sex" as safe sex suggests that no sex is safe. Elkind's good intentions are further undercut because the column's accompanying photograph of a white girl and boy kissing on the family-room sofa is captioned: "Sexually active teens should be informed about the risks associated with AIDS." Not only does the caption fail to suggest that being "informed" should mean knowing about safe sex, but it quite clearly limits any discussion of "risk" (rather than compassion) to the sexually active.

Adolescents and AIDS

A steady trickle of articles about young people's risk for contracting HIV began in 1985 and accelerated with the Global Program on AIDS' 1989 focus on youth. But the pre-1990 shriek-and-hush cycles about the risk of HIV infection among teens meant that few young people recognized themselves as the subjects of risk-reduction messages. Educators had a hard time figuring out how to alert young people without producing them as sexual subjects, potentially defeating their purpose by "causing" youth to have sex prematurely. This was not merely individual adults' anxiety, but government policy, as described by Dr. James Mason, assistant secretary of Health and Human Services in the Bush Administration:

> There is no way that the federal government can condone [condom-distribution programs]. It supersedes parental rights, and it sends a direct message to youth that having sex is OK when all our evidence is showing that it isn't. We have to take responsibility for adolescent AIDS. (Brownworth 1992: 44)

Sadly, for many young people who might eagerly have taken nonjudgmental and straightforward information under advisement, adult equivocation about their "responsibility" proved fatal. The obsession with deviance (whether homosexuality or "premature" sex) *produced* the HIV epidemic among youth. Government focus was on placing young people into categories enumerating their route of transmission (condomless intercourse, needle-sharing, use of blood products), instead of recognizing the reason they became infected: a lack of clear, applicable *risk-reduction* information. Only a few projects were directed at shifting group norms toward practicing safe sex, and these were largely with out-of-school youth. Studies throughout the late 1980s and into the 1990s showed that young people were knowledgeable about routes of transmission, but still lacked information about prevention, especially the value of condoms.

One reason for delayed concern about young people had to do with the evolution of scientific understanding: it was not until the mid-1980s, following the discovery of HIV (then called LAV or HTLV-III) and the review of epidemiologic data—especially the stored serum from a cohort of thousands of gay men in San Francisco who had been enrolled in a hepatitis B study, begun in 1978—that researchers were able to realize that time from infection to AIDS diagnosis might be a decade or more. *People diagnosed in their early-to-mid-twenties must have been infected as teenagers.* However, the framing of early concerns about young people suggests that additional cultural factors prevented the constitution of teens *as a class* of people at special risk for contracting HIV through sex.

The initial articles addressing young people and HIV infection took two forms, the first being reports about whether infected "school children" should be allowed in the classroom (McGrath 1985: Church 1985; Wallis 1986). Represented as having been infected through transfusions or from blood products used in conjunction with clotting disorders, these teenagers were lumped together with elementary-school children because their route of transmission was the same (nonsexual, iatrogenic, *innocent*). The most tenured figure in the school debates was Ryan White, whose life with HIV was widely documented in the popular press. White initially came to the nation's attention at the age of thirteen because he was banned from school and was hooked up to his classrooms via telephone. We watched White grow into early manhood—including having a girlfriend and becoming friends with rock stars Michael Jackson and Elton John. While there were a few comments in interviews that he was being sexually cautious with his girlfriend, White, like men with clotting disorders generally, was never represented as a sexual being. His already medicalized body stayed in a perpetual prepubescence: the terror of sexually active HIV-infected teenagers never emerged in the media coverage of his life.

The second area of coverage of teens and HIV concerned whether or not college students were "practicing safe sex" (Bruno 1985; Van Gelder and Brandt 1986). The age of students seemed a lesser concern than "heterosexual AIDS" in the wake of Rock Hudson's illness and death. In all of these cases, concerns centered on white, working-/middle-class young people, most of whom were "straight." Articles occasionally included token accounts of a now infected (white) gay-boy-next-door, who had not realized that he, too, might be at risk, presumably from older, more culpable gay men. The *New York Times* (Gross 1987) produced a rare article about the plight of "gay teen-agers," but rescued these teens—"many of them are not yet sexually active"—from the category "deviant" by considering them a subcategory of students, i.e., part of the mainstream.

If stories that allowed nice (white) gay boys to be counted as innocent teens suggested that homophobia had diminished, racism and sexism quickly filled the vacated space. Mass media had long represented African Americans as pathological, if only by virtue of their status as inhabitants of the decaying inner cities; and women had been represented as dangerous to men (as prostitutes) or themselves (single women searching for "Mr. Goodbar"). It was easy to transpose these earlier images of people of color and single women into specters of dangerous sexuality; it was through this lens that mid-1980s epidemiology was read. A review of epidemiologic data clearly shows that among men of color reported as having AIDS, many of whom reported sex with men, a large percentage must have contracted HIV while still in their teens. Likewise, a significant percentage of women—especially women of color—diagnosed with AIDS must have contracted HIV as teens.

Even in this early coverage, the connotative meanings associated with terms for the young began to divide into the same rough categories of deviants and general public that first emerged for adults. The adolescent of AIDS discourse is the "normally abnormal" hormone-besieged body of the white working-/middle-class male youth. The "at-risk" teen was either gay or a member of the black/Latin urban underclass, and presumably involved in drugs and prostitution. Understood as autonomous groups of young people, the "at-risk"teens were represented as a source of danger, not a site of innocence.

"Normally Abnormal"

Anna Freud (1958) described the period between childhood and adulthood as a time of upheaval and liminality in which the normal and abnormal are reversed: "To be normal during the adolescent period is by itself abnormal" (275). The stress-and-storm theory splits the adolescent into a body and a moral reasoning capacity, in order to argue that the physiological changes occurring in the adolescent body are in danger of outstripping the young person's reasoning capacities. Sexuality becomes a source of tremendous anxiety. If the prepubescent body is thought capable of innocent pleasure, the adolescent body, on the cusp of adult pleasure but without the capacity to be responsible, is thought incapable of distinguishing innocent from carnal pleasure. Even the simple pleasures of innocent childhood (kissing, hugging, etc.) become dangerous petting on the slippery slide to premature intercourse. Best to prevent the adolescent body from experiencing pleasure at all. The adolescent body is thus barely visible as the container of dangerous hormones and ill-formed moral capacity.

There is an interesting gender difference in what makes an adolescent body ready for sex. For boys, the punishment for premature intercourse is STDs; it is the wayward member, and not the entire body, that is afflicted with troublesome chancres and sores. By contrast, the pregnant adolescent, the child having a child, makes visible the result of female pleasure in the absence of moral capacity. But even if they are "big" enough, girls' "early physical development"—the visibility of their own gendered bodies—puts them at risk by attracting the dangerous attention of unscrupulous penises.

A typical pamphlet *about* (but not directed *to*) adolescents, produced by Burroughs-Wellcome, manufacturers of zidovudine (AZT) and one of the largest providers of information to non-gay-oriented clinics providing HIV counseling and testing, warns counselors that:

> Several factors make counseling of adolescents, especially early adolescents, unusually complicated. People in their early teens tend to have a low level of comfort and familiarity with their bodies, a poor capability to plan for future events and their consequences, and are unable to consider the viewpoints of others. Thus, they are prone to greater risk-taking and are therefore vulnerable to the consequences of their behavior. The development of "moral reasoning" occurs during adolescence. (Burroughs-Wellcome 1989: 17–18)

According to this ethos, desire and practice for adolescents are radically split: adolescents are presumed to engage in the "risk behaviors" only if they chance to discover such behaviors exist. They may naturally be subject to changing desires, but these desires are acorporeal and insufficient to result in sexual behavior. In Gina Kolata's (1990) account of David Kamens, adolescents are "thinking about SATs and college," and wouldn't engage in sex unless they were subjected to "pressure" by older people:

> Feeling pressured to fit in with an older crowd, he had unprotected sex with people he did not know well. (149)

But peers may also exert pressure. According to Carnegie Council on Adolescent Development policy advocates Jackson and Hornbeck (1989), the real danger of adolescence is that:

> . . . although increased identification with peers serves to fulfill needs for autonomy from parents, it also provides increased pressure to engage in adult behavior such as sex, smoking, and drinking. (833)

Subject to emerging desires that have yet to find a physical form, adolescents should be educated about AIDS (as opposed to safe sex) *before* they happen upon the dangerous practices. Unlike black or gay youth, white heterosexuals can be *fully* prevented from starting on the slippery slope to risk behaviors if they can be prevented from "experimenting."

A major strategy is to keep adolescents away from "bad influences" who might prematurely introduce the idea of sex, thus providing a concrete outlet for the raging hormones. The emphasis on the improbability of contracting HIV through casual contact and the calls to civility toward people living with AIDS, together with the lack of clear information on *prevention* (especially condom use), position the adolescent consumer of AIDS education as someone who *can't* get infected because what little contact he or she has with people subject to AIDS is limited to casual contact. Intrinsic to the notion of civility toward people living with AIDS is the idea that they are *different*—not white, not straight. In this move toward the adult mainstream "avoidance" approach to risk reduction, the white working/middle class recuperates its own youth by associating HIV only with "other" youth who are a threat to their children, now quite overtly through sex.

By 1990, the heterosexual, white working-/middle-class adolescent was defined and addressed by education that emphasized abstinence and advocated developing personal systems for weeding out risky partners. A *Good Housekeeping* editorial (1990) suggested that parents "make it clear how important it is to select a partner carefully" (257).

The unspoken strategy of keeping "nice" kids away from bad influences was hard to shake. Hot on the heels of twenty-three-year-old Ali Geertz's announcement that she had contracted HIV from a "nice" boy of her same (upper middle) class, controversial "AIDS journalist" Gina Kolata opened her 1990 *Seventeen* magazine article with Geertz's story, attempting to argue that the mistaken association of HIV infection with deviance—even with deviant teenagers—has misled readers (largely teenage girls):

> A year and a half ago more than half the teenagers Dr. Hein saw at her clinic were street kids who had run away from home or had been kicked out and who often were prostitutes, having sex with many partners and injecting drugs. Now more than half the teenagers she sees are living at home with their families or are college students. (150)

The article also includes the story of infected-teenager-turned-safe-sex-educator W. David Kamens, a "nice boy" who went to dance school. He embodies the teen whose physical precociousness results in joining an "older crowd" that pressures him to do "adult" things for which he is morally inequipped:

> Other teenagers think that if they know the person they're having sex with, they don't have to worry about AIDS. But your sex partner may not tell you about all of his previous sexual experiences. David Kamens . . . started hanging around with an older crowd and didn't ask much about his new friends' sexual experiences. . . . "I really didn't know what I was doing." (151)

Kamens operates as both the positive figure of the unsuspecting teen and, although the gender of his sexual partners is unstated, as the negative figure of the danger. We aren't sure whether he is describing himself or his sexual conquests when he says: "You really can't tell whether a boy is infected with HIV by looking at him or even talking to him" (151). Elsewhere, Kamens is treated as a member of a subculture, the more common construction of gay teenagers: for example, Brownworth (1990) describes him as openly gay and also actively involved in advocating for gay teenagers.

In AIDS discourse, the generalized adolescent emerges as the bodiless vessel of troubling desires, whose cognitive capacities are not sufficient to allow for "safe" decisions, only for avoidance. This adolescent must have no body: to prematurely emerge as a male or female body is to begin the slide into deviance. Alongside this dimly etched container emerge the deviant bodies—one racial and one homosexual—that constitute the visible boundaries of the "normal" adolescent by both inscribing and being inscribed by their need to have "knowledge" of "safe sex." In this map of positive and negative bodily spaces, gay teens make their deviance visible through their placement in a subculture. Black and Hispanic youth etch the mainstream through their *dis*placement in an urban ghetto ruled by premodern values.

Youth-as-Subculture

A second important concept of youth emerged in relation to both anthropological investigations of adolescence in other cultures (Mead 1935; Turner 1967) and the postwar popular-culture image of the "rebel without a cause." By the time of the 1960s student movements, it was easy to press the claim that young people were not merely pre-adults driven by hormones, but an inexplicable (to the "older generation") class unto themselves. Based on her comparative work, which showed great variety in the amount and existence of "stress" in the teenage years, Margaret Mead argued that "adolescence" is a cultural construct: the meanings attached to this "phase" were shown to be variable. Victor Turner's later work analyzed the symbols and practices of child-to-adult transformation in Ndembu culture to argue that the transition was understood less as a developmental phase than as a temporary and separate state of being—what Turner called a liminal phase. The influence of the two works extended far beyond the academy: Mead was widely read and extremely popular among lay readers, and Turner, though perhaps less widely read by the public, was something of a cult figure among young intellectuals in the academic boom years of the mid-sixties to mid-seventies.

These trends in popular intellectual circles converged neatly with a more general shift in post–World War II society. The persistent fear of and nostalgia for rebellious lads helped establish the new category of "teenager." Teen culture in part operated to mediate the massive economic shifts that occurred when returning veterans displaced the women and black men who had migrated to cities to work in factories during the war effort. Women were in part redomesticated through the invention of (largely white) suburban life, while unemployed blacks repopulated the inner cities. Rebellious white boys (*Rebel[s] without a Cause*) were seen as a threat to small-town life as it had been (e.g., *The Wild One*), while their black age-peers were simply considered men without jobs. Black teenage girls soon emerged as emblems of the fatherless welfare family.

Youth began to be viewed as a social role tied less importantly to physiology than to the economic and social mandates that govern entry into adult social role. The "teenager" was associated with "modern" society and "modern attitudes"; teens consumed "new" cultural products like rock and roll and automobiles, and rebelled against prevailing norms of sexual and substance-use austerity. Sexuality was still a central concern in this emerging youth subculture, but not because there was a mind (morality)/body (hormonal) mismatch. Rather, sexual neuroses stemmed from the perceived repressiveness of the 1950s: society was the source of sexual danger. Among youths and their advocates, the sexual body must be protected, not from the unreasoning adolescent mind, but from the moralizing attitudes of an adult culture that hated instead of celebrated the "natural" (i.e., sexual, pleasurable) body.

Even though psychology and social psychology had largely debunked the stress-and-storm model on empirical grounds (Powers, Hauser, and Kilner 1989), they still treated the individual young person as the unit of analysis. By contrast, the work associated with the quarterly *Youth and Society,* which emerged as and continues to be a major locus for academic work from the subculture theoretical frame, uses the methods associated with anthropology and qualitative sociology to study either clusters of youths or the concept of youth itself.

Theories that see youth as a subculture often also see youth culture as a space of resistance (Hebdige 1979). The widely distributed English version of the Danish youth movement's *The Little Red Schoolbook* (Hansen and Jensen 1971) gives a flavor of the times in this preface to the American edition:

> This is an Americanized version of a British translation of a Danish book. That it was possible to take a book written about the schools of one nation and that had been rewritten and translated to describe the schools of a second nation and use it to describe the schools of a third nation, without either distorting reality or changing the content or tone of the book substantially, illustrates an important and disturbing fact about schools in Europe and North America: they are basically the same, and the sameness is a poor quality education and an unjust educational system.
>
> The purpose of the book is to provide you with some basic information that will help you deal with that fact . . . [and] to provide you with information which you can't usually get in school or from most grown-ups—about sex and drugs, for instance—and to give you some ideas about how you can use the schools for your purposes so that you can really have what schools have always promised the young but hardly ever delivered: power over their own lives. (15–16)

This sentiment is echoed among more radical youth advocates today. Gabe Kruks, director of public policy and planning for the Gay and Lesbian Community Services Center of Los Angeles, is among those who place the blame for youth's plight on society, not on hormones:

> We really have to start looking at kids' needs instead of our own. We have the highest rate of teenage pregnancy, teenage STDs, teenage AIDS, and

> teenage date rape in the Western world here in the U.S. We are, as a society, uncomfortable with our bodies, with sex, with sexuality. We're in denial about a lot of issues associated with these subjects. We like to pretend they don't exist. That denial is killing our kids. The U.S. is a very child-hating society. If we were not aware of that before, we can see it very clearly in our response to AIDS. (Brownworth 1992: 43)

In challenging the "normally abnormal" definition used in psycho-physiologically based adolescent theory, however, youth-as-subculture theory still views these years as a phase during which young people attempt to acquire the knowledge necessary to becoming an adult. Those who apply subculture theory believe that adult culture actually tries to prevent youth from learning what they need to know about sex. Youth culture needs to be protected so that youth can share "accurate" and "appropriate" knowledge among themselves. A youth-as-subculture-inflected article about "Teen Sex: America's Worst-Kept Secret" (Brownworth 1992), which appeared in the national gay newsmagazine *The Advocate,* reinforced youth's solidarity in the face of a hostile adult society by using personal profiles of teenagers infected with HIV. The "perfect California Girl, infected with HIV at age 16," describes her family as "almost too strict about sex"; "the kid I got this from didn't know any more than I did" (44). A stress-and-storm theorist could only have interpreted these statements as fearful evidence of young people's incapacity to recognize the misfit between their bodies and their moral capacities. Says the "perfect California Girl, infected with HIV at age 16":

> Kids aren't embarrassed to have sex, so we should stop making them embarrassed to have safe sex. I want the words *sex* and *condoms* to go together like cereal and milk or peanut butter and jelly. (44)

"Gay" Teenagers

In the context of AIDS, the media retreated from any general depiction of youth as members of a subculture, reserving the notion for gay youth. Breaking from the usual policy (until about 1991) of refusing to use the term "gay," the *New York Times* ran two prominent articles (Gross 1987; Eckholm 1990) on "gay teenagers," which constructed as an autonomous group the young (white) men who were naturally discovering their unnatural sexualities in the age of AIDS. Oddly, adults in the article are still referred to as "homosexuals": evidently the *Times* wished to evade the sexuality of the "teens," who might have desires and feelings, but must not be spoken of as engaging in practices, despite the fact that key figures in each story have already contracted HIV.

The term *gay teen* suggests a self-conscious group bonded together by common experience, rather than isolated individuals who merely come together for the purpose of having sex. The young men's special experience of their deviant bodies seems to form the basis for group identification. Gay teens are:

> . . . a population in turmoil, bearing all the problems common to adolescence and another set all their own.
>
> These gay teen-agers run the gamut from inconspicuous youngsters in high school who are isolated and frightened by their stirring sexuality to runaways on the piers of Greenwich Village who openly sell themselves as prostitutes. (Gross 1987: B1)

However diverse, their community, or rather communit*ies* are separate from mainstream society: the line between gay and non-gay youth is irrevocable:

> Experts agree that the fear of AIDS will not convert a gay youngster to heterosexuality, but they might slow a decision to have intercourse. (B1)

Although not always yet engaging in sex, gay teenagers are figured as "deviants," who are obliged to know how to prevent HIV transmission. Paradoxically, the gay teenager is granted the "right" to sex, but largely to specify *which* type of teen is at risk for HIV infection. This construction of the gay teenager inscribes the heterosexual teen as safe-from-AIDS. By placing gay teens in a subculture, separate from both mainstream heterosexual and adult gay "communities," the homosexuality of the gay teenage boy is fixed, separate from the activity of his heterosexual peers.

The *Times* concedes an almost natural category of "gay teens" but implicitly suggests that the specter of HIV infection for such young men lies in liaison with adults, not in "experiments" among themselves:

> Many of them are not yet sexually active, but others wind up in physical relationships they are not prepared for. . . . Social workers say the men who buy their favors, and sometimes carry infection home to their wives, are seeking progressively young partners in the hope they will be "clean." (Gross 1987: B1)

That young men might as easily learn about safe sex from their encounters with older men is neatly denied, despite an admission that "huge numbers [of adult homosexuals] have adopted 'safer sex' " (Eckholm 1990: A1). As I have argued elsewhere (1992b), older men must always be the specter of perverted sex.

It is not in their homes or schools but in the "grim, nether world" in which "smoothfaced youngsters . . . sell themselves for $10" that the most education for gay teens has occurred, even according to the *Times*. The invisible community of the "less troubled" who have " 'passed' as heterosexual" "have no contact with counselors and receive limited sex education" (Gross 1987). Gay teens can expect no education in the information-sparse domain of the mainstream, where "choosing carefully" is the main risk-reduction strategy. If they learn anything, it is that in the already stigmatized (adult) community for which they are destined "[AIDS] makes the stakes higher" (Gross 1987).

The apparent "right to sex" granted "gay teens" is evident in the growing ease with which journalists allude to their sexuality. But this "right" is won only at the cost of considering gay youth responsible for their own fate. A sidebar to *Newsweek's* 1990 "The Future of Gay America" article describes young gays as engaging in a "pro-sex" campaign very reminiscent of the "night life and sexual freedom that defined the disco and bathhouse hedonism of the '70s" (22) and which is implicitly or explicitly situated as the "cause" of AIDS. Gay youth, like their older brothers for whom devastation is considered a done deed, are seen as inevitably on the verge of sliding into dangerous practices.

> Even young men who are better adjusted and well informed about acquired immune deficiency syndrome may not fully understand that almost any unprotected sex may be suicidal. (Eckholm 1990: B20)

"Where the Trouble Is": The Premodern Body

The stress-and-storm theory provided the most comfortable framework for explaining away the potentially risky behavior of mainstream heterosexual youth by putting blame on the volatile collision of raging hormones and unsavory types, or on explicit safe-sex education itself. Thinking about teenagers as a subculture allowed mainstream society to acknowledge gay youth without taking responsibility for educating them. Though considered geographically separate from the white mainstream, youth of color presented a more terrifying prospect than the potentially proximate gay youth—"inconspicuous youngsters" (Gross 1987)—whose identity as firmly, if fatally, gay could forestall widespread fear of bisexuality as a mode of youthful category crossover.

Creating distance from the growing HIV-related morbidity and mortality among youth of color required yet another frame. The grim vision of Third-World-like AIDS was merely incorporated into the new racist imagination, which situated rampant drug use and random killing as proper to the ghetto but threatening to explode out of it. As I have argued elsewhere in relation to western discourse about AIDS in Africa (1990, 1992a), the constitution of a displaced geographical space in which the existence of "primitive" or "premodern" heterosexual practice can be asserted as the cause of a natural disaster both legitimates the failure to provide education and deceptively distances (in this case) white heterosexual youth from risk: *their* heterosexuality is "ordinary," "civilized." "Heterosexual risk" comes loose from penile-vaginal intercourse and attaches to premodern sexuality, and, working ideologically in tandem with Western discourse on Africa, to the sexuality of black youths. The following description of "the critical transition of adolescence" illustrates the way in which stress-and-storm theory—and arguably all psychoanalytic theories that situate sexual neurosis in relation to "civilization and its discontents"— leaves open the possibility of banishing *some* bodies if they can be counted as premodern:

> In premodern times, preparation for adulthood typically extended over much of childhood. Young people had abundant opportunities for directly observing their parents and other adults performing the adult roles that they would eventually adopt when the changes of puberty endowed them with an adult body and capabilities. The skills necessary for adult life were gradually acquired and fully available, or nearly so, by the end of puberty. (Hamburg and Takanishi 1989: 825)

This apparently innocuous description suggests that in earlier times (or other places) the duration or timing of adolescence was not extended forward or backward but, in fact, *never occurred at all.* Childhood, with its early training in adult roles, blended invisibly into adulthood, with virtually everything learned by the time a now inconsequential puberty ended. The problem, argue these authors, is that this older system has not been fully supplanted:

> In contemporary societies, these social supports have eroded considerably through extensive geographical mobility, scattering of extended families, and the rise of single-parent families, especially those involving very young, very poor, and socially isolated mothers. (825)

The mainstream reader would most certainly associate this and similar invocations of "broken family" with urban blacks, already constituted as premodern by virtue of their alleged matriarchal structures, a devastatingly erroneous racist truism since the 1965 publication of *The Negro Family: The Case for National Action* (collected in Bromley and Longino 1972), more commonly known as the Moynihan report:

> In essence, the Negro community has been forced into a matriarchal structure which, because it is so out of line with the rest of the American society, seriously retards the progress of the group as a whole, and imposes a crushing burden on the Negro male and, in consequence, on a great many Negro women as well. (197)

The evocation of families whose children are on a premodern track that circumvents adolescence emerges again in *Psychology Today*'s 1988 cover feature, with its dangerously polysemous title, "The Runaway Epidemic: AIDS on the Streets." The poor family operates as a metonym for the ghetto, situating pathology in the family environment:

> Contrary to a lingering perception of runaways as adolescent adventurers, most are victims of dysfunctional families and are fleeing from stressful environments. . . . A picture emerges of deficient youth, lacking the normal states of child growth and development but acting like hardened adults. (Hersch 1988: 31–32)

While stories of street youth include descriptions of white youth, the persistent evocation of place, environment—"the extreme danger and stress of street life" (31)—defines the decayed urban core, the supposed domain of communities of color, as the geographically isolated site of a premodern AIDS epidemic. Their risk is apparently unavoidable; the environment of the ghetto, of the urban streets, the dysfunctional (i.e., nonbourgeois) family "puts them directly in the path of the disease" (35).

> If geography is destiny, runaway and homeless kids gravitate to the very locations around the country where their risk is greatest. Not only are these kids at higher risk with every sexual contact than their suburban counterparts, but they also have higher levels of drug use and sexually transmitted diseases. Often, their immune systems are already compromised by repeated exposure to infections. All of these conditions may increase the risk for developing AIDS. (Hersch 1988: 35)

Missing both childhood and adolescence in their premodern trek, street kids and teenagers of color in general are considered "hardened" adults. There is no question of "bad outsiders" influencing these young people; even if bad parenting is responsible for their diversion from "normally abnormal adolescence," they are now the youth that the white mainstream is warning their children against. Despite its utter pathos, the *Psychology Today* story in part serves to fill in the youthful faces of the category "people to avoid contact with" left vague by the surgeon general's report. Similarly, in a *Good Housekeeping* (1990) advice column in response to the question, "Is Your Child at Risk?", the columnist provides "markers" as "guides" "to identify youngsters who are at risk." In addition to the long-standing middle-class problem children—"girls with early physical development" and "adolescents with serious school problems"—the article identifies "[y]oungsters living in poverty in inner cities [who] are at very high risk. They are often under great social pressures to have sex early, and IV drugs may be rampant in their environment" (257).

Young people of color are held partially culpable for their fate. They are not *innocent* victims, either because they are said to act like (to "be") adults at much younger ages, or because they live in the harsh inner city, a premodern world in which "primitive" behaviors are expected.

An initially innocent-sounding *New York Times* article that describes a new series of AIDS-education films for African American and Hispanic audiences invokes the damning geographic metonym in its title: "Putting the AIDS Message Where the Trouble Is" (Yarrow 1989). The article is principally a service feature about the upcoming projects of AIDS Film, whose 1987 "AIDS: Changing the Rules" drew fire from the banana industry because Ruben Blades used the now defamed fruit to demonstrate condom use. (After a year of rumors that he was gay, or that his dance career had been stalled because he had AIDS, Ron Reagan is also featured discussing safe sex . . . for heterosexuals.) The writer collapses the upcoming films' cultural sensitivity with spectatorial danger by quoting the company's executive director as saying he "hoped television broadcasters will be 'bold enough' to show them." The issue of conflict between minority communities' "need to know" and the mainstream's desire not to hear about safe sex serves once again to locate risk in the geographic Other of the ghetto and in the racially Other bodies that live there. Implicitly excluded as a potential TV audience for the broadcast, the minority subjects of the film also become a visual Other. The article goes on to decribe black teenagers as in "a different milieu with different issues": different, again, reconstitutes the mainstream as the "here" and indexes the collection of concerns about being protected *from* as the central issues.

Unlike mainstream, white adolescents, who are to be protected from frank discussion of sex at all costs, youth of color are believed to already possess sexual knowledge—whether they have engaged in "actual" acts or not—so, bringing up sex and talking explicitly about drug use is seen as necessary, not culturally insensitive, and certainly not a potential "cause" of their initiation into sex or drug use. As "where the trouble is" becomes geographically situated, the risk faced by youth of color becomes a public, collective, rather than a private, individual phenomenon. Viewed as hard to reach, potentially already lost—"living on the razor's edge with drugs and sex" (Yarrow 1989)—and perhaps, not that important to save, youth of color might as well be subjected to the horror of safe-sex and safe-drug-use messages.

If adolescents can be prevented from engaging in risk behaviors if only they can be prevented from inhabiting their bodies, youth of color are intrinsically physical and unlikely to change their behavior. Where adolescents need neutral "information" to protect them from initiating drug use and sex out of marriage (i.e., unsafe sex), youth of color are thought to have the nearly impossible task of learning to make sex and drug use saf*er* (but never *safe*). AIDS education directed at young people dovetails with the "war on drugs" and the new sexual austerity: the white adolescent serves as the figure of that to be protected, while young people of color are both the figure of the lower echelon of the drug-selling world and the example of social welfare gone awry, both the victims and perpetuators of the "broken [black] family."

AIDS Education: Battling the Narratives of Deviant Bodies

As I have shown, larger cultural narratives about the developmentally appropriate transition to adult sexuality view white, working-/middle-class youth as "normally abnormal" for the duration of their adolescence. In popular AIDS discourse, they are at risk if they engage in sex before they achieve the wisdom of conventional adulthood. In stark contrast to this central trope of teen AIDS risk are the gay teenager and youth of color: neither can achieve the safety of adult sexuality. Where white working-/middle-class adolescents can be protected from HIV if they can be prevented from having sex or experimenting with drugs, gay teens and youth of color are irreversibly and persistently at risk by their nature (the homosexual desires of gay teens) or from their natural environment (the ghetto, supposed to harbor all youth of color).

White, working-/middle-class youth are presumed uninfected; gay youth, yet to be infected; and youth of color, already infected. The informational and counseling needs of the three groups are perceived to be radically different. Adolescents need to be steered away from knowledge about safe sex, lest it put ideas into their heads about having sex at all. While the media suggest that gay youth can learn about safe sex before it is too late, neither the media nor the schools seem prepared to provide such information. Gay teens are left with the sexual equivalent of on-the-job training, simultaneously inaugurating their sexual careers and learning about safe sex. Gay sex is covertly policed by making it a kind of crapshoot. If a young man is lucky, he will find an older man to initiate him into the mysteries of safe sex. If not, well, that is the price of membership. The equivocal acknowledgment of gay teens arises only because the confused, uncomfortably queer boy could become something worse: the repressed bisexual who threatens to bring HIV into heterosexual, mainstream society.

Young people of color are considered virtual adults, ironically loosening the grip of propriety that has strangled most sexually explicit educational efforts, but, because they are so "hard to reach," diminishing the likelihood of money going to such projects. Gay teens and youth of color might be able to make their sex safer, but it will never be wholly safe. The future possibility of risk (their nature, their environment) always circumscribes these deviant bodies' sexuality.

White, mainstream, heterosexual sexuality is always safe until intruded upon by prostitution or bisexuality. In a state of temporary insanity, besieged by raging hormones, at risk from elsewhere and from Others, the adolescent is, the right wing overtly argues, also at risk by the perversion of safe-sex education itself.

This logic produces a tension between reforming or circumscribing deviant bodies' practice of sex, while protecting the "normal" body through anxious silence. And this tension, with its necessary assumptions about the relative dispensability of categories of youth, continues to result in both government inaction and media misrepresentations that distract our attention from "where the trouble is": not in the black or gay ghetto, but in the hearts and minds of mainstream America.

Bibliography

Antonio, Gene. 1986. *The AIDS Cover-Up?: The Real and Alarming Facts about AIDS.* San Francisco: Ignatius Press.

Athey, Jean L. 1991. "HIV Infection and Homeless Adolescents." *Child Welfare* 70(5): 517–28.

Boyer, Cherrie B., and Susan M. Kegeles. 1991. "AIDS Risk and Prevention Among Adolescents." *Social Science and Medicine* 33, no. 1, 11–23.

Bromley, David G., and Charles F. Longino, Jr., eds. 1972. *White Racism and Black Americans.* Cambridge, Mass.: Schenkman.

Brownworth, Victoria A. 1992. "Teen Sex: America's Worst-Kept Secret." *The Adovocate,* March, pp. 38–46.

Bruno, Mary. 1985. "Campus Sex: New Fears." *Newsweek,* October 28, pp. 81–82.

Burroughs-Wellcome. 1989. *HIV Counselling.* Research Triangle Park, N.C.

Church, George. 1985. "The New Untouchables." *Time,* September 23, pp. 24–27.

Eckholm, Erik. 1990. "Cut Down as They Grow Up: AIDS Stalks Gay Teen-Agers." *New York Times,* December 13, A1, B20.

Elkind, David. 1990. "What Teens Know about AIDS." *Parents,* March, p. 193.

Freud, Anna. 1958. "Adolescence." In *The Psychoanalytic Study of the Child.* New York: International Universities Press.

Good Housekeeping. 1990. "Teenagers and AIDS." May, p. 257.

Gross, Jane. 1987. "AIDS Threat Brings New Turmoil for Gay Teen-Agers." *New York Times,* October 21, sec. B, pp. 1, 8.

Hamburg, David A., and Ruby Takanishi. 1989. "Preparing for Life: The Critical Transition of Adolescence." *American Psychologist* 44, no. 5, 825–27.

Hansen, Soren, and Jesper Jensen. 1971. *The Little Red Schoolbook.* Trans. Berit Thornberry. New York: Pocket Books.

Hebdige, Dick. 1979. *Subculture: The Meaning of Style.* New York: Routledge.
Hersch, Patricia. 1988. "Coming of Age on City Streets." *Psychology Today,* January, pp. 28–37.
Kolata, Gina. 1990. "Teenagers and AIDS." *Seventeen,* May, pp. 148–51.
McGrath, Ellie. 1985. "The AIDS Issue Hits the Schools." *Time,* September 9, p. 61.
Mead, Margaret. 1935. *Sex and Temperament in Three Primitive Societies.* New York: William Morrow.
Patton, Cindy. "Heterosexual AIDS Panic: A Queer Paradigm," *Gay Community News,* 9 February 1985.
———. 1990. *Inventing AIDS.* New York: Routledge.
———. 1992a. "From Nation to Family: Containing African AIDS." *Nationalisms and Sexualities,* Ed. Andrew Parker et al. New York: Routledge. pp. 218–234.
———. 1992b. "Fear of AIDS: The Erotics of Innocence and Ingenuity." *American Imago,* Winter 1993, pp. 123–42.
———. 1993. " 'With Champagne and Roses': Women at Risk from/in AIDS Discourse." Ed. Corinne Squire. *Women and AIDS: Psychological Perspectives.* London: SAGE, pp. 165–87.
Powers, Sally I., Stuart T. Hauser, and Linda A. Kilner. 1989. "Adolescent Mental Health." *American Psychologist* 44(2): 200–203.
Salholz, Eloise. 1990. "The Future of Gay America." *Newsweek,* March 12, pp. 20–25.
Turner, Victor. 1967. *The Forest of Symbols: Aspects of Ndembu Ritual.* Ithaca: Cornell University Press.
United States. Office of the Assistant Secretary for Health and Surgeon General. Centers for Disease Control. 1988. *Understanding AIDS.* Rockville, Md.: Centers for Disease Control.
Van Gelder, Lindsy, and Pam Brandt. 1986. "AIDS on Campus." *Rolling Stone,* September 25.
Wallis, Claudia. 1986. "Lessening Fears." *Time,* February 17, p. 90.
Yarrow, Andrew L. 1989. "Putting the AIDS Message Where the Trouble Is." *New York Times,* September 13, sec. B, p. 1.

13

The Hen That Can't Lay an Egg (*"Bu Xia Dan de Mu Ji"*)

Conceptions of Female Infertility in Modern China

Lisa Handwerker

Introduction

IN MARCH 1988 the People's Republic of China announced the birth of the first test-tube baby, born to a thirty-nine-year-old, infertile peasant woman. The official birth announcement appearing in the *Beijing Review* reads:

> This healthy little girl 3900 gms in weight and 53 cms in height was born at Beijing Medical University at 8:56 a.m. She has a ruddy complexion. When Zou, a peasant from a rural southern province and father for the first time at 42, saw his lovely daughter, he clapped his hands and wiped his tears with joy. The twelve members of his family had already arrived in Beijing several days earlier to await this happy moment. Professor Zhang, the famous scientist and head of the in-vitro research program, took the baby in her arms and happily said, "I am a grandmother again." (March 1988: 11)

On the one hand, this surprise announcement appeared in strong contradiction to China's well-publicized population control policy. On the other hand, the media attention given to this medical achievement would seem to be consistent with political, social, and economic changes, including technological innovation in science and medicine, taking place in the past decade (Lampton 1987; Simon and Goldman 1989; Suttmeier 1982; Zhao 1985). In short, this announcement speaks to the tensions facing China as it simultaneously attempts to modernize within a transnational economy, control population, and maintain "traditional" Confucian family values within a rapidly changing context.[1]

In this paper I examine the ways in which these contradictions inscribe themselves on female bodies, especially infertile female bodies. I highlight the paradox of the "problem" of female infertility within the context of China's population "explosion." How is it that the test-tube baby and not the childless woman is celebrated? Why is it that, counter-intuitively, we see a great deal of concern for infertility at the same time that population control is a top state priority? The particular paradoxical meaning of infertility in China today arises out of the confluence of those Confucian patrilineal values that place high value on women's ability to produce large families, especially male heirs, and official government policies aimed at making China more "modern" through technological advancement. I argue that these seemingly contrastive ideologies (one "traditional" and one "modern") figure women primarily as reproducers and have implications for female bodies that they do not have for male bodies. The simultaneous pursuit of specific Confucian values and modernity converge in policing women's reproductive and sexual choices, and in making their bodies special objects of knowledge and surveillance. Both the state's population policy and Confucian gender ideals assume that the normative female body is, and should be, fertile. In what follows, I examine the ways in which infertile female bodies are construed as deviant and are implicated in the power matrices of the state, the family, and medicine in modern China.

My argument draws on a larger ethnographic study conducted in 1990 on female infertility among mainly Han Chinese.[2] I selected Beijing as my primary research site because it is a city rich in infertility clinics of both *xiyi* ("Western") and *zhongyi* ("traditional Chinese") medicine, and also is the birthplace of the first Chinese test-tube baby. As the administrative capital from which family planning decisions disseminate, Beijing is a lens through which to observe some of the tensions between policies discouraging and those unintentionally encouraging births. During three prior trips to Beijing, dating from 1980, I developed extensive *guanxi* ("social relations") that were invaluable for carrying out my research at a time when, in the aftermath of the Tiananmen Square incident, many Chinese were wary of contacting foreigners.[3]

My infertility study draws on a number of different anthropological data-collecting methods and sources, including participant observation in clinics to learn about the administrative structure, semistructured interviews with questionnaires to obtain demographic information on women, informal in-depth interviews with patients and doctors to elicit beliefs about and the experience of infertility, and textual analysis of popular material (e.g., newspaper and magazine articles, letters of inquiry sent to doctors from infertile women and men, and folklore expressions) to analyze contradictory feminine gender ideology. Due to the present dearth of material on female infertility and the historical limitations of conducting fieldwork in China, this study is the first comprehensive infertility study based on ethnographic research in modern China. As such, it provides insights about new micropractices of regulating women's bodies in the 1990s.

The "Crisis" of Overpopulation

"I hope you aren't going to solve our infertility problem. China has too many people. We need less people, not more." Beijing taxi drivers, shop owners, sales personnel, students, friends, and colleagues repeated these admonishing words to me whenever I explained that I was conducting research on infertility. Today in China *renkou tai duo* ("the population is too large") is the catchall phrase used by people from diverse backgrounds to explain a host of social deficiencies, including long waits and lines, lack of supplies, crowded buses, high inflation and unemployment, and inequitable health care services. Additionally, with the possibility of impending economic, moral, and political collapse during the 1990s, China's aging national leaders, seeking to distract citizens from other issues, have focused on the now highly visible "crisis" of overpopulation.

The idea of *renkou* ("population") as a negative force to be managed, monitored, and surveilled is a modern concept linked to the emergence of Chinese socialist "development" policies. In 1954, with the first national census results reported, Mao and other leaders relaxed their opposition toward birth control in response to mounting population pressures and economic concerns (Wang 1988: 54). Nevertheless, partially due to the political atmosphere of the Cultural Revolution, mandatory population control measures were not pursued for another twenty years.

Only in the mid-1970s did demography, a modern tool for making and interpreting government population policies (Lavely 1990) take firm hold in China, coinciding with international health agencies, which in 1974 made population control a top priority. Alarmed by surveys in the 1970s, which indicated a disproportionate number of women of childbearing age (ten million women annually were expected to marry and potentially bear children) (Davin 1987: 1) and a rapidly growing population, leaders argued vehemently for fertility reduction as the essential ingredient for a "modern" Chinese nation-state.[4] Socialist concerns with social organization, economic development, and modernity gave rise to the population, especially the female body, as an entity to be regulated.[5] "Overpopulation" was now understood as the imminent threat to national development (Duden 1992: 146). By the 1970s *renkou* was an officially recognized moral, social, and economic problem.

China's Birth Planning Policy: Contradictions and Resistance

In 1979 China instituted the world's first nationwide compulsory one-child policy. The government, anticipating noncompliance because of traditional valuing of sons for old-age insurance and lineage continuity, implemented both incentives and disincentives to encourage people to limit their family sizes. Those couples that agreed to follow the one-child policy were rewarded with a job promotion, bonus money, better housing, or access to better child-care and educational facilities. Conversely, couples that refused to comply with the policy faced punitive measures, such as large fines and job transfers or demotions (e.g., Banister 1987; Croll 1985; Greenhalgh and Bongaarts 1985; Kane 1987). Additionally, the Chinese Communist Party (CCP) engaged in a dramatic ideological campaign equating large extended families with "traditional" values and outmoded patriarchal feudalism, and small nuclear families with "modern" values. An intensive media campaign—disseminated through posters, speeches, flyers, newspapers, magazines, study sessions, radio, stamps, television, theater, and film—emphasized each person's responsibility to practice birth planning toward the modernization of socialist China. Images portrayed the one-child family, and especially those with a daughter, as "modern," happy, and prosperous. Overall, materials reflected and supported a belief in an overpopulation "crisis."

To sustain a popular perception of population control as necessary and inevitable, the CCP not only proliferated data to demonstrate the severity of the population problem, but also relied on silence as a tactical strategy. For example, the 1988 CCP decision to allow peasants with one daughter to have another child was never directly reported to the Chinese people (e.g., Greenhalgh 1990a, 1990b; Potter and Potter 1990). Moreover, when changes were reported they were expressed differently in Chinese and Western newspapers (Aird 1990; Potter and Potter 1990).[6]

Another tactic was cultural isolationism. Most of my informants, especially those that had never been abroad, were unaware that China is the only country to have a mandatory one-child policy. By highlighting a population problem in geographic isolation, the CCP justified its strict enforcement of policies (Duden 1992: 153).

Policy enforcement was made possible by both vertical and horizontal administrative bodies, which facilitated unprecedented fertility surveillance through the management of every aspect of ordinary life, including registration of marriages, births, household, deaths, adoption, and residence. In the urban areas, the mechanisms for surveillance of women's lives were in place primarily through the *danwei* ("state collective") and *jiedao weiyuanhui* ("neighborhood street committees") and in rural areas through the *jiti* ("collective").[7] Specifically, family planning workers' close surveillance of female fertility was achieved through the charting of menstrual cycles; granting permission for and documenting pregnancies, births, and abortions; and the free distribution of and regulation of contraceptives, especially IUDs, which require implantation and removal by a medical professional (Banister 1987; Greenhalgh 1986, 1990b).

Despite the state's commitment to strictly control population, there is great variation in official attitudes and policy enforcement. First, even the national leadership is increasingly divided over the wisdom and mandatory enforcement of the one-child policy. Some leaders are worried that parents with an only male child will not let their son risk his life in military service (Zhang and Yang 1981: 29). Others, including demographers, are concerned about China's transition to an aging population and the subsequent long-term consequences of the one-child policy on the Chinese family (Banister 1987: 371–76).

Second, Chinese regularly violate regulations, and bureaucrats busily make adjustments responding to conditions of production and reproduction demanded by local and global capital. Under Deng Xiaoping's leadership, major political, scientific, industrial, and economic reforms have been introduced and linked to the creation of a "modern" state. The most dramatic economic reforms began in the late 1970s when rural collectives were dismantled, shifting to a system of individual profit incentives and plots worked by peasant families, who comprised approximately 79 percent of China's population (Banister 1987: 329).

With the commune system's collapse, collectivized social welfare benefits and health care suffered (e.g., Handwerker 1988; Henderson and Cohen 1982; Hsiao 1984), and as a result family planning workers' close surveillance of female fertility was disrupted in many agricultural areas. The CCP undermined its own birth planning policy by unintentionally encouraging peasant families, the primary means of controlling women, to have more, not fewer, children, to prosper economically (Anagnost 1989a, 1989b; Greenhalgh 1990: Potter and Potter 1990; Smith 1991). During my interviews with women from several regions in China, it was clear that the term "one-child policy", as it is commonly known in the West, is misleading. And in fact, in China it is referred to as the *jihua shengyu* or "birth planning policy" because there is greater variability than the term implies (Greenhalgh 1990a, 1990b; Potter and Potter 1990). In some areas women were allowed two births as long as they complied with the birth spacing; otherwise they faced stiff monetary penalties. Elsewhere, local officials allowed women three attempts without penalty to produce a male heir. Even where the one-child limit was strictly enforced, women resisted policies by giving birth in nondesignated hospitals outside their provinces. Overcoming regulations is, nevertheless, difficult, and some women were coerced by state officials to undergo abortion or sterilization.

Family-planning officials find it easier to enforce family planning in urban areas through the *danwei* than in rural agricultural areas, where the large extended peasant family has both important symbolic meanings and also enables a kind of resistance to population-control policies that families resembling the modern urban nuclear concept cannot express. State policing of the family has had a hand in restructuring the modern Chinese family on a more Western model, as couples, and particularly women, come to be blamed for *both* overpopulation and infertility problems. Overall, the CCP fails at total control because within a self-proclaimed coherent policy exist contradictions, allowing for resistance, fissures, and even transformations (e.g., Chow 1991; Greenhalgh 1990b; Ng 1990; Shue 1988).

Implications of State Policies for Female Bodies

Viewed in its entirety, the CCP's attempt to regulate population offers a lens through which to understand the construction of normative gender prescriptions. Certain forms of nationalism and state power require exercise of control over the body but are expressed differently for women and men (Foucault 1980; Parker, Russo, and Sommer 1992; Sawicki 1991). Today, in urban areas, female graduate students, not male graduate students, are expelled from the university for having children; population statistics reflect every aspect of women's, not men's, reproductive and sexual practices;[8] and women, not men, continue to be blamed for the birth of a baby girl. Policy practices and ideology coincide with and reinforce already existing beliefs about women's culpability. As Anagnost suggests: women's bodies, not men's, bear the burden of noncompliance (1989b: 334). Thus, they bear the burden of surveillance, too.

Ironically, such surveillance is often done by women, who comprise the majority of family-planning workers and obstetrician-gynecologists (*fuchanke*). Under the aegis of population control, contraceptive promotion became a veritable growth sector on its own, providing jobs for female lay organizers at the village level, who tried to induce popular acceptance of freely distributed contraceptives. The birth-control policy is one way the CCP has involved women in developing the economy. While educated urban men make decisions at the highest administrative levels, information is disseminated to women by other women. There is no equivalent of men talking to men to encourage birth control use or to implement mass-scale birth control to male workers.[9] Family-planning work is seen as "women's work" because of its childcare and childbirth concerns

(Zhang Mincai 1991: 36). Overall, it is women who are held responsible to maintain the will and ability to reproduce as directed by the CCP leadership and funded by international assistance. Although the CCP has professed to favor women's equal rights and family reforms (e.g., equal pay, older marriage age), its policies have undermined such reforms and reinforced normative gender roles of women as wives and mothers to further the Party's predominant economic aims (Robinson 1985; Stacey 1983; Weeks 1989).

By primarily focusing on economic relations and relegating gender relations to a secondary position, the CCP fails to understand the ways in which it undermines its own one-child policy through the perpetuation of control over women's reproductive and sexual lives. The same institutions that prescribe population reduction prescribe a normative feminine gender identity in terms of procreative ability. Also, the government does nothing to shift the patrilineal family values that place priority on male children—another major reason peasant women want to have more children when they have only a daughter. We have seen how this continued valuation of women as reproducers has led to the challenge of and resistance to the birth-planning policy by the peasant family. Contradictions are also reflected in beliefs toward and treatment of women without children. This next section addresses how the "problem" of infertility figures women's bodies in particular ways.

Definitions and Meanings of Infertility

Since infertility is the dynamic state of being unable to conceive or carry a pregnancy to term, defining infertility always presents a problem: Exactly when should a couple consider themselves infertile? In the United States, conventional medicine defines infertility as the inability to conceive after one year of unprotected intercourse (Berkow 1982: 1641). Textbooks further distinguish between primary infertility (the inability to become pregnant or conceive) and secondary infertility (the inability to become pregnant or carry a pregnancy to term after one or more previous successful birth(s)) (Harkness 1987: 74).

In China, by contrast, up until the mid-1980s one standardized obstetrics and gynecology textbook, *Fuchankexue,* used in "Western" medical schools nationwide, defined infertility (*bu yu zheng*) as the inability of a couple to conceive after two years of married life (Deng 1985: 374–78). By contrast, a Chinese medicine handbook defined infertility as the failure to conceive after three years even though both parties are seemingly healthy (Zhang 1991). While this labeling delay is surprising in view of Chinese continued valuation of children for old-age insurance and as a form of lineage continuity, it also reflects the highly stigmatizing nature of infertility. This definition, however, is undergoing change;

as a result of increased funding from international agencies, infertility was recently redefined in Chinese medical textbooks in accordance with World Health Organization (WHO) standards, as the inability to conceive after one year of marriage. Within the context of China's official one-child policy, primary and secondary infertility classifications are substituted by the terms *bu yu* (never given birth) or *bu yun* (never been pregnant). Male infertility is also referred to as *bu yu* in medical texts.

Despite attempts to move toward standardized medical definitions, there remains considerable variation in labeling practices. The definitions of individual practitioners (many of whom are urban-educated) and patients' self-definitions do not always concur because infertility takes on different meanings according to women's life circumstances within their broader cultural and familial contexts. Some people outright reject the label, as in the case of one woman who hadn't conceived in five years but, nevertheless, did not consider herself infertile. By accepting the label of infertility, this woman also must acknowledge the potential social implications of the label (e.g., stigma, ridicule, and the possibility of divorce).

For others, the label offered hope for treatment, as in the case of one informant, a peasant woman and mother of four young girls, who had an infected fallopian tube. According to her, many doctors had refused treatment, arguing that she was not infertile and that, furthermore, she had already produced more than enough children. But due to her inability to conceive a son, she perceived herself as infertile. She protested that her reproductive task as a woman was incomplete. The doctors, on the other hand, refused to recognize her medical condition as infertility and labeled her desire for a son "feudalistic."[10]

Other women, especially peasant women or women who married late, who had not conceived within a few months of marriage sought treatment immediately.[11] Infertile women cited five major reasons for their strong desire to have a child and to seek treatment. These included the importance of "lineage continuity" (*houdai*); someone to take care of them in their "old age" (*yang lao*); the feeling that "my life is uninteresting or too lonely without a child" ("*Wode shenghuo mei you yisi, mei you haizi de shenghuo, tai jimo*"); the belief that if "I don't have the ability to give birth, I am not a woman" ("*Ruguo wo bu neng sheng haizi, wo bu shi nu ren*"); and "societal pressure" (*shihui yali*) from family, friends, colleagues, and neighbors. In general, a couple—especially a couple from agricultural areas—who has not conceived within one year is suspect by others. A sixty-year-old unmarried woman joked with a couple who had no children after two years of married life. She said to them, "I'm regarded as a model for late marriage, and you may have set the trend for having children late" (Yuan 1991: 8). Couples, but especially women, who are without a child encounter hostility and prejudice. In popular discourse the idea persists that a woman or man who doesn't have a child is suspect and *cha yi dian* (not good enough). One of the worst insults a rural woman can give her neighbor is to call her daughter-in-law a "*bu xia dan de mu ji*". This expression equates a childless woman to a "hen that can't lay an egg" and is therefore useless. During twelve months of fieldwork in China, I discovered that, while almost all of my informants, regardless of their geographic or class background, had heard this expression, none had heard of an equivalent expression about men. One or two people said that one existed, but that they couldn't remember it. This is illustrative of the ways in which normative feminine gender ideology equating womanhood with motherhood circulates in popular culture.

While existing statistical information suggests no increase in the incidence of infertility over the past decade, there is, nonetheless, a public perception that infertility is on the rise in China. There is also the belief that female infertility is greater than male infertility, despite recent statistics suggesting an equal proportion of male and female infertility problems.[12] This perception is both created and sustained through the popular media. Within the 1980s, childlessness, both voluntary and involuntary, has become the focus of increasing numbers of television news programs and dramas and professional and popular books and articles (e.g., *Jiating Zazhi, Jiankang Baozhi, Zhongguo Funnu*). Comments about childless women in one magazine include: "Although she looks smart and healthy, she is pitiably useless," "Her belly can't bring her a baby and win her credit," and "Keep a dog and it can watch the door, keep a cat and it can catch mice. But what's the use of such a wife?" (Chen 1990: 53). Such remarks continue to construe "normal" women as reproducers and contribute to the further stigmatization of infertile women.

Comparative Medical Systems: Medical Understandings of Infertility

In addition to popular understandings of the infertile female body, two major medical systems—traditional Chinese medicine (*zhongyi*) and "Western" biomedicine (*xiyi*) as practiced in China—inscribe the female body with meaning. This next section examines the distinct ways in which Chinese and Western medical sciences construe female bodies as a culpable and controllable object of knowledge. In particular, the diagnosis, examination, and treatment of each reveal distinct ways of viewing bodies and their impact on conceptions of the infertile female body. While the therapeutic spectrum of each is by no means unified in its conceptual orientation, and patients and practitioners often draw on both practices (Unschuld 1985), here I treat these as separate systems and attempt to contrast some general principles of each approach.

While responding to different historical interpretations and individual practitioners' unique experiences, the concept of the healthy body and the clinical practice of *zhongyi* draws on *The Yellow Emperor's Classic of Internal Medicine,* which states that individual health depends on the balanced complementarity of ancient Chinese yin-yang cosmology, with the entire cosmos oscillating between yang as masculine, light, and hot and yin as feminine, dark, and cold (Veith 1966). To conceptualize health and disease, traditional physicians speak of the five yin "organs" and six yang "viscera" more in terms of function than anatomy (Furth 1987: 12). Similarly, they imagine *qi* ("air," "breathe") and *xue* (blood) as vital essences rather than gases and fluids. Since individual health depends on balance, there is constant vigilance toward harmonizing every aspect of regime and conduct, including control over one's environment and emotions (Furth 1987, 1992; Unschuld 1985; Zhang Ting-liang 1991).

Historically, Chinese medicine had the important role of ensuring the continuity of the patrilineal family by guaranteeing women's reproductive health through *fuke* (female medicine). If medical theory taught that disorder was commonplace, women learned that they were especially vulnerable. Menstruation, gestation, and childbirth subjected women to serious depletions of blood, making them chronically susceptible to disorders accompanying such bodily loss (Furth 1987: 12, 1992: 31). Female infertility problems are often linked to kidney (*shen*) deficiency, with many of the syndromes characterized by irregular or absent menses (Flaws 1989; Furth 1987, 1992; Zhang Ting-liang 1991). This accounts for the importance given to regulating menstrual periods in the infertility clinic. The Chinese term *bing* refers to a condition of imbalance resulting from excess or deficiency and is best translated as "disorder" than "disease" (Furth 1987: 13) Infertility is viewed as a process of disorder that needs balance restored—not as a fixed disease entity (Unschuld 1985).

While Chinese medicine focuses on "subjective" bodily processes, including patients' self-perceptions, "western" biomedicine relies on "objective" or technologically mediated diagnostics and test results. Modern diagnostics construe the body as distinct, divisible parts, organs and functions that can be isolated and treated separately from a person as a whole. The human body and organs are treated like clockworks; if one does not open it and take it apart, there is no way to know how it functions or to know the reasons for its failure (Unschuld 1985: 237). Biomedical explanations rely on visualizing, dissecting, and testing to identify any anatomical irregularities and hormonal dysfunctions. Explanations for infertility include tubal obstruction or removal (resulting from, for example, tuberculosis, pelvic inflammatory disease, sexually transmitted diseases, and ectopic pregnancy), congenital anomalies, endometriosis, and endocrinological dysfunctions. Modern etiology seeks the sources of illness from external pathogens rather than internal disorder (Flaws 1989; Furth 1992; Unschuld 1985). These conceptual and practical differences between Chinese and Western medicine can be observed in the clinics.

The Clinics

I observed in one Chinese medical clinic that operates within the *fuke* department of a well-known *zhongyi* hospital. While most "traditional" Chinese medical doctors of *fuke* have been male, I observed in a hospital clinic with mainly female *fuke* practitioners. The patients, on the other hand, are all female. The clinic accepts no advance appointments; women must arrive early on the one or two mornings per week that the *fuke* clinic operates, and they may wait up to a few hours to see the doctor, depending on their number. In one morning the doctor may treat up to thirty women, the majority of whom are seeking treatment for infertility. The room where I conducted observations was small, not more than seven feet by seven feet. Four of us (the doctor, her assistant, the patient, and myself) were seated on hard, backless wooden stools and crowded around a small wooden table. There was a large, colorful *nongli* (lunar) calendar hanging on the wall, for easy reference to menstrual dates. The doctor-patient interaction is generally brief, lasting less than ten minutes. The doctor, a well-known woman in her seventies, feels the patient's pulse while simultaneously observing, listening, and smelling (Flaws 1990: 9). She observes the patient's body, including the face, lips, hair, and tongue, for unusual color, texture, or thickness. She also listens to the voice quality, as she questions the patient about her symptoms, including details about her menstrual cycle (color, length, timing, pain), bowel movements, and urination. Less frequently, the doctor inquires about specific sexual behavior (e.g., frequency, timing, position). Occasionally, she interrupts the patient to elicit more information or to highlight some clinical information to the assistant-in-training. Finally, she smells for and takes note of any unusual odors. The combined efforts of an active patient—who provides vital information about how she feels—and an observant doctor—who utilizes sensory perception, prescribes lifestyle changes concerning sex, diet, and exercise, and provides emotional advice—are seen as necessary in *zhongyi* to stimulate the patient's inner defenses to *zhi bing* (cure disorder) (Unschuld 1985: 255). The doctor told me that a woman must have the *naixin* (patience) to return weekly for a minimum of four months to *zhi bing*. There is the expectation that the woman is committed to the process of restoring balance and not merely the desired product—a baby (Flaws 1989: 100).

After the pulse reading and evaluation, the doctor prescribes Chinese herbal medicine to each *bingren* (patient), while her assistant records the prescription in the woman's chart. Contrary to popular belief, some Chinese medicine is more costly than Western medicine. Additionally, it is stronger tasting and smelling. One patient complained that she could not keep her infertility "problem" a secret from neighbors because they complained about the frequent and strong smell from the boiling herbs and inquired about her condition.

While drawing heavily on traditional medicine techniques, the *zhongyi* clinic, in an effort to gain legitimacy and compete with the lure of biomedical technologies, recently has incorporated some "modern" diagnostic techniques, such as basal body temperature charts, to confirm a diagnosis. While not a central diagnostic tool, pelvic exams may also be done. Although there is a longstanding perception that Chinese medicine is suitable for chronic problems, especially gynecological problems (Flaws 1989; Furth 1992: 29), equating Western medicine with modernity and prestige has necessitated the incorporation of such techniques in an increasingly competitive market economy. In reality, most patients use both systems, viewing them as complementary.[13]

By contrast, on the two days per week that the "Western" medicine clinic operates, numerous female patients arrive before 8 A.M., when the clinic opens, to *gua hao* (take a number) in hopes of avoiding the long waits. In one morning as many as twenty women may receive medical care. The women, anxious to receive the best possible care from an "infertility specialist," often purchase the slightly more expensive ticket for three yuan, instead of the one-yuan ticket to see a general practitioner.[14] Since there is only one specialist, trained in the United States, all infertility cases are referred to her, regardless of the ticket price. However, it is an indicator of how important women view this problem and how much hope they have invested in the "miracles" of Western medical specialists and scientific technology. By the time they come to the Western medicine clinic, many of the women, especially those from rural areas, have already been to Chinese medical practitioners and other traditional healers (e.g., *qigong* masters).

The clinic, located on the second floor, has a sign *fuchanke* (Obstetrics and Gynecology) on the door. Another sign, reflecting gender segregation reads, "Men not permitted beyond this point." In general, this clinic, where women seek contraceptives, maternity care, abortions, and treatment for gynecological problems, is regarded as a women's sphere, where the majority of obstetricians/gynecologists (*Fuchanke daifu*) are female. (As is the case in the *zhongyi* clinic.) Female friends or relatives accompanying a woman may remain seated on wooden benches in the long hallway off the examination rooms but are not permitted to enter these rooms. Occasionally, a patient's anxious mother-in-law or mother enters the examination room, only to be reprimanded and sent away by the doctor. These hallways, painted white and lime green, are usually permeated with discernible odors from the clinic, including steam from the autoclave, menstrual blood, urine, and the occasional smell of burning Chinese herbs. Hanging on the wall, there is a plastic model hand with a condom-looking device covering two fingers. Ironically, this model—the only educational material in the clinic—is so abstract that it fails to educate patients about the male anatomy or the proper use of condoms. Reflecting biomedicine's compartmentalization of bodies into isolated organs and bodily pathologies, different rooms are designated for specific medical procedures, such as ultrasound, abortions, prenatal checkups, and hysterosalpingogram (HSG). The room for general examinations, separated from the hallway by a curtain, is a large room. After a woman drops off her number and medical chart at the doctor's desk, she waits in the hallway for her number to be called. The treatment of women as entities to be examined is reinforced by the inattention to names, with individuals reduced to mere numbers.

Once called, a woman sits in the examination room at a wooden desk across from or next to the doctor, who takes her medical history. The female patient answers her questions about her age, marital status, date of last menstrual period, number of previous pregnancies, miscarriages or births, history of childhood illness, and other nonspecific medical history. While these questions are also often asked in the *zhongyi* clinic to gather information, in the *xiyi* clinic doctors use this information to judge patients' specific behavior. For example, if an older woman delays childbearing, she is often reprimanded for so doing. The language used, including the framing of questions, is powerful in both shaping and reinforcing gender norms. The narrow range of questions constructs women in reproductive terms by highlighting what "normal" female bodies can do and by contrasting this to the deviant nature of infertile female bodies. It also constructs reproductive health as separate from other aspects of bodily experience. Furthermore, the terminology expresses a negative body image, with menarche referred to as *bu ganjing de shihou* ("the unclean time"); laboratory tests and examinations (Pap smear, serum progesterone, endometrial biopsy, laparoscopy, HSG) are scheduled according to the timing of a woman's menses.

The doctor's wooden desk is covered by a piece of glass with a paper calendar displaying both Gregorian (*gongli*) and lunar (*nongli*) dates. The doctor often refers to this calendar when women from agricultural areas tell the doctor the date of their last menstrual period in terms of the lunar calendar. This calendar highlights one of many differences between the urban-educated doctor and her rural patients—a different conception of time. Such differences have led to misunderstandings in scheduling examinations and operations and prescribing medications, such as clomiphene citrate, a drug administered cyclically by the Gregorian calendar.

The history-taking process is fraught with multiple interruptions and much noise, as other doctors cut ahead to consult on behalf of another patient or an out-of-town relative; or friends appear from nowhere, urgently in need of a medical excuse or prescription; or a former patient, in gratitude, drops off a gift or tickets to some exclusive event. After taking the medical history, the doctor routinely does a pelvic exam. In contrast to traditional Chinese medicine, the pelvic exam is a central diagnostic tool in Western biomedicine. Biomedicine and its "modern" tools privilege vision, especially looking inside the female body via pelvics, postcoital tests, laparoscopy (the direct visualization of the fallopian tubes with an optic instrument inserted into the abdomen), x-rays, which require the introduction of gas or dye into the uterus and tubes to check for blockage, cultures to detect any infections that may prevent conception, biopsies of the uterine lining during phases of the menstrual cycle to discern its responsiveness to hormones, a daily temperature reading to determine whether and when a woman ovulates, and blood tests to check for abnormal female hormone levels. These techniques result in both the depersonalization and the dismemberment of women into distinct reproductive body parts, e.g., *zigong* (uterus), *luanzi* (ovaries), and *shuluanguan* (fallopian tubes). This is in contrast to *zhongyi's* attempt to integrate the whole body into a balanced state. These methods of biomedicine also imply a distinction between the body's inner and outer surfaces that *zhongyi* does not emphasize.

Overall, Western biomedicine (*xiyi*) sets itself apart from Chinese traditional medicine by drawing on different notions of the body, manifest in the routines of examinations, diagnostics, and treatment of infertility. While Chinese medical treatment is based on the combined efforts of the doctor and the female patient to stimulate inner defenses in a struggle against disease, biomedical treatment uses external measures to produce changes or destroy pathogenic agents that have penetrated from the outside through cutting, removing, and attacking (Unschuld 1985: 248), with a passive female patient. Nevertheless, despite these differences, both Chinese and Western medicine continue to view woman's health status as reflected in her reproductive health, with special attention to regular menstruation, pregnancy, and childbirth. The result is that both construe the fertile female body as "normal" and the infertile female body as "deviant."

The Infertile Male

In both the Western and Chinese medical clinics, infertile women and men are scrutinized differently with respect to reproduction and sexuality. Despite the fact that the investigation of female infertility is more time-consuming, more expensive, and increasingly more invasive for women than it is for men, until recently the evaluation of males has been overlooked. Previously, it was assumed that the female body was culpable for an infertility "problem." While the male body has attracted some interest, neither *zhongyi* nor *xiyi* has had a separate male reproductive health specialty. Rather, men have been scrutinized by *zhongyi's* internal medicine and Western medicine's urology (*minniaoke*) departments. Only in the 1980s did "andrology" (*nanke*), or the field of male reproductive science, emerge in China (Yuan 1991: 8).

Although I did not observe men in the clinics and do not have comparative data, I did interview men about infertility beliefs. While some men willingly seek care in the hospital's urology department, others are reluctant to do so because the Western medicine exam includes producing a semen specimen. Within a cultural context in which ejaculation is associated with a loss of vital energy, and masturbation is viewed negatively—and where hospital clinics lack individual bathrooms or private medical rooms—many men hesitate to provide a semen specimen. Additionally, many men believe that adequate sexual performance is an indicator of adequate fertility. One young man I interviewed was convinced that his wife had the "problem," because he could perform sexually, only to find out that he suffered from azoospermia, or lack of sperm in his semen. Afraid of gossip by family members or colleagues, he, like most men, hid his infertility. Many men thus fear that an infertility problem will be viewed as a sexual problem, such as impotence. This fear is not unwarranted in a culture that closely associates reproduction with sexuality and in a climate where popular journals have begun to openly discuss men's sexual problems. Recent estimates suggest that as many as ten percent of Chinese men suffer from sexual dysfunction (Yuan 1991: 10). Whereas previously public discussions on sexuality were not permitted, it now appears legitimate to address sexuality, specifically male sexuality, under the guise of "modern" science in the 1990s. For example, on February 1, 1993, a new store called "Adam and Eve" opened in Beijing with government support under the auspices of the Beijing Science and Technology Center. While other stores carry sex objects and products to improve sex, this is the only store in China devoted solely to their sales. Under the slogan "Get rid of ignorance; head toward science," such devices as "The Physiotherapeutic Ring" for men needing help maintaining erections, "Well Men Gem" spray for increasing sexual pleasure, and "Men's Cheerful Tissues," which promises to enhance sexual performance if rubbed in the appropriate region, are displayed and sold (*San Francisco Chronicle* 1993). Such stores further conflate sexuality and reproduction and reinscribe reproduction as central to women's lives and sexual pleasure to men's lives.

This conflation of sexuality and reproduction creates a context in which men are afraid to acknowledge an infertility problem. One man was so fearful of ridicule that he changed jobs, destroyed his medical chart, avoided his friends, and even considered moving to another city. While infertile men also suffer from the stigma of infertility, they are ultimately protected from public ridicule in ways that women are not. Even after a male diagnosis of infertility, women often continue to seek risky medical procedures for what they still perceive as their problem (and in cultural terms it is!). And while men never hide their wives' infertility problems, women often veil their husbands' infertility in secrecy. One woman in this study told me that, even if she publicly acknowledged her husband's problem, no one would believe her. In another case, in which I wondered out loud if protecting her husband might lead to further stigmatization of female infertility, the woman remarked, "It's my wifely duty to protect him. As the household head, he stands to lose face."

Shrouding male infertility in secrecy is further motivated by the medicalization of artificial insemination with the creation of sperm banks in the mid-1980s (Chao 1988). A couple considering the artificial insemination by husband (AIH) or by donor (AID) procedure points to the female as the "problem" source. Assuming the woman conceives after undergoing AID or AIH, friends and family members can believe she was successfully treated, and furthermore, they would never suspect a male infertility problem. In one case from Shanghai a couple secretly underwent AID treatment. After the child was born, the paternal grandparents became suspicious on the grounds that the child did not look like their son. They accused the wife of having an extramarital affair. Under pressure, the man admitted that, as a result of his sterility, his wife had undergone artificial insemination by donor. The parents refused to believe their son was sterile and rejected the grandson as one of their own. The husband, forced to choose between his parents and his wife, eventually divorced [personal communication]. This story and others suggest how shrouding male infertility in silence perpetuates a cultural insistence that female bodies are the site of infertility. Thus, while some physicians may decry the cultural insistence on blaming only women for sterility, they seldom view men as thwarting "nature's plan" or society's mandate to reproduce in the same way that they accuse women.

Female Culpability: Infertility as a Disease of Civilization

While *xiyi* and *zhongyi* systems scrutinize the bodies in different ways for signs of deviancy, infertile women are increasingly perceived as culpable, bringing this upon themselves through modern lifestyles and choices. In the China of the 1990s, the origins of female infertility are linked to individual choices or actions, suggesting that, even if a woman is not responsible for her infertility "problem," she can be blamed for actions that predisposed her to such a condition (Sandelowski 1990). In both medical diagnostics and women's subjective explanations there is a similar tendency to view a woman's body and, especially, her reproductive capacity as direct indices of women's gender conformity or nonconformity. For example, a young urban woman with whom I spoke had to delay childbearing until her graduate studies were completed. During graduate school, after she accidentally became pregnant, she had an abortion. Now, having completed her studies, she was trying, but unable, to get pregnant. The doctor yelled at her because she had undergone an abortion and continued her studies rather than fulfill her one-child quota. While there is no evidence to suggest that sterility is more common among urban-educated women, there is, nevertheless, an increasing perception in the 1990s that a woman's involvement in intellectual pursuits can potentially divert her energy away from reproduction.[15]

This idea of holding women accountable has reappeared at a particular historical moment, when mainly middle-class physicians, pressured for explanations, still cannot account for numerous infertility cases. Physicians' disagreement about the causes of sterility allow for new explanations to emerge. Several practitioners suggested that the introduction of tampons to China has led to increased incidence of endometriosis and that increased sexual permissiveness leading to "excessive" abortions has led to higher infertility rates. Moreover, some of the practitioners' explanations converge with patients' popular explanations. Women believed their own actions that defied cultural norms, such as engaging in sex during menstruation, being active in sports, or drinking cold liquids, caused their infertility problems. Additionally, there is an emerging public tendency to think of peasant women as more fertile and urban women, especially heavy-set women, as less fertile. Popular opinion attributes excessive weight to laziness or "bourgeois" decadence. These new explanations for female culpability are reflected in folklore variations, such as "the *fat* hen that can't lay an egg" (emphasis added).

Both practitioners and patients have increasingly constructed infertility as a dis-ease, a disorder of civilization and modern living, involving culpable female acts. In their view, the preoccupation with the internal management of the body is produced by instabilities at the macro-level of society. All of this material points to a view of the female physical body as reproducing the vulnerabilities and anxieties of the social body (Horn 1991; Jacobus, Keller, and Shuttleworth 1990). It appears to be women's freedom—in intellectual pursuits and freedom to violate cultural norms about marriage, sexuality, and diet—that is pointed to as a source of physical and social instability, precisely at a time when women are gaining more access to education, social power, and lifestyle options. Accordingly, the disruption of the "natural" differences between masculine and feminine gender roles must be diagnosed and treated.

The Modern Dis-ease, the Modern Cure: In vitro Fertilization

While women are held responsible for their "problem," the authority to diagnose is invested in the expert, and the cure is linked ironically to the cause. Infertility, the disease of "modernity," is seen as curable by "modern" solutions. Recent articles and television programs disseminate glowing success stories of test-tube babies, suggesting these "miracle babies" are the answers to infertility problems. Presented as a solution to a variety of medical and social problems, the marketing of in vitro fertilization creates the social context in which normative female gender constructs are (re)produced, even within the context of "modern" China. This is best expressed by one test-tube-baby mom, a thirty-nine-year-old peasant woman. She recounts:

> I have been a primary-school teacher all my life. I work with children daily and I love kids, so I couldn't understand how this could have happened to me. You have no idea what a terrible disease this is, this infertility. It's the worst disease (*bing*) possible in China. I suffered so much in my life. My parents died when I was young, so I was raised by my grandparents. After my grandmother died, my grandfather continued to take care of me, and now, he is in his eighties and my only surviving relative. I must care for him. When my husband and I married, his family agreed to certain conditions due to my unusual family circumstances. First, rather than I move in with his family, he would move in with me and my grandfather. Second, our first child would be named after my grandfather and any subsequent children after his name. Who could have predicted I would be infertile? For ten years we tried every treatment in our area. We were poor and couldn't seek medical care in the larger cities. Then, in 1988, after we had already given up hope, my relative heard an interview with Professor Zhang on television and told me about the test-tube baby. At first, I didn't understand it—I thought you go and pick out a baby from a tube and bring it home. We saved our money, as it costs almost 2,000–3,000 *yuan* to do the operation.[16] Due to the economic reforms, my family is so much wealthier now, so it wasn't like before when we had no money. When I became pregnant I was so happy. After the baby was born I named her after Professor Zhang and my grandfather. When my husband found out he was furious. Before this birth, even though I was infertile, he was always good to me, but now he has started beating me. It is terrible and I don't know what to do. I came here to ask the doctor for a chance for a second baby and this time, I'll name the baby after him.

In contrast to the celebratory account that appeared in the popular press, this version of the test-tube woman's story shows that even unconventional routes to motherhood, potentially liberating, have not solved the "problem," but rather, appear to reproduce female culpability and normative gender prescriptions. Whereas Chinese medicine has historically been implicated in the perpetuation of Confucian normative gender ideology by ensuring a women's fundamental obligation to produce babies and continue the patrilineal family, biomedicine—alongside Chinese medicine—has taken up that role in the 1990s.

Contradictions

My own observations and interviews in the clinics confirm that infertility specialists—promoting birth control and abortion, on the one hand, and aggressively assisting women, on the other hand—perceive no conflict. While the obligatory call for women to reproduce is in stark contrast to the call for women to control family size in response to modernization, these female practitioners argue that their efforts to help infertile women are both humanitarian and beneficial to the development of the field of reproductive science.

With respect to the humanitarian dimensions, biomedical doctors perceived that they were fulfilling a social need. As one doctor stated:

> Too many births cause havoc, but no births at all produces greater anxiety. The treatment of infertility remains a worthy topic as long as the Chinese family remains. . . . In China, a country in its initial stages of development, infertile patients are more overwhelmed with anxieties and stronger in their appeals for the test-tube baby.

One infertility specialist showed me numerous letters that she had received from women all over China begging for the chance to have a child and detailing their suffering. One letter reads:

> For over ten years I have been trudging along for great distances in search of medical assistance in order to acquire the right of childbirth. I have traveled the greater part of China and visited scores of herbalists, swallowed down hundreds of doses of bitter medicine and exhausted almost all secret prescriptions under the sun. I narrowly escaped the destiny of death but I remain without a child. . . .

Another letter reads:

> Have pity on my family, please, my dear doctor. My family has just had one successor for three generations. When I was six months old, my father died a violent death. When I was only one and a half, my brother breathed

> his last breath, leaving me the only child to take care of my aging grandparents. Later my grandmother died. My grandfather, now almost 90 years old, has been waiting for me to bear a child for twenty years but still I am barren. I am now almost 40 years old, my husband is a peasant. Without a child, my family is at a loss about our future. . . .

Confronted with these realities, the doctors argue that they cannot easily dismiss the desire for a child as "feudalistic" or "traditional." Rather, as one doctor said, "They have come to view childbirth as an innate right of all women as well as their sacred duty." These comments further discredit the decision of many educated women to voluntarily remain child-free and leads to the conflation of voluntary and involuntary childlessness.[17] The doctors pointed out that even the word for female uterus (*zigong*) literally means the "palace or room for children." One doctor summarized, "Any woman who has the ability to give birth wants at least one child." I was warned by her that, even if a childless woman told me she didn't want a child or didn't care, this was only because she couldn't have one. Almost all of the doctors with whom I spoke shared the idea that it was "natural" for all Chinese women to want a child. Ironically, such "naturalness" is asserted even in the face of intense medical intervention of women!

To defend the aggressive treatment of infertile women within the population policy context, one doctor stated:

> In order to facilitate the implementation of the state family planning policy, in our capacity as obstetricians and gynecologists, we need not only be adept in the field of contraception and birth control but also adept at reversing women's reproductive function in cases of reverse sterilization because of the accidental death of a child.

Another physician also argued:

> Programs such as in-vitro fertilization and gamete intrafallopian transfer serve not only as new techniques to help infertility but also contribute to the understanding of the human reproductive process. At the same time they open up a new vista to promote basic medical research in genetics, immunology, and early embryology. Therefore, they are of major importance to the implementation of the policy of family planning and eugenics (Ni 1988: 303).

In other words, women's fertility and infertility are situated as critical markers of national "progress." In the context of Chinese socialism, they can be signs of feudalism (as in the case of wanting too many children) or bourgeois values (as in the case of not wanting children). In the 1990s, high-tech medicine for infertility treatment has also become a signifier of Chinese modernity. This plays a role in state support for infertility research and development in a country ostensibly committed to population control.

While the CCP neither officially recognizes infertility as a "social problem" (in the same way that it recognizes a "population crisis") nor endorses the reimbursement of infertility treatment, it has unwittingly encouraged the growth of an infertility industry.[18] As one international population official told me:

> We provide money annually towards the development of contraceptive research and technology in China. We presently fund five major factories for contraceptive production in various parts of China. During a recent site visit we discovered that officials have been using our money to secretly manufacture medicines to treat the infertile and sexually impotent because of the potential for large private profits. (personal communication)

This industry has been stimulated through new economic policies that increase dependency on international funding and technological transfers in science and medicine, encourage medical privatization and the marketing of medicines and technological products, and make infertility a very lucrative business. As a result, it is not surprising that China, and in particular Beijing, is witnessing a growing number of physicians and clinics of both "Western medicine" (*xiyi*) and "Chinese medicine" (*zhongyi*) willing to provide services to women seeking infertility treatment and men seeking to improve their sex lives, further conflating sexuality and reproduction. Doctors' failure to understand the ways in which these contradictions may impact on women, ironically, results in further stigmatizing the condition of childlessness for women.

Summary

Within the context of the official population reduction rhetoric, one would anticipate that childless couples, either through voluntary or involuntary means, would be rewarded. To the contrary, childless couples complain that state policies perpetuate outright discrimination against them. Childless women observe mothers with one child receiving such benefits as better housing with more bedrooms, monetary bonuses, child-care support, and a vacation on Children's Day, while they receive nothing. They wonder why a government seeking to reduce population does not reward childless couples as exemplars of population reduction goals.

The CCP sees its role as merely managing women by controlling the desirable number, quality, and spacing of births. The focus on women, not men, perpetuates the idea that the ability and will to reproduce and maintain the integrity of the family and the nation is vested in women. It is important to note, however, that different classes of women are called upon to serve the nation in distinct ways. While a poor female peasant or worker has the duty to restrict her fertility, a female intellectual or cadre is encouraged to fulfill her/their one-child quota. According to this logic, poor, overly reproductive (defined as women who exceed the birth quota) female bodies threaten national security and economic development, while nonreproducing female bodies threaten moral values of patriarchal control over women's reproduction and sexuality. Despite population reduction rhetoric, then, normative feminine gender prescriptions equating womanhood with motherhood are reinforced, never challenged. As one woman in the study aptly noted, "The one-child policy is really the 'you must have one-child policy' " (Handwerker 1990:10). Even in the 1990s, voluntary childlessness is rarely an acceptable option.

This study has explored the paradox of infertility, how women's bodies are situated in it, and how the new modes of conceptualizing and treating them are instituted. I argue that the treatment of infertility in China involves a complex hybrid of medical systems and cultural frameworks that makes it difficult to come up with any simple summary statements. The "objectivity" or "modernity" of Western medicine is contrasted to the "subjectivity" or "traditionality" of Chinese medicine. Although *zhongyi* and *xiyi* operate in distinct ways, practitioners of both medical systems uphold the idea that women are primarily responsible for and destined to bear babies. The treatment of infertile female bodies recirculates and reconsolidates specific cultural values that link gender identity and fertility. Amidst a population "explosion" in late-twentieth-century China, medicine continues to endorse women's special biological function as mothers. Within the context of the hyperpopulation discourse, it is not surprising then that infertile female bodies have been constructed as deviant and an infertility industry has unwittingly been created. This ideology fits well with the political agenda of the Party and the increasing reliance on technology as the answer to China's problems. Women's bodies are at the vortex of these seemingly competing systems of "modernity" and "tradition"—both cause of and solution to China's national progress and welfare.

Notes

Acknowledgments. This article is based on twelve months of research conducted in the People's Republic of China in 1990, which was generously supported by a joint grant from the Fulbright-Hays Doctoral Dissertation Award and the Committee on Scholarly Communication with the People's Republic of China. Equipment and material assistance were also provided by the Wenner-Gren Anthropological Association Dissertation Fund, a National Science Foundation Doctoral Dissertation Improvement Grant (#BNS-13347), and the Association for Women in Science. Additionally, the Soroptomist International Award (1991–92) and University of California San Francisco Humanities Award (1992–93) provided dissertation writing support. Jacqueline Urla and Jennifer Terry's vision was essential for the completion of this chapter. I am extremely grateful for their detailed readings, insightful comments, and encouragement. My work has also benefited from comments at earlier stages by J. Ablon, G. Becker, M. Inhorn, A. Ong, L. A. Rebhun, N. Scheper-Hughes, and S. White. For their support, I thank T. Gold, the Handwerkers, Dr. H. Yuan, E. Kim, S. Louie, A. Rinawi, J. Urla, and Y. Verdoner. Finally, I thank the women and men of this study who shared with me both their joys and sorrows. While I have conveyed their stories to the best of my ability, all interpretations are my own.

1. In this paper I use "tradition" and "modern," for lack of better terms. Rather than a static condition, I am referring to a complex and dynamic set of historical practices that are constantly invented and reinvented by the Chinese nation-state and people. Here the word refers to selective dominant cultural ideology about women that has been reworked over space and time.

2. My female-centered focus reflects both my commitment to and long-standing interest in women's health and a gender-hierarchical medical and cultural system that limited my access to male patients. Despite my clinical experience in the United States, which included educating, diagnosing, and treating men with sexually transmitted diseases at a free clinic, I was unable to act as a participant-observer in the urology department, where men were examined and treated. On the contrary, I was able to observe women, mainly Han Chinese women, from divergent socioeconomic backgrounds and a wide geographic area (with some women traveling long distances to come to the clinic), being examined, diagnosed, and treated by physicians in various clinical settings with few restrictions. This accounts in large part for this chapter's focus.

3. The importance of *guanxi* in conducting social science research in China has been well documented. For example see Turner-Gottschang (1987).

4. Despite the numerous problems in implementing large-scale surveys and censuses in China (including misreporting, undercounting, overcounting, residence, migration, and national security concerns due to geography, cultural resistance, and lack of trained demographers), the resultant data was extracted and presented to the public as unproblematic, objective "facts."

5. Thomas R. Malthus, a world-renowned demographer who articulated the connection between population and national welfare, mentioned the culpability of the Chinese female body when he stated that the Chinese population must be curbed because, among other conditions, ". . . the climate of China was favorable to the production of children, and that *women* [my emphasis] were more prolific than in any other part of the world." See Meeks (1971).

6. This was, in part, a consequence of the international controversy over China's birth-planning policy, which led the United States Agency for International development (USAID) to suspend contributions for the United Nations Fund for Population Activities (UNFPA) for China.

7. The state *danwei,* or work unit, is an elaborate social institution in urban China, which potentially can manage every aspect of state employees' daily lives, including food, health care, housing, money, education, child-care, and pension. As Sydney White (personal communication) has pointed out to me, it is important to distinguish between the *danwei,* or state work unit, in urban areas of China and the *jiti,* or collective work unit, in rural areas. For an excellent description of a *danwei,* see Henderson and Cohen (1984).

8. Statistics on women's fertility and sexual and reproductive practices are gathered and circulated at every possible opportunity. Information is collected on the number of women of childbearing age, live births, miscarriages, abortions, pregnancies, and contraceptive use and knowledge. By contrast, statistics on men's fertility and sexual practices are rare.

9. Women in China are primarily responsible for contraceptive use. According to the national fertility study of 1982, 70 percent of the 170 million women of reproductive age were practicing some form of birth control (Banister 1987). After the first birth or more, women are likely to use the IUD or pill, or undergo sterilization to prevent additional births. Among men, condom usage is rare, although some doctors told me that recently there appears to be an increase in the numbers of men in urban areas willing to use condoms (personal communication). For additional information on birth control methods, see, e.g., Banister 1987; Croll 1985; Greenhalgh 1986, 1990; Kane 1987; and Potter 1985.

10. Terms like "feudalistic" and "bourgeois decadence" are frequently used by the CCP to refer to persons or ideas that undermine the government's latest moral, social, economic, and political prescriptions.

11. In China, late marriage generally refers to an unmarried woman from an urban area who is twenty-five years of age or older or an unmarried woman from a rural area who is twenty years of age or older. Although this criteria is changing because of the large numbers of women who are choosing to marry later or not marry at all, there is, nevertheless, still extreme social pressure for early marriage.

12. A recent article appearing in *Women in China* (Yuan 1991: 8–10) suggests a ten percent infertility rate in large cities in China. Prior to this survey, there were no available national statistics and only a few available local or regional statistics suggesting the problem of infertility is 40 percent male and 40 percent female and 10 percent couple or an unidentified problem (personal communication).

13. Although patients utilize both medical systems, I observed a tendency for rural patients to first consult *zhongyi* practitioners and, conversely, for urban patients to first consult *xiyi* doctors. Nevertheless, in urban areas some individuals first seek treatment at a designated *hetong yiyuan* ("matching hospital") of either "Chinese" or "Western" medicine. If they seek care at the designated hospital, the *danwei* may reimburse them for some preliminary exams and treatment.

14. The *yuan* is the Chinese monetary unit. During the time my research was conducted, from December 1990 to December 1991, five *yuan* were the equivalent of one U.S. dollar.

15. Both Chinese and Western medicine doctors expressed this view to me. At the time of my research, I was 32 years old, and they warned me to return home quickly to finish my Ph.D. so that I could give birth before I became too old. The social construction of the "appropriate age" to have children is also an interesting topic and one that I pursue in greater depth in my dissertation (Handwerker 1993).

16. The reimbursement of infertility treatment is extremely complex and depends on many factors, including the specific *danwei*, geographic region, *guanxi*, or social network, type of treatment or care, one's social position, diagnosis, etc. Beginning in the late 1960s and extending until the economic reforms of the 1980s, peasants relied on a collectivized health-care system. According to this system, they were referred first to the local level for primary health care, and only later, if the problem could not be solved, were they referred to secondary and tertiary facilities at the provincial and city levels. For the most part, medical referrals were reimbursed. Now, with the collapse of a rural health-care system, infertile peasants must pay out-of-pocket for treatment. By contrast, urban residents may or may not be reimbursed for infertility treatment. In 1990, during my fieldwork in China, a state doctrine circulated that forbade the reimbursement of infertility treatment. But when I questioned doctors about this document, they argued that it could not be enforced, but rather could be interpreted in highly variable ways. They did agree that in vitro fertilization, costing anywhere from 2,000–3,000 *yuan* ($1.00 = 5 *yuan*) and requiring anywhere from one month to two years to save this amount, depending on a person's work, would not be reimbursed by the state *danwei*. Nevertheless, during my fieldwork, I met one woman who thought she might be able to receive reimbursement from her *danwei*.

17. As Ann Anagnost has pointed out to me (personal communication), not having chil-

dren may be seen by some as a sign of modernity because *"waiguo mei you women ye mei you"* ("Foreigners don't have, we also don't have"). On the other hand, there is growing societal anxiety about the numbers of middle-class, educated women choosing to remain unmarried and child-free because it falls outside the bounds of a newly normalized reproduction and a newly reconstituted household as the primary unit of production. Couples that choose not to have a child are most fearful that other people will mistakenly believe they are sterile, leading to the further stigmatization of their child-free lifestyle. For more information on voluntary childlessness in China, see Handwerker (1993).

18. For the purposes of this chapter, it has been necessary to simplify what is, in fact, a very complex discussion. The official position toward infertility as a "social problem" varies according to historical time, geographic locale, and to whom one speaks. For example, officials from the Ministry of Public Health and from the Ministry of Family Planning are likely to present different positions in both the same and different geographic regions. This reflects the variations in political regulation throughout China.

References

Aird, John S.

1990 *Slaughter of the Innocents: Coercive Birth Control in China.* AIE Studies, 498. Washington, D.C.: American Enterprise Institute for Public Policy Research.

Anagnost, Ann

1989a "Family Violence and Magical Violence: Woman as Victim in China's One-Child Family Policy." *Women and Language* 11(2): 16.

1989b "Transformations of Gender in Modern China." In *Gender and Anthropology.* Ed. Sandra Morgen, pp. 313–42. Washington, D.C.: American Anthropological Association.

Banister, Judith

1987 *China's Changing Population.* Stanford: Stanford University Press.

Beijing Review

1988 First Test-tube Baby on Mainland. March 21–27, 31(12): 11. (Chinese announcement appeared in the *Renmin Ribao,* March 1988.)

Berkow, Robert, ed.

1982 *The Merck Manual of Diagnosis and Therapy.* 14th ed. Rahway, N.J.: Merck Sharp and Dohme Research Laboratories.

Chao, Jingshen

1988 "More Than Three Hundred Women in Tianjin Have Received Artificial Insemination." *Renmin Ribao* (overseas edition), October 24.

Chen, Huihe

1990 "China's Childless Couples," *Nexus: China in Focus.* Autumn, 52–55.

Chow, Rey

1991 *Woman and Chinese Modernity: The Politics of Reading between West and East.* Minneapolis: University of Minnesota Press.

Croll, Elisabeth, Delia Davin, and Penny Kane, eds.
1985 *China's One-Child Family Policy.* London: Macmillan.

Davin, Delia
1987 "Gender and Population in the People's Republic of China." In *Women, State, and Ideology: Studies from Africa and Asia.* Haleh Afshar, ed. p. 1. Albany: State University of New York Press.

Deng, Huai Mei
1985 *Fuchankexue.* 2d ed. Shanghai Di Yi Xue Yuan. Shanghai: Ren Min Wei Sheng Chu Ban Shi.

Duden, Barbara
1992 "Population." In *The Development Dictionary: A Guide to Knowledge as Power.* Ed. Wolfgang Sachs, pp. 146–57. London: Zed Books.

Flaws, Bob
1989 *Endometriosis and Infertility, and Traditional Chinese Medicine: A Laywoman's Guide.* Boulder, Col.: Blue Poppy Press.

Foucault, Michel
1980 *The History of Sexuality.* Vol. 1: *An Introduction.* Trans. Robert Hurley. New York: Vintage.

Furth, Charlotte
1987 "Concepts of Pregnancy, Childbirth, and Infancy in Ch'ing Dynasty China." *Journal of Asian Studies* 46(1): 7–35.
1992 "Chinese Medicine and the Anthropology of Menstruation in Contemporary Taiwan." *Journal of Medical Anthropology Quarterly,* 6(1): 27–48.

Greenhalgh, Susan
1986 "Shifts in China's Population Policy, 1984–86: Views from the Central, Provincial, and Local Levels." *Population and Development Review* 12(3): 491–515.
1990a "The Evolution of the One-Child Policy in Shanxi 1979–1988." *The China Quarterly* 122: 191–229.
1990b *The Peasantization of Population Policy in Shaanxi: Cadre Mediation of the State-Society Conflict.* Working Paper, no. 21. New York: The Population Council.

Greenhalgh, Susan, and J. Bongaarts
1985 "An Alternative to the One-Child Policy in China." *Population and Development Review* 11(4): 585–617.

Handwerker, Lisa
1988 "Rural Health Care in China: Recent Trends." *Synapse, University of California at San Francisco Medical School Newspaper,* special ed., February 25.
1990 The Hen That Can't Lay an Egg: Preliminary Thoughts on Infertility Research in China. Paper presented at the Center for Chinese Studies, Fall Regional Seminar. Berkeley: University of California at Berkeley, 1990.
1993 The Hen That Can't Lay an Egg (Bu Xia Dan de Mu Ji): The Stigmatization of Female Infertility in Late-Twentieth-Century People's Re-

public of China. Ph.D. diss. University of California, San Francisco and Berkeley.

Harkness, Carla
1987 *The Infertility Book: A Comprehensive Medical and Emotional Guide.* San Francisco: Volcano Press.

Henderson, Gail E., and Myron S. Cohen
1982 "Health Care in the People's Republic of China: A View from inside the System." *American Journal of Public Health* 72(11): 1238–45.
1984 *The Chinese Hospital: A Chinese Socialist Work Unit.* New Haven: Yale University Press.

Horn, David
1991 "Constructing the Sterile City: Pronatalism and Social Science in Interwar Italy." *American Ethnologist* 18(3): 581–601.

Hsiao, William C.
1984 "Transformation of Health Care in China." *New England Journal of Medicine* 310: 932–36.

Huang, Ti
1966 *The Yellow Emperor's Classic of Internal Medicine.* Trans. Ilza Veith. Berkeley: University of California Press.

Jacobus, Mary, Evelyn Fox Keller, and Sally Shuttleworth, eds.
1990 *Body/Politics: Women and the Discourses of Science.* New York: Routledge.

Kane, Penny
1987 *The Second Billion: Population and Family Planning in China.* Ringwood, Victoria: Penguin Books.

Lampton, David M. (ed.)
1987 *Policy Implementation in Post-Mao China.* Berkeley: University of California Press.

Lavely, William, et al.
1990 "Chinese Demography: The State of the Field." *The Journal of Asian Studies* 49(4): 807–34.

Marx, Karl
1971 *Marx and Engels on the Population Bomb.* 2d ed. Ed. Ronald L. Meek. Trans. Dorothea L. Meek. Ramparts Press.

Ng, Vivien W.
1990 *Madness in Late Imperial China: From Illness to Deviance.* Norman: University of Oklahoma Press.

Parker, Andrew et al.
1992 *Nationalisms and Sexualities.* New York: Routledge.

Potter Sulamith Heins
1985 "Birth Planning in Rural China: A Cultural Account." Reprinted in *Child Survival: Anthropological Perspectives on the Treatment and Maltreatment of Children.* Nancy Scheper-Hughes, ed. p. 33. Boston: D. Reidel, 1987.

Potter, Sulamith Heins, and Jack M. Potter
1990 *China's Peasants: The Anthropology of a Revolution.* Cambridge: Cambridge University Press.

Robinson, Jean C.
1985 "Of Women and Washing Machines: Employment, Housework, and the Reproduction of Motherhood in Socialist China." *China Quarterly* 101: 32–57.

Sandelowski, Margaret J.
1990 "Failures of Volition: Female Agency and Infertility in Historical Perspective." *Signs* 15, no. 3, 475–99.

San Francisco Chronicle
1993 "Lotions and Sprays at China Sex Shop: Ways to Excite 'Cheerful Tissues,' " February 19, sec. B1.

Sawacki, Jana
1991 *Disciplining Foucault: Feminism, Power, and the Body.* London: Routledge.

Simon, Denis F., and Merle Goldman, eds.
1989 *Science and Technology in Post-Mao China.* Harvard Contemporary China Series, 5. Cambridge: Harvard University Press.

Shue, Vivienne
1988 *The Reach of the State: Sketches of the Chinese Body Politic.* Stanford: Stanford University Press.

Smith, Christopher J.
1991 *China: People and Places in the Land of One Billion.* Boulder, Col.: Westview Press.

Stacey, Judith
1983 *Patriarchy and Socialist Revolution in China.* Berkeley: University of California Press.

Suttmeier, Richard P.
1982 Science and Technology in China's Socialist and Economic Development. Paper presented at the World Bank Science and Technology Unit, Project Advisory Staff, January 1982.

Turner-Gottschang, Karen
1987 *China Bound: A Guide to Academic Life and Work in the PRC: for the Committee on Scholarly Communication with the People's Republic of China.* . . . Washington, D.C.: National Academy Press.

Unschuld, Paul U.
1985 *Medicine in China: A History of Ideas.* Berkeley: University of California Press.

Wang, Feng
1988 "Historical Demography in China." *Review and Perspective* 236: 53–69.

Weeks, Margaret R.
1989 "Virtuous Wives and Kind Mothers: Concepts of Women in Urban China." *Women's Studies International Forum* 12(5): 505–18.

Yuan, Lili
1991 "Reproduction and Happiness." *Women of China*, February, pp. 8–10.

Zhang, Mincai
1991 "Male Family Planning Worker." Trans. Lin Guanxing. *Women of China* 1 (January): 36–37.

Zhang, Ting-liang
1991 *A Handbook of Traditional Chinese Gynecology.* Compiled by Song Guang-ji and Yu Xiao-zhen of the Zhejiang College of Traditional Chinese Medicine. Ed. Bob Flaws. Boulder, Col.: Blue Poppy. (*Zhong Yi Fu Ke Shou Ce.* Original publication date 1984. 1985 edition used for translation.)

Zhang, X., and C. Yang
1981 "Qianjin zhong de xin wenti" ("New problems in the forward march"). *Guangming Ribao* (*Bright Daily*) (Beijing) 2, 29 September.

Zhao, Z.
1985 "Gaige keji tizhi, tuidong keji he jingji, shehui xietiao" ("Reform the science and technology system, and promote its coordination with the economy and society"), *People's Daily* 12 (March): 1–3.

14

The Media-ted Gene

Stories of Gender and Race

Dorothy Nelkin and M. Susan Lindee

"Mother nature is a bigot" who discriminates against men, according to a 1991 article in *Health* magazine.[1] The writer recited some genetic differences between men and women: women live longer than men; they have sharper senses of taste and smell and greater sexual capacities; they are psychologically and physically more resilient and better at distinguishing colors. He included some differences that are presumably not genetic: twice as many women as men clean their navels every week, and more women than men believe in reincarnation. The implication, however, was that these too were innate: "Every living thing is assembled according to instructions in its chromosomes."

Increasingly in the 1990s, differences between men and women and between racial groups are appearing in popular culture as genetically driven. Such genetic images encourage stereotypes of the nurturing female, the studious Asian, and the violent African American male. But the images of pathology have moved from gross to hidden body systems. Once blacks were portrayed with large genitalia and women with small brains; today the differences lie in their genes.

We are studying the form and significance of this growing popular interest in the genetic basis of difference as expressed in media reports, popular fiction, television programming, advertising and other sources. The gene in popular culture is not a biological entity. Though it refers to a biological construct and draws its cultural power from science, its symbolic meaning is independent of biological definition. In this chapter, we draw from our larger study to explore

Research for this paper was supported by the National Institutes of Health, National Center for Human Genome Research, Grant 1R01 HG0047-01. It appears in expanded form in Dorothy Nelkin and Susan Lindee, *The DNA Mystique: The Gene as a Cultural Icon,* New York: W. H. Freeman, 1995.

how genes are used to construct gender and race as individual qualities that are central to personal identity.

This use of "nature" is of course an old story—ideas about biology and heredity have encoded social arrangements at least since the baroque claims of craniometry in the nineteenth century. In the contemporary context, such claims are fueled by the high status of research on human genetics. But the emphasis on genetic difference reflects cultural priorities rather than the specific details of molecular biology, for one of the fundamental insights of contemporary genomics is that all human beings—indeed, all species—share a great deal of DNA. Genetic differences—diseases, talents, and presumably even personality traits—exist, but humans have so much DNA in common that the social message of molecular genetics could be that we are all fundamentally the same. While the science of genetics focuses on biological differences between individuals (because such differences provide insight into the processes of heredity), it has also revealed striking similarities both within and between species.

Why, then, has the narrative of difference played such an important role in popular interpretations of DNA? The power of the gene as a cultural icon reflects the appeal of scientific explanations that reinforce and legitimate existing social categories. In a context of anxiety over persistent racial tensions and threats posed by the changing social status of women, genetic arguments seem especially persuasive. The gene appears to be a solid and predictable marker, an unambigious sign of natural human difference.

Popular belief in essential differences has been reinforced by scientific studies of body parts such as genes or neurons that seem to explain behavior, and by scientific theories about evolution that seem to ground social practices in biological imperatives. Molecular genetics, behavioral genetics, neurobiology, and sociobiology have provided a language through which group differences that are culturally desirable can be interpreted as biologically determined. These sciences have encouraged the increasing acceptability of genetic explanations and their strategic role in the continuing debates over gender and race.

Genes and Gender

The January 20, 1992, cover of *Time* featured a photograph of two small children: the girl looked admiringly and flirtatiously at the boy, who was flexing his muscles. The caption read: "Why are men and women different? It isn't just upbringing. New studies show they are born that way." The cover story, called "Sizing up the Sexes," reported on scientific research purporting to show that "gender differences have as much to do with the biology of the brain as with the way we are raised."[2] The article began from a study of the responses of young children playing with a variety of toys. The researcher found that boys systematically favored sports cars and fire trucks, while girls were drawn to kitchen toys and dolls. Those girls who defied these expectations and actually preferred male toys, the researcher claimed, had higher than normal testosterone levels.

Time went on to report a general public conversion to the idea that differences between men and women are genetically determined. Even skeptics in the professions, the article said, had come to see that nature dictated social place and behavior. It quoted psychologist Jerre Levy to illustrate a typical intellectual shift: "When I was younger, I believed that 100% of sex differences were due to the environment." But later, observing her 15-month-old daughter, she believed that skill in flirting must be inborn. The *Time* writer editorialized: "During the feminist revolution of the 1970s, talk of inborn differences in the behavior of men and women was distinctly unfashionable, even taboo. Men dominated fields like architecture and engineering, it was argued, because of social, not hormonal, pressures. Women did the vast majority of society's child rearing because few other options were available to them." Feminists, said the author, expect that the "end of sexism" will change all this, "but biology has a funny way of confounding expectations. . . . Perhaps nature is more important than nurture after all."[3]

The 1992 *Time* story about the origins of gender variation differed in striking ways from a *Newsweek* cover story from a decade earlier on the same subject. Its title, "Just How the Sexes Differ,"[4] was similar, and, like the story in *Time,* it attempted to convey prevailing scientific wisdom about the relative role of nature and nurture in shaping differences in the behavior of men and women. It even drew on some of the same authorities and referred to some of the same scientific data. The 1981 article, however, stressed the overriding importance of different social experiences and expectations on behavior, citing several experts, including anthropologist Sarah Blaffer Hrdy and Michael Lewis of the Institute for the Study of Exceptional Children, on the influence of culture on gender differences: "As early as you can show me a sex difference, I can show you culture at work." And it cited psychologist Eleanor Maccoby: "Sex typing and the different set of expectations that society thrusts on men and women have far more to do with any differences that exist . . . than do genes or blood chemistry."

The *Newsweek* journalist pointed to the broad variation among individuals of each sex: "Not all males in a given group are more aggressive or better at math than all females . . . women and men both fall along the whole continuum of test results." This 1981 article concluded with a statement on the nature-nurture debate: "It is clear that sex differences are not set in stone. . . . By processes still not understood, biology seems susceptible to social stimuli." Ten years later,

Time conveyed a different message—that sex differences are innate and unchangeable, and that it is just a matter of time until scientists will be able to prove it.

These two approaches to the same subject matter, over such a short period, capture changes in the social meaning of science. Neither popular article was about scientific data per se. Rather, both were manifestations of popular expectations about gender and genetics. By 1992 the public consensus that had dominated public policy since the 1950s—that "nurture" was more important than "nature"—had changed.

The media have speculated on the causes of gender differences since the early 1970s when the women's movement became a threat to traditional sex roles. A 1972 article in *Time* described the terms of a growing debate: "Women's Liberationists believe that any differences—other than anatomical—are a result of conditioning by society. The opposing view is that all of the differences are fixed in the genes."[5]

Popular interest in gender differences during the 1970s became apparent in the response to explanations offered by the emerging field of sociobiology. The media were attracted to claims that differences between the sexes were "genetically programmed" into male and female behavior to assure species survival. According to *Cosmopolitan,* "Authorities now say nature not nurture makes him thump and thunder while you rescue lost kittens and crimp. . . . Recent research has established beyond a doubt that males and females are born with a different set of instructions built into their genetic code."[6] Machismo, said *Time,* is biologically based. It "says in effect: I have good genes, let me mate."[7] The *Boston Globe* commented on innate differences in intelligence: "On the towel rack that we call our anatomy, nature appears to have hung his-and-hers brains."[8] Drawing on the writings of sociobiologists E. O. Wilson and Richard Dawkins, journalists described examples of male behavior—aggression, promiscuity, rape, and bravado—as evolving because they were useful reproductive strategies.

At a time when women were competing in the workplace, media interest also turned to differences in natural abilities. In 1980 two psychologists, Camille Benbow and Julian Stanley, published a research paper in *Science*[9] on the differences between boys and girls in mathematical reasoning. Examining the correlation between SAT scores and classroom work, they found that differences in school preparation did not account for differences in test performance. The *Science* article provoked so many popular articles that, in 1980, the *Reader's Guide* added a new heading called "Math Ability" to its listings. While their original *Science* article contained the standard qualifications about the limits of the research, Benbow and Stanley themselves encouraged more brazen and thus more provocative interpretations in their interviews with the press. In a *New*

York Times interview, they urged educators to "accept the possibility that something more than social factors may be responsible. . . . You can't brush the differences under the rug and ignore them."[10] *Time,* writing of the "gender factor in math," summarized the findings: "Males might be naturally abler than females."[11] *Discover* reported that male superiority (in math) is so pronounced that "to some extent, it must be inborn."[12]

But the media also published critiques that questioned the methodology of the research, the limited nature of the evidence, and its relevance to predicting performance of individual women. Journalists assumed that the message was controversial: "People are so eager not to believe there is a difference in mathematical reasoning ability . . . that all kinds of people are taking pot shots," said *Time* in 1982.[13]

By the late 1980s, however, sociobiological explanations of gender differences increasingly appeared in news articles and commentaries as givens, evoking little discussion or debate. For example, they entered discussions about whether women should be drafted. In 1986, neoconservative economist George Gilder wrote that hard evidence showing that men are more aggressive means that women should not go into combat.[14] Similarly, a letter to the *New York Times* used sociobiology to oppose the drafting of women. The male is "hardwired through evolution to defend the tribe. . . . A woman can have 10 to 15 surviving babies in her reproductive lifetime," so survival of societies has depended on the fact that males rather than females are predestined biologically to go to war.[15] And *Time,* in 1992, suggested that men have better spatial perception than women because theirs evolved from tracking game.[16] Women, as an evolutionary result of gathering berries, are better on the details: "He can read a map blindfolded, but can he find his socks?" To some readers the metaphor could suggest that husbands should read maps, while their wives, better foragers, pick up their socks.[17]

Popular books and articles on brain research also presume the existence of innate and intrinsic differences between the sexes. In *The Brain: The Last Frontier,* psychologist Dr. Richard Restak, who also wrote the script for the popular TV series called *The Brain,* stated that "these differences are innate, biologically determined, and relatively resistant to change."[18] Girls, he said, are better in language; boys are better in math. They have distinct thought processes and use their brains in different ways. Restak also provided a biological explanation for women's intuition, suggesting that women are predisposed to allow their emotions to influence their thought processes, while men compartmentalize their emotions and thoughts and are therefore better able to think along purely rational lines.

Purported differences in the brain were also the focus of Ann Moir and

David Jessel's *Brain Sex: The Real Differences between Men and Women.* They popularized research on nonhuman primates, concluding that we should accept differences and that they should be built into early childhood education.[19] An article in *Mirabella* magazine describing this book showed a photo of a beautiful naked woman. The writer asked: "Are women genetically programmed to enjoy pleasure? A new book . . . argues, yes. . . . Women hear better, are more tactile, even see better in the dark. . . ."[20] The women's magazine *Elle* described these books on the brain in an article called "Mind over Gender,"[21] asserting that: "It just isn't the case that child rearing practices are the only influence on how we grow up. . . . The primary determinant of what and who we will turn out to be is to be found in our bodies. In other words, the old nature-versus-nurture quibble has finally been settled—and nature now appears to be the winner."

The preoccupation with gender differences sometimes becomes a source of humor and irony, reflecting the tensions that pervade the subject of male and female roles in contemporary society. A social critic, Maureen Dowd, suggests that "women are genetically engineered to turn into bitches at a certain age." If you are successful and stand out or speak your mind, you become a bitch—"you stop crusading to please guys at work or at home."[22] Elsewhere she responds to studies of the relationship between testosterone and aggression, concluding that "men may be biologically unsuited to hold political office and leadership positions."[23] Capturing the tension, a cartoon portrays a man, lost on a country road, who turns in irritation to his nagging wife: "Because my genetic programming prevents me from stopping to ask directions—that's why!"[24]

One important focus of the popular debate is whether women are uniquely suited to raise children. The idea that women are "ordained by their nature to spend time meeting the needs of others" is an old and oft-repeated theme.[25] It was used in the nineteenth century by physicians who disapproved of intellectual or professional interests for women. More recently, it has appeared in the narratives of sociobiology, which posit that women emotionally prefer caring labor to intellectual labor.[26] Sociobiologists have insisted that the bias toward women staying at home has prevailed in most societies because, as Wilson puts it, "the genetic bias is intense enough to cause a substantial division of labor even in the most free and most egalitarian of future societies."[27] Sociobiology and behavioral genetics provide a means to define the spheres assigned to men and women as dictated by nature. And this defining is sometimes used by feminist critics. Camille Paglia, for example, finds a scientific truth in sexual stereotypes: "I can declare that what is female in me comes from nature and not from nurture. . . . The traditional association of assertion and action with masculinity, and receptivity and passivity with femininity, seems to me scientifically justified. . . . Man is contoured for invasion, while woman remains the hidden, a cave of archaic darkness."[28] Some feminists have deployed genetic arguments to celebrate the "creative power that is associated with female biology" and the "natural talent and superiority of women."[29] But most feminists try to validate female identity without perpetuating biologically based stereotypes. For example, Carol Tavris, in her 1991 book, *The Mismeasure of Women,* argued that biological differences are only a small and insignificant variable influencing women's behavior, and that differences between men and women relate to their social context and expectations.[30] Carol Gilligan's *In a Different Voice* located differences in the way that male and female children interpret and explain moral dilemmas more in socialization than in biology.[31] But the message resonated with popular stereotypes that women and men are fundamentally different, and this idea was taken up and embraced by the media, which transformed her account into evidence of women's biological predisposition to caring. *Self* magazine, for example, asserted that women are biologically predisposed to being more sensitive than men to the feelings of others.[32] *Time* wrote about the biological basis of female intuition, suggesting that women naturally excel in interpreting emotions and thus have greater skill in caring for toddlers.[33]

Such biological explanations can be used to argue that independence and professional work, while natural for men, are unnatural for women. They can also be used to promote the idea that all women biologically prefer to devote themselves full-time to caring for their children, and that the social changes linked to feminism have created a feminine identity crisis.

The message that gender differences are biologically determined sometimes appears in more subtle ways. The press describes studies of sperm activity, linking motility to competitiveness and superiority in language that reinforces the idea of biological difference. A *New York Times* science article, for example, assigns human attributes to sperm: "Even from conception, males apparently enjoy an advantage over females. . . . [T]here seems to be discrimination against women even before they are born."[34] *Discover,* reporting on studies of sperm motility, writes of "competitive sperm . . . the basic source of maleness." There are "supersperm" that engage in a "kamikaze mission," in which some barge ahead to "further the success of their brothers." In contrast, eggs are described as passive and receptive. They "send alluring chemical clues"; they are "passive spectators."[35]

At a time when changing opportunities for women are a frequent source of resentment, when women are seen as competing with men for scarce jobs, such biological explanations for social stereotypes appear increasingly acceptable. The 1992 cover article in *Time* captured the changes that have taken place in popular culture when its author observed that the subject of genetic differences was once "taboo," but that "now it is O.K. to admit the possibility."[36] Similarly,

Richard Restak tells stories of being "hooted down" by feminists who objected to hearing him speak on the subject of brain differences, but now he finds that audiences are eager to hear more.[37]

Genes and Race

At the 1990 annual meeting of the American Psychological Association, psychologist Philippe Rushton presented a paper arguing that blacks have smaller brains than whites, and that this explained the differences in their educational performance. Rushton, apparently oblivious to the "elephant problem" (if brain size is all, why are not elephants the most intelligent creatures on earth?) suggests that variation in brain size in different races is a consequence of evolutionary pressures.[38]

There are few subjects more controversial than the identification of genetic differences in intelligence between racial groups. The American ideological commitment to equality as a social value has, since World War II, favored cultural explanations of the differences in IQ test scores or crime rates between races. Environmental explanations offer the hope that social problems can be solved through public action and government policy. This was the theme of a 1989 National Academy of Sciences report called *Common Destiny,* on the status of blacks in American society.[39] But in 1990, social psychologist Richard Herrnstein published a strong critique of *Common Destiny,* in which he called the authors to task for failing to consider the different "average endowments of people in the two races.... We can ignore the iron law of selection but we cannot elude it."[40]

The race IQ controversy appeared in the press as early as 1922, when Walter Lippmann published a series of critiques of the efforts to measure intelligence. He focused his attack on the views of I. Lothrop Stoddard, who claimed, as did Herrnstein some seventy years later, that inequality was an "iron law" of nature. Stoddard argued that intelligence testing demonstrated that people in lower social classes were less intelligent and, therefore, that socialist revolutions aimed at equalizing class differences violated nature.[41]

In 1969, psychologist Arthur Jensen revived the debate over the causes of group differences in performance on IQ tests, focusing particularly on race. In an essay in the *Harvard Educational Review,* "How Much Can We Boost IQ and Scholastic Achievement?"[42] Jensen argued that there was a significant (15-point) average difference in IQ between blacks and whites on the conventional IQ test.[43] He acknowledged that environment played some role in creating this difference, particularly since the scores of poor, white, urban children were similar to those of poor, black, urban children. But he claimed that between one-half to one-fourth of the mean difference in IQ scores—7.5 points to 3.8 points—between blacks and whites could be attributed to genes. Jensen even used the fact that prosperous blacks tend to have lighter skin as evidence of their genetic superiority, rather than of their greater social acceptability in mainstream white culture. He concluded, as he put it in an interview with *Newsweek,* that the school system needed to learn to accommodate "large numbers of children who have limited aptitudes for traditional academic achievement."[44]

The basic arguments in this explosive paper became known, pejoratively, as "Jensenism," a play on "Jansenism," the doctrine of the 17th-century bishop Cornelis Jansen, who was considered a heretic by the Roman Catholic Church because he rejected the doctrine of free will.[45] The term has since come to mean a scandalous doctrine, or one that is immoral, and certainly Jensen's critics interpreted his account as both scandalous and immoral.

Jensen, however, was not alone. In his 1971 analysis of the IQ controversy, Hans Eysenck agreed with Jensen and expressed regrets at the cruelty of nature in dividing human beings into intelligent and unintelligent groups. The biological inequality of blacks, he said, is a tragedy for the moral belief in the equality of men, as opposed to the "scientific fact of hereditary inequality." Eysenck called up the nineteenth-century notion of neoteny (prolonged childhood as a sign of higher evolutionary status) to claim that black children mature faster and are therefore inferior.[46]

Jensen and his supporting chorus received extensive media coverage, partly because of the controversy they provoked in the scientific community. Conservative writers portrayed Jensen as a progressive and authoritative source of knowledge whose studies confirmed common sense. *US News and World Report* warned its readers that the findings about Negro scores on IQ tests must be taken seriously as "evidence of genetic differences."[47] Other magazines took a critical position. *Science News* called Jensen a "pyrotechnic scholar,"[48] and *Newsweek* called his work a "potential social hydrogen bomb."[49]

In an August 1971 essay in *Atlantic Monthly,* Richard Herrnstein made the ambitious claim that intelligence was eighty percent genetic. He argued that, as social programs equalized environmental forces—that is, as all children came to have access to high-quality education, good nutrition, a stable home environment, and cultural opportunities—genetic ability would become even more important in determining levels of achievement. Under such conditions of equal opportunity, inherited ability would be the only variable shaping IQ, success, and social standing.[50] And since blacks were genetically inferior in intelligence, he said, they would suffer the most from a complete meritocracy.

In 1994 Herrnstein reiterated these arguments in his controversial book, with Charles Murray, *The Bell Curve.* They asserted that IQ, genetically determined and differing in different races, explained the state of American society: "Success and failure in the American economy and all that goes with it, are increasingly a matter of the genes that people inherit." Using such genetic arguments to buttress their attack on the welfare state, they argued that welfare and affirmative action policies "subsidize births among poor women at the lower end of the intelligence distribution," causing a general "decline in intelligence in America" and leading inevitably to growing social problems of crime, illegitimacy, and poverty.[51]

Herrnstein blamed the media for the reluctance to recognize the genetic basis of racial differences in IQ. "It takes great pains to get a balanced and reasonably complete account of intelligence testing into the public forum where it might inform public policy."[52] Increasingly defensive, he accused the media of trivializing the research and systematically favoring sociological over genetic explanations.[53]

Similarly, in 1988 media critics Mark Snyderman and Stanley Rothman also accused the media of opposing hereditary views in favor of environmental explanations of group differences.[54] They blamed the liberal "metro-Americans," who feel estranged from American values, yet who influence the public through the media. Yet, by publishing their claims in popular magazines, it was Jensen and Herrnstein who had used the media to establish the veracity of scientific research.

Speculation about the influence of genetics on IQ was further encouraged by the striking success of Asian students at a time of anxiety about America's waning economic clout, as compared to the Japanese. Popular writer and editor at *Fortune,* Daniel Seligman, wrote in 1982 that the Japanese are genetically more intelligent because of "outbreeding" or "heterosis" after World War II. Seligman took issue with explanations that attributed increased IQ to nutrition, health, and postwar environmental improvements. Rather, he wrote, during the war a lot of young men got out of their villages, so that "Japan was producing more intelligent kids."[55]

Another remarkable explanation of Japanese success credited samurai warriors as the real heroes of modern Japan. W. David Kubiak, writing in the *Whole Earth Review* in 1990, said that "the taste for organizational life" is genetic, and, like most genetic traits, it is "randomly distributed." In Japan, independent types who abhorred hierarchy and authoritarianism were decimated by the "joiner" types. As a consequence, the independents' "gene pool slowly began to bleed away." During the Edo period, the samurais killed those deemed dangerous "with the same impunity that a breeder culls his flocks of undesired traits." Thus, some fifteen generations of the population were "genetically pruned" for "assertive and egalitarian DNA." This carefully bred population, Kubiak says, explains the economic success of Japan today.[56]

Economic anxieties have converged with concerns about welfare and the changing ethnic composition of major U.S. cities to encourage racial stereotypes. These appear, often indirectly, in such code words such "welfare mothers," "teenage pregnancy," "inner-city crime," and the "urban underclass." Such labels articulate the resentments that foster biological generalizations. Thus, Marianne Mele Hall, a Reagan administration appointee, could describe African Americans as "conditioned by 10,000 years of selective breeding for personal combat and the anti-work ethic of jungle freedoms."[57]

In his 1991 column in *Newsweek,* George Will articulated similar assumptions about the biological basis of violence among blacks. Will was reviewing a speech that James Q. Wilson, author of *Crime and Human Nature,* had presented at the American Political Science Association. Seldom are speeches to professional associations considered newsworthy, but Wilson had explained the increasingly visible problems of the underclass in terms of the fundamental difficulties in socializing certain males. He argued that men are innately uncivilized, reckless, and assaultive: "Nature blundered badly in designing males." Socialization, however, has successfully constrained biology, turning wild males into gentlemen. Today, however, "two epochal events" have changed this picture: "the great migration of Southern rural blacks to Northern cities and the creation of a welfare state that made survival not dependent on work or charity." These events, he argued, countered the constraining effects of civilization, creating the historical conditions for a warrior class.[58]

These assumptions also appeared to underlie the Bush administration's "violence prevention initiative," as its director, Dr. Frederick Goodwin, described it in a 1992 speech to the American Psychiatric Association. This NIH program included research, mainly on animals, that was intended to identify those people who may be prone to violent behavior. Goodwin described primate studies indicating that violence is a "natural way for males to knock each other off." This led him to speculate that the loss of social structure in American society, particularly within inner-city areas, has removed some of the "civilizing evolutionary things that we have built up. . . . Maybe it isn't just the careless use of the word when people call certain areas of certain cities 'jungles', that we may have gone back to what might be more natural, without all of the social controls that we have imposed upon ourselves."[59] The speech evoked a strong reaction from the Black Congressional Caucus, who called it racist, and Goodwin later apologized for the comment, claiming to be unaware that it was offensive.[60]

Genetic language appears in a more benign, but no less stereotyped, form in the context of sports. Sportscasters have been cautious ever since CBS commentator Jimmy the Greek was reprimanded for observing, on the air, that blacks are better athletes than whites because they were bred that way. Their bigger thighs, he said, go back to slave trading days, "when the slave owner would breed his big black to his big woman so that he would have a big black kid."[61] Today, most references to genetic characteristics are oblique, as when black atheletes are described as having natural grace or gifted bodies. For example, Ken Griffey, outfielder for the Seattle Mariners, has "genetic talent," according to *Sports Illustrated*.[62] Black ballplayer Reggie Jackson expressed his cynical perspective about such images, as he explained why blacks are seldom starting pitchers: The subtle message, he said, is that "we have genetic talent, but we're just not intelligent."[63] But the message from Al Campanis, former general manager of the L.A. Dodgers, was far from subtle. Offering his view on the lack of minorities in upper management in sports, he said: "Blacks lack the necessities to take part in the organizational aspects to run a sports team."[64]

In the 1990s, race theorists are more and more willing to publicly express their views about genetic differences between blacks and whites, and to suggest their significance for social policy. Anthropologist Vince Sarich at Berkeley lectures on the genetic basis of racial differences to hundreds of undergraduates.[65] Michael Levin of City College in New York has argued that blacks are less intelligent than whites and has used his theories to oppose affirmative action. He has asserted that differences in average test scores are self-evident proof of genetic differences—as though the SAT provided direct access to DNA. And he has placed the burden of proof on those who would deny genetic differences: "To show that no average race differences in intelligence exist one would need to show there is no systematic performance gap on standard tests . . . or that these tests predict black performance less accurately than white performance." He attributes differences to genetics, believing it amply confirmed over the last several decades that, on average, blacks are significantly less intelligent than whites.[66] At the same time, Leonard Jeffries, an African American teacher at City College, argues that the skin pigment melanin affords blacks physical and intellectual advantages over whites. Distinguishing "sun people" (blacks) from "ice people" (whites), he is interested in stirring up collective identity on the basis of a biological construction of "race."[67]

He is not alone in this goal. Just as some feminists use biological explanations to affirm the special characteristics of women, many African Americans want to affirm the essential characteristics that define their unique identity, while rejecting stereotypes imposed by others. Purity of blood is a frequent theme in literature dealing with racial difference in American society. Dep, the hero of Spike Lee's 1990 film, "Mo Better Blues," chanted during a game of "playing the dozens": "No White Blood in me. My stock is 100% pure." His girlfriend replied: "Master was in your ancestor's slave tent just like everyone else."[68] Novelist Toni Morrison, in her 1981 book *Tar Baby,* wrote about the plight of Jadine, a "yalla," suggesting that having large amounts of white blood creates a biological tug of war in the psyche.[69] Genetic images appeal to these writers as a way of resisting cultural imperialism and establishing collective identity on the basis of shared identification with a common ethnic heritage.

Ideas about the biological inferiority of whites also appear in this literature. Exploring racial differences, Morrison presents one perspective in her 1977 *Song of Solomon.* The character Guitar says that "white people are unnatural. As a race they are unnatural. They have a biological predisposition to violence against people of color. The disease they have is in their blood, in the structure of their chromosomes."[70] The character's comments are reflective of the claims of some leaders of the Afrocentric movement. Michael Bradley's 1991 book *The Iceman Inheritance: Prehistoric Sources of Western Man's Racism, Sexism, and Aggression*—popular with Afrocentrists—argued that white men are brutish because they are descended from Neanderthals.[71]

Afrocentrists are effectively attempting to transform their differences into positive biological strengths. But they share an assumption with racist critics: that race is a biological reality with some meaning for this debate. As this rhetoric demonstrates, the biological construction of race has rhetorical value for anyone negotiating the social place of racial groups, either as inferior or superior. Indeed, the fact that African Americans, contesting the mainstream assessment of blacks, also choose to embrace genetic images suggests their shrewd assessment of the discursive power of explanations that establish an apparently scientific basis for human differences.

Conclusion

In the public debates over human differences—over, for example, the meanings of gender and race—genetic images are strategically employed in an effort to delineate boundaries, justify rights, or legitimate inequalities. Genetic causation can be used by those who believe education will make no difference in the social status of black Americans; by those who promote equality of the sexes; and by those who oppose equality in general. Genes can be understood in this debate as rhetorical devices that can be utilized in many different ways. Genes seem to cause the bodily differences that matter most in this cultural debate. But bodily difference is historically specific, written not in the body but in the culture that defines what aspects of the body are most important when one begins to sort people into groups.

Gender and racial stereotypes, purportedly rooted in the chromosomes, both reflect and perpetuate popular attitudes. A 1990 nationwide survey of 1,372 adults found that most white people continue to interpret the behavior of African Americans through stereotypes built upon particular interpretations of biology. Three out of four believe that blacks prefer to live on welfare, and most believe that blacks are more likely to be lazy, violent, and less intelligent than whites.[72]

Scientific claims become a way to reinforce such stereotypes. While science is a form of cultural knowledge, it is often seen to represent a natural reality, an unbiased, objective approximation of truth. Thus, to say that "everything is assembled according to instructions in its chromosomes" seems more acceptable than saying "blacks are less intelligent," or "women cannot do math." But explanations based on "natural" or inherent abilities serve the same social purposes. They place people in desired contexts—women in the home, blacks in sports—and exclude them from other contexts—mathematics departments or managerial positions. They are, in effect, a way to construct the body in ways that will legitimate existing social categories.

Notes

1. Edward Dolnick, "Superwomen," *Health*, July/August 1991, pp. 42–48.
2. Christine Gorman, "Sizing Up the Sexes," *Time*, January 20, 1992, pp. 42–45.
3. Ibid.
4. Daniel Gelman, "Just How the Sexes Differ," *Newsweek*, May 18, 1981, pp. 72–78.
5. "The New Woman, 1972," *Time*, March 20, 1972, p. 25.
6. Tim Hacklin, "Is Anatomy Destiny?" *Cosmopolitan*, March 1982. See discussion in Dorothy Nelkin, *Selling Science* rev. ed. (New York: W. H. Freeman, 1995).
7. "Sociobiology and Sex," *Time*, August 1, 1977, p. 63.
8. Allan R. Andrews, "Studies Find Differences in Male, Female Brains," *Boston Globe*, June 19, 1980.
9. Camille Benbow and Julian Stanley, "Sex Differences in Mathematical Reasoning: Fact or Artifact?" *Science* 210 (December 12, 1980): 1262–64.
10. "Are Boys Better at Math?" *New York Times*, December 7, 1990.
11. "The Gender Factor in Math," *Time*, December 15, 1980.
12. Pamela Weintraub, "The Brain: His and Hers," *Discover*, April 1981, pp. 15–20.
13. "Who Is Really Better at Math?" *Time*, March 22, 1982, p. 64.
14. George Gilder, "The Case against Women in Combat," *New York Times Magazine*, January 28, 1989, p. 44.
15. *New York Times*, March 2, 1990. Letter to the editor from Jesse D. Sheinwald.

16. Gorman, "Sizing Up the Sexes," p. 45. Also Sandra Blakeslee, "Why Don't Men Ask Directions? They Don't Feel Lost," *New York Times,* May 26, 1992.

17. I acknowledge Betsy Hanson for this observation. The man-the-hunter, women-the-gatherer thesis has been rigorously critiqued by anthropologists. See, for example, Donna Haraway, *Primate Visions: Gender, Race, and Nature in the World of Modern Science* (New York: Routledge, 1989).

18. Richard Restak, *The Brain: The Last Frontier* (Garden City: Doubleday, 1979), p. 9.

19. Anne Moir and David Jessel, *Brain Sex: The Real Difference between Men and Women* (New York: Carol, 1991).

20. Advertisement, *Mirabella,* April 1991, p. 135.

21. Meme Black and Jed Springarn, "Mind over Gender," *Elle,* March 1992, pp. 158–62.

22. Maureen Dowd, "The Bitch Factor," *Working Woman,* June 1991, pp. 78+.

23. Maureen Dowd, "When Men Get a Case of the Vapors," *New York Times,* June 30, 1991.

24. *New Yorker,* September 23, 1991.

25. Martha T. Mednick, "On the Politics of Psychological Contracts: Stop the Band Wagon, I Want to Get Off," *American Psychologist* 44 (1989). 1118–23.

26. Anne Fausto-Sterling, *The Myths of Gender* (New York: Basic Books, 1985), p. 4.

27. E. O. Wilson, "Human Decency Is Animal," *New York Times Magazine,* October 12, 1975, pp. 39+.

28. Camille Paglia, *Sex, Art, and American Culture* (New York: Vintage, 1992), p. 108.

29. See, for example, Alison Jaggar, *Feminist Politics and Human Nature* (Sussex: Harvester Press, 1983).

30. Carol Tavris, *The Mismeasure of Woman* (New York: Simon and Schuster, 1992).

31. Carol Gilligan, *In a Different Voice* (Cambridge: Harvard University Press, 1982).

32. Madeline Chinnici, "Are Moods Contagious?" *Self,* December 1990, p. 58.

33. Gorman, "Sizing Up the Sexes," p. 42

34. Natalie Angier, "In Sperm, Men Have Prenatal Advantage," *New York Times,* October 30, 1990, p. C3.

35. Meredith Small, "Sperm," *Discover,* July 1991, pp. 48–53.

36. Gorman, "Sizing Up the Sexes."

37. Quoted in Black and Springarn, "Mind over Gender," p. 162.

38. Stephen J. Gould, *The Mismeasure of Man* (New York: Norton, 1981). Gould explores early efforts to measure brain size, either by weighing brains after death or by filling skulls with lead shot. He shows that attempts to find any measure showing brain size to be a crucial variable failed. If brain size is corrected for body size, women have larger brains than men. Some criminals have large brains, and some esteemed men of French science were found to have had small brains.

39. Gerald M. Jaynes and Robin Williams, Jr. (eds.), *A Common Destiny: Blacks and American Society* (Washington, D.C.: National Academy of Sciences Press, 1989).

40. R. J. Herrnstein, "Still an American Dilemma," *The Public Interest,* no. 98 (Winter 1990), pp. 3–17.

41. See Walter Lippman's articles in *The New Republic,* 1922 through 1923. As cited by David Livesay, *The Race/I.Q. Controversy in the Popular Press,* paper, Cornell University, May 6, 1983.

42. Arthur Jensen, "How Much Can We Boost IQ and Scholastic Achievement?" *Harvard Educational Review* 39 (Winter 1969), pp. 1–123.

43. In the *Harvard Educational Review,* a student-edited publication reaching about 12,000 readers. Ibid.

44. Hans Eysenck, *The IQ Argument: Race, Intelligence, and Education* (New York: Library Press, 1971).

45. It is not irrelevant that Jansen also believed in predestination—and he claimed that Christ's death redeemed only a part of humanity, not all of it.

46. Eysenck, *The IQ Argument.*

47. "Can Negroes Learn the Way Whites Do?" *U.S. News and World Report,* March 10, 1969, pp. 48–51.

48. "Genetics vs. Headstart," *Science News* 95 (April 1969): 326–27.

49. "Is Intelligence Racial?" *Newsweek,* May 10, 1971, p. 69.

50. Richard Herrnstein, "IQ," *Atlantic Monthly,* September 1971, pp. 43–58.

51. Richard Herrnstein and Charles Murray, *The Bell Curve* (New York: Free Press, 1994).

52. Richard Herrnstein, "In Defense of Intelligence Tests," *Commentary,* February 1982, pp. 40–50.

53. R. J. Herrnstein, "IQ Testing and The Media," *The Atlantic Monthly,* August 1982, pp. 68–74.

54. Mark Snyderman and Stanley Rothman, *The IQ Controversy, The Media and Public Policy* (New Brunswick, N.J.: Transaction Books, 1988).

55. Daniel Seligman, "The Drama Backstage: How Japan Got Smarter," *Fortune,* July 12, 1982, pp. 37–40.

56. W. David Kubiak, "E Pluribus Yamato—The Culture of Corporate Beings," *Whole Earth Review,* Winter 1990, pp. 4–10.

57. Cited in Micaela di Leonardo, "White Lies, Black Myths," *Village Voice,* September 22, 1992, p. 31.

58. George F. Will, "Nature and the Male Sex," *Newsweek,* June 17, 1991, p. 70.

59. Fredrick Goodwin, "Conduct Disorder as a Precursor to Adult Violence," *American Psychiatric Association Annual Meeting,* May 1989, p. 9.

60. Warren Leary, "Struggle Continues over Remarks by Mental Health Officials," *New York Times,* March 8, 1992.

61. Sarah Ballard, "An Oddsmaker's ODD Views," *Sports Illustrated,* January 25, 1988, p. 7.

62. Tim Kurkjian, "West A1; In Baseball's Best Division, Kansas City and California Are Still Trying to Catch Oakland," *Sports Illustrated,* April 16, 1990, p. 66.

63. Reggie Jackson, "We Have a Serious Problem That Isn't Going Away," *Sports Illustrated,* May 11, 1987, p. 40.

64. *Nightline,* ABC, April 6, 1987.

65. Paul Selvin, "The Raging Bull at Berkeley," *Science* 251 (January 25, 1991): 368–71.

66. Michael Levin, "Affirmative Action Bears Great Moral Costs" (letter to the editor), *New York Times,* June 9, 1990.

67. James Traub, "Professor Whiff," *Village Voice,* October 1, 1991.

68. Spike Lee, *Mo' Better Blues* (Universal City: Universal Pictures, 1990).

69. Toni Morrison, *Tar Baby* (New York: New American Library, 1981).

70. Toni Morrison, *Song of Solomon* (New York: New American Library, 1977), p. 155.

71. Michael Bradley, *Iceman Inheritance: Prehistoric Sources of Western Man's Racism, Sexism and Aggression* (New York: Kayode, 1991).

72. "Survey Finds Whites Retain Stereotypes of Minority Groups," *New York Times,* January 10, 1991.

Notes on Contributors

Anne Fausto-Sterling is Professor of Medical Science at Brown University. The author of *Myths of Gender: Biological Theories about Women and Men,* she is writing a book on biology and the social construction of sexuality.

Joseph Grigely's recent exhibitions include *Body Signs: Deviance, Difference, and Eugenics* at Washington Project for the Arts, *Conversations with the Hearing* at White Columns, New York, and installations at the 1995 Venice Biennale and the Musée d'Art Moderne de la Ville de Paris/ARC. His publications include *Textualterity: Art, Theory, and Textual Criticism.*

Carol Groneman is Professor of History and Chair of the interdisciplinary Thematic Studies Department at John Jay College of Criminal Justice, City University of New York. She writes on working-class and immigrant women and is the associate editor for the section on immigration in *Encyclopedia of New York City.* She is working on a study of the history of nymphomania in Western culture from the late eighteenth century to the present.

Lisa Handwerker holds a Ph.D. in anthropology from the University of California at Berkeley and an M.P.A. from the University of California at San Francisco. Her areas of research include reproductive health and technologies, bioethics, infertility, and gender studies, with a special focus on China.

David G. Horn is Assistant Professor of Comparative Studies at Ohio State University and the author of *Social Bodies: Science, Reproduction and Italian Modernity.*

Janice M. Irvine is Assistant Professor in the Department of Sociology at the University of Massachusetts, Amherst. She is the author of *Disorders of Desire: Sex and Gender in Modern American Sexology* and coeditor of *Sexual Cultures and the Construction of Adolescent Identities.*

Susan Jahoda is Associate Professor in the Art Department at the University of Massachusetts, Amherst, and co-art editor of the journal *Rethinking Marxism.*

Her work has been exhibited in both Europe and the United States and has appeared in journals such as *Afterimage, Heresies, Arts Magazine,* and *Creatis.*

M. Susan Lindee is Assistant Professor of the History and Sociology of Science at the University of Pennsylvania and author of *Suffering Made Real: American Science and the Survivors at Hiroshima.*

K. Tsianina Lomawaima (Creek) is Assistant Professor of American Indian Studies and Anthropology at the University of Washington. She is the author of *They Called It Prairie Light: The Story of Chilocco Indian School.*

Nicholas Mirzoeff is Assistant Professor of Art History at the University of Wisconsin, Madison. His essay included in this volume is derived from his book *Silent Poetry: Deafness, Sign and Visual Arts 1750–1920.*

Dorothy Nelkin is University Professor at New York University. She writes about the intersection of science and culture and is the author of *Selling Science: How the Press Covers Science and Technology* and coauthor of *Dangerous Diagnostics: The Social Power of Biological Information.* She is an elected member of the National Academy of Sciences Institute of Medicine.

Cindy Patton, Assistant Professor of Rhetoric and Communications at Temple University, has worked with the AIDS Action Committee, the World Health Organization's Global Program on AIDS, and the Centers for Disease Control on numerous HIV-related projects. She is the author of *Sex and Germs: The Politics of AIDS* and *Inventing AIDS,* and is coauthor of *Making It: A Woman's Guide to Sex in the Age of AIDS.*

Robert N. Proctor is Professor of the History of Science at Pennsylvania State University. He is the author of *Racial Hygiene: Medicine under the Nazis* and *Value-Free Science? Purity and Power in Modern Science.*

Alan C. Swedlund is Professor and Chair in the Department of Anthropology, University of Massachusetts, Amherst. His major area of research, historical epidemiology of nineteenth-century New England, has precipitated his interest in the development of thought in biological anthropology and medicine during the Victorian and Progressive eras.

Jennifer Terry, Assistant Professor of Comparative Studies at Ohio State University, has written articles on queer theory, women and medical surveillance, and the history of sexual science in the United States. She is at work on a book entitled *Siting Homosexuality: A History of Surveillance and the Scientific Production of Deviant Subjects.*

Rachel J. Tolen is a doctoral candidate in anthropology at the University of Pennsylvania. In addition to performing historical research, she carried out fieldwork in a railway colony in Madras, India. Her dissertation looks at issues of class, culture, lifestyle, and labor through an exploration of relations between the households of officers of the Indian Railways and the servant households that work for them.

Jacqueline Urla is Assistant Professor of Anthropology at the University of Massachusetts, Amherst. She does research on language minorities, social movements, alternative media, and cultural politics. She is working on a collaborative research project exploring the representation of whiteness in native peoples' art, material culture, and visual media.

Name Index

Abel, Theodor, 175
Abraham, Karl, 242
Acton, William, 240, 242
Adams, David Wallace, 201, 213, 214
Adams, Kathleen J., 21, 22, 48
Adelson, Andrea, 278, 309
Ahern, Wilbert H., 213, 214
Aherns, A., 298, 313
Aird, John S., 361, 382
Aiyappan, A., 95, 103
Albrecht, Paul, 123
Alcott, William, 241, 242
Alexander, Elizabeth, 43
Allen, Laura, 169
Allen, Woody, 323
Alloulah, Malek, 59, 75
Altick, Richard D., 20, 28, 32, 43, 44
Aly, Gotz, 193, 195
Ames, M. Ashley, 168
Anagnost, Ann, 362, 363, 381, 382
Anderson, Carole G., 334
Anderson, James D., 201, 214
Andrews, Allan R., 400
Angier, Natalie, 169, 401
Ankey, C. Davison, 41, 45
Antonio, Gene, 356
Apfelbaum, Bernard, 336
Appadurai, Arjun, 79, 84, 88, 89, 103
Appel, Toby A., 23–26, 42, 44
Aristotle, 116
Arnold, David, 79, 102, 103, 213, 215
Arnold, Thomas, 66, 76
Astruc, Jean, 227, 242
Athanassoglou-Kallmyer, Nina, 75
Athey, Jean L., 356
Augustine, Saint, 317
Ayyar, K. N. Krishnaswami, 89, 104

Bailey, F. G., 82, 104
Bailey, J. Michael, 169
Baillie, Mathieu, 37, 44
Ballard, Sarah, 402
Balsamo, Anne, 301, 305, 306, 309
Banister, Judith, 361, 362, 381, 382
Banner, Lois, 308, 309
Barker-Benfield, Graham, 240, 242
Barman, Jean, 212, 213, 215
Barnett, S. A., 101, 104
Barnum, P. T., 30
Baron, Naomi S., 75
Bartky, Sandra Lee, 300, 310
Bartman, Saartje. *See* Sarah Bartmann
Bartmann, Sarah, 8, 19, 20, 23, 26–36, 38–41
Batterberry, Ariane, 43, 44
Batterberry, Michael, 43, 44
Battey, Robert, 230, 231, 240, 241, 249
Beam, Lura, 166, 167
Bebian, Roch-Ambroise Auguste, 63, 75
Becker, Arnie, 327
Becker, Brent A., 41, 45
Becker, Rafael, 194
Beecher, Catherine, 201
Begbie, Harold, 91, 104
Behad, Ali, 49, 53, 73
Bell, Alexander Graham, 70, 71, 77
Belskie, Abram, 290
Benbow, Camille, 390, 400
Benishay, D. S., 169
Bennett, William, 289, 310
Bergler, Edmund, 168, 242, 245
Berkhofer, Robert F., Jr., 213, 215
Berkow, Robert, 364, 382
Bernal, Martin, 43, 45
Berthier, Ferdinand, 63, 64, 73, 74, 76
Bertillon, Alphonse, 67, 118, 125
Berube, Allan, 168
Bhabha, Homi, 64, 76
Bieber, Irving, 168
Bieneck, Edeltraut, 176, 194
Bienville, M. D. T., 227, 228, 240, 242
Bigelow, H. J., 241, 243
BillyBoy, 278, 282, 304, 307, 310
Binding, Karl, 171
Binny, John, 81, 107
Bittner, Marie, 44
Black, Meme, 401
Black-Perkins, Kitty, 285
Blackburn, Stuart H., 84, 104
Blackwell, Benjamin, 90, 104
Blades, Ruben, 354

Blakeslee, Sandra, 401
Blakey, Michael, 287, 289, 307, 310
Blanckaert, Claude, 75, 123, 125
Blankenburg, Werner, 182, 183
Bloch, Iwan, 241, 243
Bloch, Marc, 76
Bloch, Ruth, 226, 243
Block, A. J., 236, 241, 243
Blome, Kurt, 191
Blumenbach, Johann Friedrich, 37
Boas, Franz, 287, 288, 289
Boeters, Gustav, 171
Bonaparte, Napoleon, 25, 39, 65
Bongaarts, J., 361, 383
Booth, Bramwell, 93, 95
Booth, Catherine E., 90, 215
Booth, Emma, 91
Booth, William, 90
Booth-Tucker, Frederick, 91–98, 104
Bordo, Susan R., 277, 298, 300, 301, 305, 310, 331, 336
Bormann, W., 181, 195
Bouchard, P.-L., 70, 71
Bouhler, Philipp, 170, 171, 182
Bourdieu, Pierre, 78, 94, 103, 104, 202, 213, 215
Boyer, Cherrie B., 338, 356
Boyle, Harry, 59
Brack, Victor, 182–84, 195
Bradley, Michael, 399, 402
Braille, Louis, 66
Brandt, Karl, 184, 185
Brandt, Pam, 343, 357
Breckenridge, Carol A., 79, 88, 104, 307, 310
Brill, A., 136, 164
Broca, Paul, 68, 69
Bromley, David G., 352, 356
Brown, Isaac Baker, 229, 230, 243, 268, 270, 271, 276
Brownworth, Victoria A., 342, 346, 349, 356
Bruegger, Karl, 166
Brumberg, Joan Jacobs, 289, 310, 331, 336
Brumble, H. David, 212, 215
Bruno, Mary, 343, 356
Bryant, Louise Steven, 166
Bucher, Bernadette, 21, 45
Bucknill, John Charles, 232, 240, 243
Bullough, Vern, 242, 243
Burkhalter, Evelyn, 282
Burleigh, Michael, 171
Burr, Chandler, 169
Bush, George, 2, 397
Butler, Ben, 44
Byne, William, 169

Cadet de Gassicourt, Charles Louis, 124
Campanis, Al, 398
Camper, Pieter, 34
Canguilhem, Georges, 54, 55, 74, 123, 126
Cannady, Beverly, 283
Caprio, Frank S., 168
Carnes, Patrick, 317, 321, 335, 336
Carpenter, Edward, 134, 164
Carpenter, Karen, 302
Carpenter, Minnie Lindsay, 91, 104
Carrel, Alexis, 171
Cezar, Peter, 28, 29
Chancey, George, 164, 235, 241, 243
Chao, Jingshen, 373, 382
Chapman, J. Milne, 233, 234, 240, 241, 243
Charcot, Jean-Martin, 163
Chen, Huihe, 366, 382
Chernin, Kim, 301, 310
Chinnici, Madeline, 401
Chorover, S., 336
Chow, Rey, 363, 382
Chunn, William Pawson, 230, 239, 241, 243
Church, Archibald, 241, 243
Church, George, 342, 356
Churchill, Fleetwood, 240, 243, 256, 276
Clouston, T. S., 239, 240, 243
Cogan, Frances B., 308, 310
Cohen, Myron S., 362, 384
Cohn, Bernard S., 78, 79, 81, 91, 104
Coleman, William, 24, 26, 27, 45
Comaroff, Jean, 78, 96, 103, 105, 213, 215
Comaroff, John, 103, 105
Comte, Auguste, 122
Condillac, Etienne Bonnot de, 54, 56, 65, 71
Conrad, Peter, 336
Copley, Antony, 226, 243
Cott, Nancy, 226, 243, 327, 336
Crary, Jonathan, 3, 17
Crawley, Lawrence Q., 274, 276
Cressey, Paul Frederick, 83, 105
Crinis, Max de, 193
Croll, Elisabeth, 361, 381, 383
Cullen, William, 240
Cushing, E. W., 230, 241, 243
Cuvier, Frederic, 27, 32, 45
Cuvier, Georges, 8, 19, 20, 23–29, 32–45, 125
Cuxac, Christian, 74, 75, 77

Daniel, E. Valentine, 78, 103, 105
Dannemeyer, William, 162
Darwin, Charles, 110, 117
Davin, Delia, 360, 383
da Vinci, Leonardo, 143
Davis, Kathy, 301, 310
Dawkins, Richard, 390
De Blainville, Henri, 8, 23–25, 28, 32–36, 40–41, 43, 45
De Bry, Theodor, 21
Deburau, Jean-Baptiste, 50, 51, 73
De Fougeray, E. Hamon, 76
Degler, Carl, 241, 243, 327, 336
Delacroix, Eugene, 60, 61, 75
de Man, Paul, 74
D'Emilio, John, 168, 240, 243, 334
Deng, Huai Mei, 364, 383
Deng Xiaoping, 362
Derrida, Jacques, 58, 64, 69, 73–76
Desbarrolles, Adolphe, 67, 76
Descartes, M., 52
Deutsch, Albert, 168
Diaz, Kathryn E., 169
Dickinson, Robert Latou, 139, 141, 143–47, 150, 151, 154, 166–67, 290, 308
Diderot, Denis, 71
Diehl, Karl, 196
Diethelm, Oscar, 224, 225, 227, 240, 243
Dietrich, Marlene, 303
Digby, Anne, 226, 243
di Leonardo, Micaela, 402
Dirks, Nicholas B., 84, 105
Doherty, Thomas, 281, 307, 310
d'Ohsson, Ignatius Mouradgea, 61, 75
Dolce, Joe, 169
Dolnick, Edward, 400
Dorner, Gunter, 168
Dover, Harriette Shelton, 212, 215
Dowd, Maureen, 392, 401
Downey, Jennifer, 168
Dublin, Lewis, 289
DuBois, Ellen Carol, 337
Dubos, Rene J., 335
Du Camp, Maxime, 69, 76
duCille, Ann, 307, 310
Duden, Barbara, 360, 361, 383
Duff, John M., 231
Dumont, Louis, 84, 102, 105
Dunn, R., 228, 244
Duprest-Rony, A. P., 241, 244
Duranty, Edmond, 68, 76
Dwyer, Ellen, 240, 244
Dykstra, Bram, 242, 244

Ebersole, Lucinda, 302, 310
Eccles, Audrey, 225, 244
Eckholm, Erik, 349, 350, 356
Eder, Jeanne, 212, 218
Edwards, Milne, 117
Edwards, Paul, 20, 43, 45
Ehrenreich, Barbara, 202, 215
Elkind, David, 338, 341, 356
Ellis, Albert, 240, 244
Ellis, Havelock, 76, 131, 133, 134, 136, 137, 143, 163, 164, 166, 167, 241, 244
Ellis, Lee, 168
Elphick, Richard, 21, 22, 45
Engelstein, Laura, 167
Englander, Martin, 174, 193
English, Deirdre, 202, 215
English, Donald E., 76
Epee, Abbe Charles-Michel de l', 53, 54, 62, 70, 73, 74
Epstein, Barbara Leslie, 211, 215
Esquirol, J. E. D., 241, 244
Ewald, François, 110, 126
Eysenck, Hans, 395, 401, 402

Fausto-Sterling, Anne, 5, 8, 41, 45, 123, 125, 169, 287, 401
Fee, Elizabeth, 114, 126, 287, 311, 334
Feher, Michel, 3, 17
Fellman, Clair, 241, 244
Fellman, Michael, 241, 244
Fenichel, Otto, 242, 244
Fere, Charles, 241, 244
Ferenczi, Sandor, 136, 164
Ferrand, Jacques, 224, 225, 242, 244
Ferrero, Guglielmo, 112, 114–21, 125, 127, 163, 166, 235, 246
Ferri, Enrico, 110, 111, 114, 126, 127
Ferrier, David, 240, 244
Ferris, G. N., 235, 248
Figlio, Karl M., 45
Fioretti, Guilio, 127
Fischberg, Maurice, 193
Fischer, Eugen, 175, 185, 186
Flaubert, Gustave, 69, 76
Flaws, Bob, 367–69, 383
Fletcher, Robert, 111, 123, 124, 126
Flourens, P., 23, 26, 45
Flower, W. H., 44, 45

Foreest, Pieter van, 224
Forel, A., 241, 244
Foreman, Judy, 335
Forster, Ian, 335
Foucault, Michel, 3, 9, 17, 50, 63, 73, 75, 98, 101, 105, 110, 111, 122, 126, 197, 198, 207, 213, 216, 240, 244, 293, 314, 315, 333, 334, 336, 363, 383
France, Kim, 308, 311
Francez, J. P., 239, 244
Francis, W., 80, 83, 85, 86, 88, 105
Frank, Adolphe, 69, 76
Frederick, Carlton, 235, 244
Frederickson, Vera Mae, 212, 216
Freedman, Estelle, 237, 240, 242–44, 334
Freitag, Sandria, 79, 89, 105
Freud, Anna, 344, 356
Freud, Sigmund, 135–37, 163–65, 238, 241, 242, 244
Friedlander, Henry, 195
Friedli, Lynne, 226, 244
Friedman, Richard C., 164
Frith, Simon, 280, 311
Fumento, Michael, 169
Furth, Charlotte, 367, 368, 369, 383

Gabor, Zsa Zsa, 302
Gaillard, Henri, 57
Gaitskill, Deborah, 203, 216
Galassi, Peter, 302
Galbraith, Hugh, 165
Galen, 224
Gall, Franz Joseph, 34, 44, 67, 68
Gallagher, Catherine, 3, 17
Gallaudet, Thomas, 55, 74
Galton, Francis, 68, 76, 118, 119
Gandhi, Mohandas K., 79, 105
Garland, Judy, 302
Garofalo, Raffaele, 111, 123, 127
Gassendi, M., 52
Gates, Henry Louis, 75
Gautier, Theophile, 61, 73
Gebhard, Bruno, 291, 293
Geertz, Ali, 346
Gelman, Daniel, 400
Geoffroy Saint-Hilaire, Etienne, 25, 27, 29, 32, 39, 43, 45
Gernado, Joseph Marie de, 62–63, 85
Giago, Tim A., Jr., 212, 216
Gilder, George, 391, 400
Gilligan, Carol, 393, 401
Gilman, Sander, 20, 31, 37, 45, 109, 126, 173, 193, 229, 240, 241, 244
Ginzburg, Carlo, 76
Goebbels, Joseph Paul, 177
Goffman, Erving, 94
Goldberg, Ann E., 239, 240, 244
Goldin, Nan, 305, 311
Goldman, Merle, 358, 385
Goldmann, Franz, 175, 194
Goldschmidt, Richard, 187, 196
Goleman, Daniel, 334, 355
Goodwin, Frederick, 397, 402
Gordon, Linda, 337
Gordon, Stewart N., 103, 105
Gorman, Christine, 400, 401
Gorren, L. J., 169
Gorski, Roger, 169
Gould, Robert E., 169
Gould, Stephen Jay, 7, 17, 20, 31, 34, 43, 46, 123, 126, 287, 311, 401
Gould, W. A., 171
Goulemot, Jean Marie, 240, 245
Graham, John, 34, 46
Grati, F. G. de, 71
Gray, J., 298, 313
Gray, Stephen, 20, 46
Green, David, 9, 17, 118, 119, 126
Green, Richard, 164
Greenberg, David F., 241, 245
Greenhalgh, Susan, 361, 362, 381, 383
Griffey, Ken, 398
Griffiths, P. D., 168
Groneman, Carol, 5, 11
Gros, Antoine Jean, 65
Gross, Jane, 343, 349–51, 356
Gross, Walter, 173, 186, 193, 195
Grotjahn, Alfred, 190
Guenther, Mathias Georg, 22, 46
Guha, Ranajit, 79, 106
Guille, Sebastien, 76
Gunther, Hans F. K., 185, 194
Gurin, Joel, 289, 310
Gurtner, Franz, 180

Hacking, Ian, 122, 126
Hacklin, Tim, 400
Hafner, Karl-Heinz, 193
Hagen, Wilhelm, 178, 194
Hahn, Susanne, 196
Haig-Brown, Celia, 213, 216
Haikerwal, Bejoy Shanker, 82, 106

Halberstam, Judith, 309, 311
Hall, Diana Long, 167
Hall, Marianne Mele, 397
Hall, Rufus, 231
Hamburg, David A., 352, 356
Hamer, Dean H., 169
Hamilton, F. H., 241, 245
Hamilton, J. A. G., 241, 245
Hammond, William, 241, 245
Handler, Ruth, 279, 281, 309
Handwerker, Lisa, 5, 13, 14, 362, 379, 381–83
Hansen, Soren, 348, 356
Haraway, Donna, 3, 17, 22, 30, 46, 109, 126, 401
Hare, E. H., 241, 245
Harkness, Carla, 364, 384
Hatch, W. J., 86, 88, 106
Hauser, Stuart T., 348, 357
Hauy, Valentin, 66
Hawker, Robert, 74
Haynes, Todd, 302
Hebdige, Dick, 348, 357
Hebert, Yvonne, 213, 215
Hefelmann, Hans, 172, 182
Hegar, Alfred, 230, 231, 240, 241, 245, 249
Hegel, Georg Wilhelm Friedrich, 73
Heinecke, Samuel, 53
Helms, Jesse, 162
Hemingway, F. R., 84, 86, 88, 106
Henderson, Gail E., 362, 380, 384
Henry, George W., 149–52, 154, 165, 167
Herman, Judith Lewis, 337
Herrnstein, Richard, 394–96, 401, 402
Hersch, Patricia, 353, 357
Heyde, Werner, 184
Heydrich, Reinhard, 191
Hierosolimitano, Domenico, 51
Hildebrandt, Wilhelm, 175, 194
Hillary, James Jager, 234, 245
Himmler, Heinrich, 172, 180, 189, 191, 195
Hinchen, Keziah, 336
Hippocrates, 224
Hirschfeld, Magnus, 134, 136, 164, 165, 241, 245
Hitler, Adolf, 170, 171, 176, 182, 185, 192, 196
Hitschmann, Edward, 242, 245
Hoche, Alfred, 171
Hoffbauer, Johanne Christoph, 74
Hoffman, Malvina, 290
Hofman, M. A., 169
Hohmann, Delf Maria, 309, 311
Hohmann, Joachim S., 196
Hooton, Earnest, 139, 166, 287
Hoover, J. Edgar, 280
Hormann, Bernhard, 195
Horn, David, 5, 13, 110, 113, 122–24, 126, 166, 287, 374, 384
Horney, Karen, 242, 245
Howe, Joseph W., 222, 240, 245
Hrdlicka, Ales, 286–88, 295, 307, 308, 311
Hrdy, Sarah Blaffer, 389
Hsiao, William C., 362, 384
Huang, Ti, 384
Hubbard, Ruth, 169
Hudson, Rock, 343
Huhner, Max, 232, 241, 245
Hultgren, Mary Lou, 201, 213, 216
Humphries, Tom, 72, 77
Hunt, Helen, 203
Hunt, James, 63, 75
Hunt, Margaret, 337
Hunter, Nan D., 169
Hunter, Richard, 241, 245

Inden, Ronald B., 78, 79, 106
Irvine, Janice, 5, 15, 334–36
Itard, Jean-Marc-Gaspard, 55, 56, 60, 74

Jackson, Michael, 343
Jackson, Reggie, 398, 402
Jacobus, Mary, 374, 384
Jaggar, Alison, 401
Jahoda, Susan, 5, 12
James, John Angell, 24, 46
Janin, Jules, 73
Jansen, Cornelis, 395
Jaynes, Gerald M., 401
Jeffries, Leonard, 398
Jelgersma, G., 76
Jensen, Arthur, 394–96, 401
Jensen, Jesper, 348, 356
Jessel, David, 392, 401
Jimmy the Greek, 398
John, Elton, 343
Johnson, Lee, 75
Johnson, Virginia E., 35, 47, 168, 315, 316, 323, 334
Johnson, William Henry, 30
Johnston, Basil H., 212, 216
Jones, David, 81, 89, 106
Jones, Lisa, 284, 285, 297, 311

Jong, Erica, 327, 336
Jordanova, Ludmilla, 3, 17, 74, 239, 245

Kahn, Alice, 302, 311
Kamens, W. David, 345, 346
Kane, Penny, 361, 381, 383, 384
Kaplan, Helen Singer, 169, 315, 318–19, 321–23, 335
Kardiner, A., 168
Kasl, Charlotte, 325, 335, 336
Katz, Jonathan Ned, 168
Kaul, Friedrich, 193, 195
Kaw, Eugenia, 277, 311
Keaton, Diane, 323
Kegeles, Susan M., 338, 356
Keller, Evelyn Fox, 374, 384
Keltin, Kristin, 196
Kelves, Daniel, 166
Kennedy, Edward M., 197, 216
Kennedy, Foster, 171
Kennedy, Hubert C., 164
Kiernan, James G., 163, 222, 241, 245
Kilner, Linda A., 348, 357
King, A. Richard, 213, 216
Kinsey, Alfred, 130, 138, 154–60, 163, 166–68, 323
Kirby, Percival R., 20, 28, 29, 31, 46
Klare, Kurt, 196
Klee, Ernst, 193
Koch, Gerhard, 184, 193
Koch, Robert, 190
Koester, F., 190, 196
Kolata, Gina, 345, 346, 357
Kovaleskaya, Sonya, 44
Krafft-Ebing, Richard von, 131–37, 163–65, 221, 222, 232, 233, 237, 241, 245
Kretschmer, Ernst, 190
Kroger, William S., 169
Kroker, Arthur, 3, 17
Kroker, Marilouise, 3, 17
Kruger, Barbara, 306
Kruks, Gabe, 348
Krushchev, Nikita, 279
Kubiak, W. David, 396, 397, 402
Kubie, L., 168
Kurkjian, Tim, 402
Kwinter, Sanford, 3, 17

Lacan, Jacques, 58, 74
Lacassgne, Alexandre, 124
LaFlesche, Francis, 212, 216
Lamacz, Margaret, 335
Lamartine, Alphonse, 75
Lampton, David M., 358, 384
Lane, Harlan, 74, 77
Lang, Theobald, 187, 193, 196
Lange, Johannes, 179, 194, 196
Langone, John, 169
Lankester, E. Ray, 163
Laqueur, Thomas, 3, 17, 44, 46, 226, 239, 246
Latour, Bruno, 23, 24, 43, 46, 124, 127
Lavater, J. C., 34, 46, 67
Lavely, William, 360, 384
Lawson, Carol, 309, 311
Leary, Warren, 402
Leavy, Jane, 283, 311
Leblais, Madame, 44
Le Bruyn, Corneille, 49, 73
Lee, Richard B., 23, 46
Lee, Spike, 399, 402
Lehmann, J. F., 180
Leiblum, Sandra R., 318, 324, 334–36
Leif, Harold, 315, 316
Lennox, W. G., 171
Lenz, Fritz, 180, 190, 194
LeVay, Simon, 169
Levin, Michael, 398, 402
Levy, Jerre, 389
Lewes, Kenneth, 165
Lewin, Roger, 22, 46
Lewis, Michael, 389
Lindee, M. Susan, 5, 16
Lindfors, Bernth, 20, 30, 31, 43, 46
Linnaeus (Carl von Linne), 25, 46
Lippmann, Walter, 394, 401
Littlefield, Alice, 213, 216
Livesay, David, 401
Lomawaima, K. Tsianina, 5, 14, 212, 213, 216
Lombroso, Cesare, 13, 67–69, 109–25, 127, 141, 163, 166, 179, 235, 246
Lombroso-Ferrero, Gina, 111–14, 121, 127
London, Louis S., 168
Longino, Charles F., 352, 356
Lord, M. G., 307, 311
Louis XIV, 52, 65
Loustau, Leopold, 62
Louyer-Villermay, J. B., 224, 240, 241, 246
Lowe, Rob, 327, 336
Lunbeck, Elizabeth, 242, 246
Luxenberger, Hans, 190

Lydston, G. Frank, 163, 241, 246
Lynn, Richard, 41, 47

Maaskant-Kleibrink, Marianne, 240, 246
Macalpine, Ida, 241, 245
Maccoby, Eleanor, 389
Maddock, John, 41, 46
Magendie, M., 228, 246
Magian, A. C., 237, 241, 246
Magnan, Valentin, 163
Malthus, Thomas R., 380
Manitowabi, Edna, 212, 217
Manley, Ginger, 335
Mao Tsetung, 360
Maria, Tono, 29
Marriot, McKim, 78, 79, 106
Marshall, John, 44, 47
Marshall, Stuart, 9, 11, 17
Martin, Clyde E., 167
Martin, Emily, 3, 17
Martin, Felix, 70
Martin, John Rutledge, 234, 241, 246
Marx, Karl, 384
Mason, James, 342
Masters, William H., 35, 47, 168, 315, 316, 323, 334
Mathes, Valerie Sherer, 213, 217
Maudsley, Henry, 232, 233, 246
Mauss, Marcel, 78, 79, 94, 106
May, Elaine Tyler, 279, 311
Mayhew, Henry, 81, 107
Mayr, Ernst, 40, 47
McBean, Anne, 336
McBeth, Sally, 213, 217
McCaskill, Don, 213, 215
McCully, S. E., 234, 241, 246
McGrath, Ellie, 342, 357
Mead, Margaret, 347, 357
Mednick, Martha T., 401
Meigs, Charles D., 229, 246
Melosh, Barbara, 286, 308, 312
Mengele, Josef, 186
Meniere, Prosper, 57
Mennecke, Friedrich, 184, 185
Menninger, Karl, 168
Merchant, Carolyn, 22, 47, 119, 127
Mercuriale, Girolamo, 224
Meredig, Cathy, 308
Meredith, Howard L., 213, 218
Meriam, Lewis, 197, 217
Merleau-Ponty, Maurice, 74
Meyer, Adolf, 139, 166
Meyer-Bahlberg, Heino, 168
Mielke, Fred, 193, 195, 196
Miles, Catherine Cox, 166
Mill, John Stuart, 201
Millett, Kate, 201, 217
Mills, Charles K., 236, 237, 239, 246
Minton, Henry L., 165
Mirzoeff, Nicholas, 5, 9
Mitchell, Deborah, 285
Mitchell, Timothy, 75
Mitchinson, Wendy, 240, 246
Mitscherlich, Alexander, 193, 195–96
Moench, L. Mary, 141, 143
Moir, Ann, 391, 401
Molin, P., 201, 213, 216
Moll, Albert, 163
Money, John, 164, 168, 319, 321, 335
Monroe, Marilyn, 302
Montagu, Mary, 49
Montesquieu, Charles de Secondat Baron de, 58
Morris, Barry, 213, 217
Morrison, Toni, 41, 47, 399, 402
Morrison, W. Douglas, 124
Morse, Margaret, 308, 312
Moscucci, Ornella, 223, 226, 239, 240, 246
Mosimann, J., 298, 313
Mott, Frederick Walker, 261, 265, 277
Motz, Marilyn Ferris, 278, 280, 281, 283, 312
Mullaly, Frederick S., 82, 84–87, 107
Muller, Johannes, 43, 47
Muller-Hill, Benno, 193, 195, 196
Munby, Arthur J., 266, 276
Munn, Nancy, 103, 107
Murad III, 51
Murie, James, 44, 45
Murray, Charles, 395, 402

Naddaff, Romona, 3, 17
Nandy, Ashis, 79, 107
Napoleon I. *See* Bonaparte, Napoleon
Neal, Harry Edward, 90, 107
Nelkin, Dorothy, 5, 16, 400
Nerval, Gerard de, 61, 73
Ng, Vivien W., 363, 384
Niceforo, Alfredo, 112, 123, 124, 127
Nitsche, Paul, 184
Nixon, Richard, 279
Nunn, Sam, 162
Nye, Robert, 75, 76, 123, 124, 127

Onta, Pratyoush, 103, 107
Oppenheim, Janet, 240, 246
Outram, Dorinda, 25, 40, 42, 44, 47

Padden, Carol, 72, 77
Paglia, Camille, 392, 401
Parent-Duchatelet, Alexandre Jean-Baptiste, 229
Parke, J. Richardson, 232, 242, 246
Parker, Andrew, 363, 384
Parsons, Bruce, 169
Parsons, Talcott, 334
Pasquino, Pasquale, 110, 111, 127
Passeron, Jean-Claude, 202, 213, 215
Pasteur, Louis, 114, 124
Pattatucci, Angela M. L., 169
Patton, Cindy, 5, 15, 16, 357
Paul, Diane, 166
Paul, Ru, 304
Payne, R. L., 234, 239, 247
Peabody, Richard, 302, 310
Peirce, Charles, 103
Pelissier, Pierre, 56
Penzer, N. M., 73
Perrot, Michelle, 240, 241, 247
Perry, Ruth, 44, 47
Peyson, Frederic, 62
Phillips, L. M., 221, 239, 247
Pick, Daniel, 124, 127
Pieroni Bortolotti, Franca, 110, 125, 127
Pietropinto, Anthony, 334, 336
Pilcz, A., 174, 193
Pilliard, Richard, 169
Pinel, Philippe, 56, 74
Pinney, Christopher, 79, 107
Plant, Richard, 196
Plato, 54, 74
Platter, Felix, 224
Polack, John O., 230, 239, 247
Pomeroy, Wardell B., 167
Pommerin, Rainer, 194
Poovey, Mary, 234, 239, 247
Porter, Roy, 239, 247
Potter, Jack M., 361, 362, 385
Potter, Sulamith Heins, 361, 362, 381, 384, 385
Powell-Hopson, Darlene, 285
Powers, Polly I., 348, 357
Pratt, Richard H., 199, 213
Proctor, Robert, 5, 8, 169
Provine, William, 194
Prucha, Francis P., 213, 218
Puccini, Sandra, 123, 127

Quetelet, Adolphe, 112

Rabinowitz, Paula, 280, 281, 312
Racine, Jean, 52, 73
Ramanujam, T., 97, 100, 107
Ramm, Rudolf, 195
Rand, Erica, 302, 312
Raven-Hart, Rowland, 21, 47
Reagan, Ronald, 2, 397
Reagan, Ronald, Jr., 354
Reed, James, 166
Reel, Estelle, 203, 205, 206, 214
Regnard, Andre, 75
Reichler, Max, 174, 193
Restak, Richard, 391, 394, 401
Reyhner, Jon, 212, 218
Richards, F. J., 86, 107
Richards, Josephine E., 218
Rieger, Jurgen, 194
Riley, Denise, 122, 127
Ripa, Yannick, 240, 247
Ritter, Robert, 179, 189, 194, 196
Robertston, Josephine, 291, 308, 312
Robinson, Jean C., 364, 385
Robson, Ruth Ann, 169
Roehm, Ernst, 188
Rohe, George H., 230, 247
Rosen, Raymond C., 318, 324, 334–36
Rosenberg, Charles, 226, 239, 240, 248
Rothermund, Indira, 79, 107
Rothman, Stanley, 396, 402
Rousseau, G. S., 240, 247
Rousseau, Jean-Jacques, 53, 74
Routh, C. H. F., 233, 240–42, 247
Rowland, Craig, 335–37
Rubin, Gayle, 336
Rudin, Ernst, 187
Ru Paul, 304
Rushton, J. Philippe, 41, 45, 394
Russett, Cynthia Eagle, 3, 17, 34, 47, 240, 247, 287, 312
Russo, Mary, 363
Ryan, Jack, 279
Rycaut, Paul, 51, 52, 73
Rygg, Larry, 213, 218

Sagarin, Edward, 240, 244
Sahli, Nancy, 241, 247

Said, Edward, 61, 62, 75
Salholz, Eloise, 357
Salmon, Richard F., 334–36
Saltzman, Elizabeth, 308
Sandall, Robert, 91, 108
Sandelowski, Margaret J., 374, 385
Sanders, J., 187
Sargent, Dudley, 289, 290
Sarich, Vince, 398
Sawicki, Jana, 363, 385
Schaef, Anne Wilson, 325, 336
Schiebinger, Londa, 3, 17, 27, 34, 44, 47, 226, 247, 287, 307, 312
Schlossman, Steven, 242, 247
Schluter, Dolph, 41, 47
Schneider, Cynthia, 302
Schneider, Joseph W., 336
Schottky, Johannes, 180, 194
Schrenk-Notzing, A. von, 241, 247
Schulz, Edgar, 175, 194
Schwartz, Hillel, 289, 312
Schwartz, Mark, 335
Scull, Andrew, 240, 241, 247
Seidman, Steven, 237, 247
Sekula, Allan, 9, 18, 76, 112, 118, 119, 123–25, 128
Seligman, Daniel, 396, 402
Sells, Cato, 200
Selvin, Paul, 402
Shapiro, Eben, 278, 312
Shapiro, Harry, 291, 308, 312
Sharpe, F. S., 241, 247
Shaw, J. C., 235, 248
Sheinwald, Jesse D., 400
Sheridan, Alan, 197, 198, 218
Sherwood-Dunn, B., 231, 248
Shohat, Ella, 22, 47
Shorter, Edward, 239, 240, 248
Shortland, Michael, 228, 240, 248
Shortt, S. E. D., 240, 248
Showalter, Elaine, 240, 248
Shue, Vivienne, 363, 385
Shuttleworth, Sally, 374, 384
Simenauer, Jacqueline, 334, 336
Simmons, Christina, 286, 308, 312
Simon, Denis F., 358, 385
Skultans, Vieda, 242, 248
Slade, Adolphus, 62, 75
Sloan, M. G., 242, 248
Small, Meredith F., 169, 401
Smith, A. Laptham, 248
Smith, Christopher J., 362, 385
Smith, Heywood, 233, 241
Smith-Rosenberg, Carroll, 226, 239, 240, 248
Snyderman, Mark, 396, 402
Socarides, Charles W., 168
Socrates, 74
Solomon, Richard, 318, 334
Sommer, Doris, 363
Sorsby, Maurice, 173, 193
Spencer, Frank, 307, 312
Spicer, Hayley, 298
Spitzka, Edward Anthony, 44, 47
Sprague Zones, Jane, 309, 313
Springarn, Jed, 401
Squillacciotti, Massimo, 123, 126
Stacey, Judith, 364, 385
Stall, Sylvanus, 201, 202
Stammler, Martin, 176, 194
Stanley, Julian, 390, 400
Steinberg, Marvin A., 336
Stepan, Nancy Leys, 28, 48, 75
Stephen, James Fitzjames, 82
Stern, Leonore O., 164
Stevenson, Richard, 278, 313
Stewart, William, 230, 248
Stocking, George, 7, 18, 25–28, 48, 75, 112, 123, 128
Stoddard, I. Lothrop, 394
Stokoe, William, 72, 77
Stoller, Richard C., 164
Storer, Horatio, 219, 220, 232, 248
Sudanandha, Samuel, 102
Sutcliffe, J. A., 241, 248
Suttmeier, Richard P., 358, 385
Swaab, D. F., 169
Swedlund, Alan, 5, 13
Szasz, Margaret, 212, 218

Tagg, John, 76, 119, 125, 128
Taine, Hippolyte, 59, 60, 75
Tait, Robert Lawson, 241, 242, 248
Takanishi, Ruby, 352, 356
Talbot, Eugene S., 163
Talmey, Bernard, 222, 235, 241, 248
Tarde, Gabriel, 123, 124, 128
Tarnowsky, Pauline, 118, 141
Tarra, Giulio, 69
Tavris, Carol, 393, 401
Tazi, Nadia, 3, 17
Terman, Lewis, 139, 166

Terry, Jennifer, 5, 7, 125, 128, 165, 166, 168, 169, 242, 248, 288, 308, 313
Thoinot, Leon, 235, 248
Thompson, E. P., 95, 96, 108
Thornton, Alice Wandesford, 262, 276
Thurston, Edgar, 81, 82, 84, 87, 88, 108
Tiedemann, Frederick, 43, 48
Tiffany, Sharon, 21, 22, 48
Tilt, Edw. J., 240, 248
Tirala, Lothar, 187
Tolen, Rachel, 5, 10
Tomes, Nancy, 240, 248
Toomey, Colleen, 309
Tournefort, Joseph Pitton de, 52, 53, 73
Traub, James, 402
Trennert, Robert, 202, 213, 218
Trumbach, Randolf, 31, 48
Tuke, Daniel Hack, 232, 240, 243, 249
Turner, Bryan S., 3, 18
Turner, Victor, 347, 357
Turner-Gottschang, Karen, 380, 385

Ulrichs, Karl Heinreich, 134, 164
Unschuld, Paul U., 366–68, 371, 385
Urla, Jacqueline, 5, 13
Utley, Robert, 213, 218

Vaille, H. R., 239, 249
Vance, Carole S., 322, 335, 336
Van Gelder, Lindsy, 343, 357
Veith, Ilza, 367
Verschuer, Otmar Freiherr von, 174, 185–87, 189, 194–96
Vespucci, Amerigo, 21, 32
Virey, J.-J., 55
Voght, Martha, 242, 243

Wack, Mary, 225, 249
Wagner, Gerhard, 173, 174, 180, 193–95
Wald, Elijah, 169
Wallach, Stephanie, 242, 247
Wallis, Claudia, 342, 357
Walton, John Tompkins, 220, 221, 249
Walvin, James, 20, 29, 45, 48
Wang, Feng, 360, 385
Warner, Patricia Rosalind, 301, 313
Washington, Booker T., 200, 201, 205
Watney, Simon, 9, 18
Webster, Daniel, 44
Wechsler, Judith, 73
Weeks, Jeffrey, 240, 249, 320, 321, 335
Weeks, Margaret R., 364, 385
Weil, Arthur, 187, 195
Weinreich, Max, 192, 196
Weintraub, Pamela, 400
Welch, George T., 241, 249
Wells, T. Spencer, 230, 231, 240, 241, 249
Wertham, Frederic, 193
Westphal, Karl Friedrich Otto, 131, 163
White, Ryan, 343
White, Sydney, 380
Widerkehr, Joseph de, 62
Wigand, Arpad, 179, 194
Wildmon, Donald, 162
Will, George, 397, 402
Williams, John, 213, 218
Williams, Lena, 277, 313
Williams, Robin, Jr., 401
Willis, Susan, 281, 282, 285, 313
Wilson, Albert, 164
Wilson, Edward O., 390, 392, 401
Wilson, Elizabeth, 285, 313
Wilson, James Q., 397
Winau, Rolf, 193
Winslow, Caroline, 44
Wiseman, C. J., 298, 313
Wolf, Naomi, 277, 313
Wolff, Georg, 175, 194
Wortis, Benjamin, 168
Wortis, Joseph, 165
Wylie, W. Gill, 240, 249

Xiaoping, Deng. *See* Deng Xiaoping

Yang, Anand, 79, 81, 87, 102, 108
Yang, C., 362, 386
Yarrow, Andrew L., 354, 357
Young, Robert, 76
Young-Bruehl, Elisabeth, 242, 249
Yuan, Lili, 365, 372, 381, 386

Zhang, Mincai, 364, 386
Zhang Ting-liang, 367, 386
Zhang, X., 362, 386
Zhao, Z., 358, 386